AF361651

The State-of-the-Art Propagation and Breeding Techniques for Horticulture Crops

The State-of-the-Art Propagation and Breeding Techniques for Horticulture Crops

Editor

Sergio Ruffo Roberto

MDPI • Basel • Beijing • Wuhan • Barcelona • Belgrade • Manchester • Tokyo • Cluj • Tianjin

Editor
Sergio Ruffo Roberto
Londrina State University
Londrina, Brazil

Editorial Office
MDPI
St. Alban-Anlage 66
4052 Basel, Switzerland

This is a reprint of articles from the Special Issue published online in the open access journal *Horticulturae* (ISSN 2311-7524) (available at: https://www.mdpi.com/journal/horticulturae/special_issues/Propagation_Breeding).

For citation purposes, cite each article independently as indicated on the article page online and as indicated below:

LastName, A.A.; LastName, B.B.; LastName, C.C. Article Title. *Journal Name* **Year**, *Volume Number*, Page Range.

ISBN 978-3-0365-8444-7 (Hbk)
ISBN 978-3-0365-8445-4 (PDF)

Cover image courtesy of Sergio Ruffo Roberto

Contents

About the Editor

Sergio Ruffo Roberto

Sergio Ruffo Roberto received his Ph.D. degree from Sao Paulo State University, Brazil, in 1998. Since 2000, he has been a professor at the Agricultural Research Center, Food Research Institute, State University of Londrina. His research focuses on the propagation of fruit species, plant production and postharvest of fruit and vegetables. His recent interests include plant growth regulators, artificial lighting, packaging materials and fruit quality.

Preface

Horticulture has established its importance in many aspects, including innovation, improving land use, promoting crop diversification, generating employment, and providing food to the world population. Thus, innovation in plant propagation and breeding is essential to meet the challenges of global changes such as population growth and climate change. Over the years, horticulturists have developed several propagation methods which have supported breeding programs and allowed the production of high-quality nursery plants and higher-yielding crops. Traditional breeding is one of the main strategies used to improve agronomic traits. In many horticultural species, several cultivars have been developed through conventional methods, such as mutagenesis, inter- and intra-specific crosses, and clonal selection. Conventional breeding is a long-term and expensive process; a long period of time and resources are needed to obtain progenies and to evaluate their traits. In addition, sexual breeding is not always feasible because some cultivars to be used in crosses are incompatible, sterile, or polyembryonic. Moreover, in many cases, after breeding, backcrosses are required to recover the desired features of the improved cultivar, further lengthening breeding programs. Since the 1990s, new biotechnology techniques have been applied to the propagation and breeding of horticultural species, providing efficient alternatives to traditional methods for the improvement of novel cultivars. This has been possible through the development of transformation protocols, starting from many sources of explants. More recently, several new techniques have been developed and classified as new plant breeding techniques.

Sergio Ruffo Roberto
Editor

Editorial

The State-of-the-Art Propagation and Breeding Techniques for Horticulture Crops

Sergio Ruffo Roberto *, Aline Cristina de Aguiar and Viviani Vieira Marques

Agricultural Research Center, State University of Londrina, Celso Garcia Cid Road, km 380, P.O. Box 10011, Londrina 86057-970, Brazil; aline.aguiar@uel.br (A.C.d.A.); vivianimarques@yahoo.com.br (V.V.M.)
* Correspondence: sroberto@uel.br; Tel.: +55-43-3371-4774

Citation: Roberto, S.R.; de Aguiar, A.C.; Marques, V.V. The State-of-the-Art Propagation and Breeding Techniques for Horticulture Crops. *Horticulturae* **2023**, *9*, 804. https://doi.org/10.3390/horticulturae9070804

Received: 16 June 2023
Revised: 3 July 2023
Accepted: 6 July 2023
Published: 13 July 2023

Horticulture has established its importance in many aspects including innovation, improving land use, promoting crop diversification, generating employment, and providing food to the world population. Thus, innovation in plant propagation and breeding is essential to meet the challenges of global changes such as population growth and climate change [1].

Over the years, horticulturists have developed several propagation methods which have supported breeding programs and allowed the production of high-quality nursery plants and higher yielding crops. Traditional breeding is one of the main strategies used to improve agronomic traits. In many horticultural species, several cultivars have been developed through conventional methods, such as mutagenesis, inter- and intra-specific crosses, and clonal selection. Conventional breeding is a long-term and expensive process; a long period of time and resources are needed to obtain progenies and to evaluate their traits. In addition, sexual breeding is not always feasible because some cultivars to be used in crosses are incompatible, sterile, or polyembryonic. Moreover, in many cases, after breeding, backcrosses are required to recover the desired features of the improved cultivar, further lengthening breeding programs [2].

Since the 1990s, new biotechnology techniques have been applied to the propagation and breeding of horticultural species, providing efficient alternatives to traditional methods for the improvement of novel cultivars. This has been possible through the development of transformation protocols starting from many sources of explants. More recently, several new techniques have been developed and classified as new plant breeding techniques. The aim of this Special Issue was to present the latest advances in new horticultural propagation and breeding methods.

The Special Issue "The State-of-the-Art Propagation and Breeding Techniques for Horticulture Crops" brings together some of the latest research results of new techniques in this field. It presents twelve original papers, which deal with a wide range of research activities.

Vendrame et al. [3] examined the growth response of ornamental bananas, Musa 'Little Prince' and Musa 'Truly Tiny', to different light sources, including LEDs and regular fluorescent light. The authors found that shoot mass and length could be promoted by controlling light quality and intensity. However, the effect of light quality and intensity related to plant growth and development were not evident. Although not directly evaluated in this study, the number of in vitro shoots produced per explant were higher for some of the cultures grown under the LED lights, with multiple shoots produced, while no shoots were formed in cultures grown under regular fluorescent lighting. The different responses between the two banana cultivars indicated a genotype effect in combination with different light environments. LED lighting affected the relative chlorophyll content as well as stomata size in banana plantlets in vitro. LEDs at 90 μmol m^{-2} s^{-1} were identified to be a suitable selection for the micropropagation of ornamental bananas.

Some interesting aspects of the seeds of the succulent species *Suaeda aralocaspica* were found by Si et al. [4]. The authors demonstrated that the highest germination percentage of brown seeds was 100% and that of black seeds was 17%. Thus, brown seeds were more suitable for further culturing experiments than black seeds. For brown seeds, the sterilization effect of NaClO was better than that of $HgCl_2$. Rinsing with 75% ethanol for 60 s, sterilizing with NaClO for 8 min, and cultivating at pH 8.0 MS for 7 days was the best of all sterilization procedures and cultivation methods tested, which has been successfully applied to in vitro *S. aralocaspica* cultures.

Another important horticultural species, i.e., vanilla (*Vanilla planifolia*), was studied by Serrano-Fuentes et al. [5]. The authors tried to induce somaclonal variation in V. planifolia through gamma radiation and detected it by using inter-simple sequence repeat molecular markers. The results showed a hormetic effect on the explants, promoting development at a low dose (20 Gy) and showing inhibition and death at high doses (60–100 Gy). The LD50 was observed at 60 Gy. The primers UBC-808, UBC-836, and UBC-840 showed the highest % P, with 42.6%, 34.7% and 28.7%, respectively. Genetic distance analysis showed that treatments with and without irradiation produced somaclonal variation. The use of gamma rays during in vitro culture was shown to be an alternative method to broaden genetic diversity for vanilla breeding.

In vitro experiments were conducted by Hanász et al. [6] to study the responses of potato (*Solanum tuberosum* L.) genotypes to osmotic stress. In vitro shoot cultures of 27 breeding lines and their drought-tolerant parents were tested under osmotic stress induced by the addition of PEG 6000, D-mannitol, and PEG 600 to the Murashige–Skoog medium. It was demonstrated that 7.5% and 10% PEG 6000 or 0.2 M and 0.3 M D-mannitol treatments are suitable for the selection of osmotic stress-tolerant potato genotypes.

Hadi et al. [7] provided the first report on the development of an in vitro propagation protocol for seed germination and somatic embryo formation from seeds of the threatened endemic medicinal plant species, *Aconitum violaceum*. The authors found that that seeds are suitable explants for efficient multiplication and restoration of A. violaceum within a short period of time, approximately three to 5 months, starting from the initiation of seed germination or somatic embryo development to final tissue culture-raised plantlets.

Interesting results were also reported by Khuat et al. [8] using black cardamom (*Amomum tsao-ko*). They aimed to improve the seed germination rate and uniform germination through mechanical scarification, immersion in hot or cold water, acid scarification, and the application of plant growth regulators. Applying mechanical scarification treatment before sowing was shown to be the most effective for improving seed germination rates. Immersion in cold water or plant growth regulators before sowing were also recommended. Finally, the authors described that the developed in vitro propagation protocol is an effective solution for rapid multiplication of high-yielding elite plants to meet the needs of expanding cultivation of this important crop.

A high-performance in vitro propagation system for *Plectranthus amboinicus*, a medicinally important aromatic perennial herb, was investigated by Faisal and Alatar [9] through direct shoot organogenesis. The system was developed using axillary node explants cultured on MS medium augmented with various plant growth regulators. The authors found that after 8 weeks of culture, the explants cultured in full-strength MS basal medium (pH 5.7) with 5.0 µM BA and 2.5 µM NAA exhibited the highest percentage of regeneration and the maximum number of shoots per explant. Individual elongated shoots were rooted on half-strength MS basal medium containing 0.25 µM IBA after 4 weeks of culture.

Nasrat et al. [10] assessed the effect of induced calluses of the medicinal herb *Bougainvillea glabra* in vitro under different light conditions and plant growth regulators and measured their phytochemical and antioxidant activities using different extraction solvents. The maximum number of days to callus initiation were recorded when nodal explants were cultured on woody plant medium supplemented with 7.5 µM 2,4-D + 0.5 µM BAP under light conditions. On the contrary, the minimum number of days to callus initiation was obtained when nodal explants were treated with 2.5 and 5 µM 2,4-D + 1 and 1.5 µM BAP under dark conditions. In addition, an aqueous extract of conventionally propagated nodal explants exhibited the highest phenolic content and antioxidant activities.

Another interesting result was obtained by Dewir et al. [11] who compared in vitro flower induction and the formation of daughter corms of saffron (*Crocus sativus*) in gel and liquid cultures. Additionally, different concentrations of glutamine, salicylic acid, and jasmonic acid were tested with the aim to improve the formation of saffron daughter corms. It was demonstrated that saffron flowering could be induced in vitro, and the harvested stigma of these flowers could be used as a source of spice or pharmaceuticals. Compared with solid culture, liquid cultures/bioreactors improved daughter corm diameter and fresh weight. Moreover, salicylic acid at 75 mg L^{-1} and glutamine at 600 mg L^{-1} increased corm diameter and fresh weight.

Sand [12] reviewed the main results of experiments conducted on Syngonium *podophyllum*, which is recognized as a valuable ornamental and medicinal species, between 1996 and 2004 and published after 1998, to provide new insights specifically related to de novo shoot formation from callus. The author described the lessons learned from all experiments performed on Syngonium—including the principles to implement industrial-scale micropropagation—and may further support the production of phenolic compounds relevant for in vitro systems at the industrial scale.

Moving towards another crop species, *Morus alba*, also known as mulberry, Chen et al. [13] conducted a trial to clarify the inherent mechanism of cutting rooting, and to explore the relationship between growth hormones and endogenous hormones. The plant growth regulators IAA, IBA, and ABT-1 were able to promote the rooting of cuttings of vegetable mulberry and fruit mulberry, with ABT-1 exhibiting the best effect.

Finally, Nascimento et al. [14] estimated the adaptability and the temporal stability of strawberry (*Fragaria ananassa*), as well to select genotypes that are easy to propagate with lower cold requirements using a mixed linear model. Through this model, the authors showed the superiority of 11 genotypes that had the potential to be released as cultivars. The strawberry RVFS07M-34 was the most promising genotype to be registered as a new cultivar.

Author Contributions: All authors, S.R.R., A.C.d.A. and V.V.M., contributed equally to this work. All authors have read and agreed to the published version of the manuscript.

Conflicts of Interest: The authors declare no conflict of interest.

References

1. Roberto, S.R.; Colombo, R.C. Innovation in Propagation of Fruit, Vegetable and Ornamental Plants. *Horticulturae* **2020**, *6*, 23. [CrossRef]
2. Kullaj, E. Rootstocks for improved postharvest quality of fruits: Recent advances. In *Preharvest Modulation of Postharvest Fruit and Vegetable Quality*; Siddiqui, M.W., Ed.; Academic Press: Cambridge, MA, USA, 2018; pp. 189–207. [CrossRef]
3. Vendrame, W.; Feuille, C.; Beleski, D.; Bueno, P. In Vitro Growth Responses of Ornamental Bananas (*Musa* sp.) as Affected by Light Sources. *Horticulturae* **2022**, *8*, 92. [CrossRef]
4. Si, Y.; Haxim, Y.; Wang, L. Optimum Sterilization Method for In Vitro Cultivation of Dimorphic Seeds of the Succulent Halophyte *Suaeda aralocaspica*. *Horticulturae* **2022**, *8*, 289. [CrossRef]
5. Serrano-Fuentes, M.; Gómez-Merino, F.; Cruz-Izquierdo, S.; Spinoso-Castillo, J.; Bello-Bello, J. Gamma Radiation (60Co) Induces Mutation during In Vitro Multiplication of Vanilla (*Vanilla planifolia* Jacks. ex Andrews). *Horticulturae* **2022**, *8*, 503. [CrossRef]
6. Hanász, A.; Dobránszki, J.; Mendler-Drienyovszki, N.; Zsombik, L.; Magyar-Tábori, K. Responses of Potato (*Solanum tuberosum* L.) Breeding Lines to Osmotic Stress Induced in In Vitro Shoot Culture. *Horticulturae* **2022**, *8*, 591. [CrossRef]
7. Hadi, A.; Singh, S.; Rafiq, S.; Nawchoo, I.; Wagay, N.; Mahmoud, E.; El-Ansary, D.; Sharma, H.; Casini, R.; Yessoufou, K.; et al. In Vitro Propagation of Aconitum violaceum Jacq. ex Stapf through Seed Culture and Somatic Embryogenesis. *Horticulturae* **2022**, *8*, 599. [CrossRef]
8. Khuat, Q.; Kalashnikova, E.; Kirakosyan, R.; Nguyen, H.; Baranova, E.; Khaliluev, M. Improvement of In Vitro Seed Germination and Micropropagation of *Amomum tsao-ko* (Zingiberaceae Lindl.). *Horticulturae* **2022**, *8*, 640. [CrossRef]
9. Faisal, M.; Alatar, A. Establishment of an Efficient In Vitro Propagation Method for a Sustainable Supply of *Plectranthus amboinicus* (Lour.) and Genetic Homogeneity Using Flow Cytometry and SPAR Markers. *Horticulturae* **2022**, *8*, 693. [CrossRef]
10. Nasrat, M.; Sakimin, S.; Hakiman, M. Phytochemicals and Antioxidant Activities of Conventionally Propagated Nodal Segment and In Vitro-Induced Callus of *Bougainvillea glabra* Choisy Using Different Solvents. *Horticulturae* **2022**, *8*, 712. [CrossRef]
11. Dewir, Y.; Alsadon, A.; Al-Aizari, A.; Al-Mohidib, M. In Vitro Floral Emergence and Improved Formation of Saffron Daughter Corms. *Horticulturae* **2022**, *8*, 973. [CrossRef]

12. Sand, C.; Antofie, M. De Novo Shoot Development of Tropical Plants: New Insights for *Syngonium podophyllum* Schott. *Horticulturae* **2022**, *8*, 1105. [CrossRef]
13. Chen, H.; Lei, Y.; Sun, J.; Ma, M.; Deng, P.; Quan, J.; Bi, H. Effects of Different Growth Hormones on Rooting and Endogenous Hormone Content of Two *Morus alba* L. Cuttings. *Horticulturae* **2023**, *9*, 552. [CrossRef]
14. Nascimento, D.; Gomes, G.; de Oliveira, L.; de Paula Gomes, G.; Ivamoto-Suzuki, S.; Ziest, A.; Mariguele, K.; Roberto, S.; de Resende, J. Adaptability and Stability Analyses of Improved Strawberry Genotypes for Tropical Climate. *Horticulturae* **2023**, *9*, 643. [CrossRef]

 *horticulturae*

Article

Adaptability and Stability Analyses of Improved Strawberry Genotypes for Tropical Climate

Daniele Aparecida Nascimento [1], Gabriella Correia Gomes [2], Luiz Vitor Barbosa de Oliveira [2], Gabriel Francisco de Paula Gomes [2], Suzana Tiemi Ivamoto-Suzuki [2], André Ricardo Ziest [3], Keny Henrique Mariguele [4], Sergio Ruffo Roberto [2] and Juliano Tadeu Vilela de Resende [1,2,*]

[1] Campus CEDETEG, State University of Centro-Oeste, Alameda Élio Antonio Dalla Vecchia, 838, Guarapuava 85040-167, Brazil; dani123nas@gmail.com
[2] Agricultural Research Center, State University of Londrina, Celso Garcia Cid Road, km 380, Londrina 86057-970, Brazil; gabriella.correia@uel.br (G.C.G.); luizvitor.barbosa@uel.br (L.V.B.d.O.); gabriel.fdepaulagomes@gmail.com (G.F.d.P.G.); suzanatiemi@uel.br (S.T.I.-S.); sroberto@uel.br (S.R.R.)
[3] Agricultural Research Center, Federal University of Santa Catarina, R. Eng. Agronômico Andrei Cristian Ferreira, s/n—Trindade, Florianópolis 88040-900, Brazil; andre.zeist@ufsc.br
[4] Santa Catarina State Rural Extension and Agricultural Research Corporation, Admar Gonzaga Street, 1347, Itajaí 88034-901, Brazil; kenymariguele@epagri.sc.gov.br
* Correspondence: jvresende@uel.br; Tel.: +55-(43)-3376-5990

Abstract: Strawberries are grown worldwide, and the fruit is known for its flavor, pleasant aroma, and the presence of important nutraceutical compounds. Under temperate conditions, the species is octaploid and presents a complex inheritance. Exploring polyploidy in varietal crosses is the main alternative to developing genotypes of high-temperature regions; thus, breeding programs must evaluate the interaction based on parameters, such as the heritability, stability, easy propagation, and adaptability to different soil and climate variations. To estimate the stability and temporal adaptability of pre-selected triple hybrids of day-neutral strawberries, thirty-six experimental genotypes, three commercial genotypes ('Albion,' 'Monterey,' and 'Dover'), and four single hybrids ('RVFS07,' 'RVFS06,' 'RVDA11,' and 'RVCA16') were evaluated in a protected cultivation from August to February under tropical climate conditions (southern hemisphere) using the mixed linear model (MLM). The genotypes RVFS07M-34, RVFS07M-24, RCDA11M-04, RVFS07M-154, RVFS07M-36, RVFS07M-33, RVFS07M-80, RVFS07M-10, RVDA11M-21, RVDA11M-13, and RVFS06AL-132 had the highest values of total fruit mass, adaptability, and stability. The mean predicted genotypic values of the selected genotypes was 138% higher than the mean of the controls. Therefore, these genotypes have the potential to be released as cultivars.

Keywords: *Fragaria* × *ananassa* Duch; heritability; propagation; low cold requirement; neutral photoperiod

Citation: Nascimento, D.A.; Gomes, G.C.; de Oliveira, L.V.B.; de Paula Gomes, G.F.; Ivamoto-Suzuki, S.T.; Ziest, A.R.; Mariguele, K.H.; Roberto, S.R.; de Resende, J.T.V. Adaptability and Stability Analyses of Improved Strawberry Genotypes for Tropical Climate. *Horticulturae* **2023**, *9*, 643. https://doi.org/10.3390/horticulturae9060643

Academic Editor: Yuepeng Han

Received: 19 April 2023
Revised: 23 May 2023
Accepted: 25 May 2023
Published: 30 May 2023

1. Introduction

Strawberry production in the southern hemisphere is totally dependent on cultivars developed by breeding programs in the United States and Spain [1]. In addition to low adaptability and stability in tropical regions, these cultivars increase production costs, since commercial nursery plants are multiplied and imported in dollars [2]. Another recurring problem concerns the physiological and phytosanitary quality, where several crops were affected by pathogens that did not exist in the producing regions, and they were probably introduced by imported nursery plants.

Climate change has promoted an increase in the temperature and in the concentration of carbon dioxide (CO_2) in the atmosphere [3,4]. High temperatures limit the culture, as strawberry plants require cold hours for floral induction. Thus, it can negatively affect the yield of cultivated plants, interfering with the balance of morphophysiological and hormonal processes [5,6].

The development of adaptable and stable tropical cultivars can be a solution to overcome this issue. Therefore, it is important that genetic breeding programs are stimulated to develop productive cultivars and adapted to a wide range of latitudes, allowing cultivation area expansion [7]. Several studies in tropical regions have demonstrated the potential of new genotypes with low chilling requirements and higher yields, compared to well-established cultivars in the market [2,8–10].

The environment interferes with agronomic characteristics [11] and the quality of the fruit post-harvest [12]. Considering the wide latitude of strawberry cultivation worldwide, studies of adaptability and stability should be carried out both in relation to the location and throughout the harvest period [1].

Studying the genotype × environment interaction aims to identify the behavioral variation that genotypes undergo when exposed to varied environmental conditions [13]. One way to measure these interaction effects is through correlation studies between the characteristics of interest [14]. Correlations estimated among variables provide strategic information for studies to improve adaptability and stability in strawberries [15]. In addition, these studies allow genotype identification with predictable behavior, as well as those responsive to environmental variations, under specific conditions [16,17].

The variability of the genus Fragaria is wide, and it is classified according to the ploidy level, in which the basic number of chromosomes is equal to seven ($x = 7$) [18]. Among the twenty-five known species, thirteen are diploid ($2n = 2x = 14$), five are tetraploid ($2n = 4x = 28$), one is petaploid ($2n = 5x = 35$), one is hexaploid ($2n = 6x = 42$), three are octaploid ($2n = 8x = 56$), and two are decaploids [18]. The cultivated species (*Fragaria × ananassa*), classified as octaploid, is derived from immediate ancestors (*F. chiloensis* and *F. virginiana*), considered allopolyploids [19], which, in turn, are derived from two or more different diploid ancestors [20]. The octaploid commercial strawberry genome comprises 813.4 megabases (Mb) that are distributed across 28 pseudochromosomes, with 108,087 protein-coding genes and 30,703 RNA-coding genes [21]. Therefore, this wide variability generates expectations of obtaining genotypes adapted to a low chilling requirement.

Most polyploids obtained by chromosome duplication have characteristics of vigorous plants and larger fruits [22]. Eukaryotic polyploids also show strong resistance to biotic and abiotic factors [23].

Mixed linear models consider genotype effects as modifications to estimate adaptability and stability, allowing a genetic effects analysis to be made using best linear unbiased predictors (BLUPs) [24]. However, mixed linear models are routinely prone to experimental inconstancy and the heterogeneity of environmental variations [25]. Therefore, in this method, the genotypic values (e.g., productivity), adaptability, and genotypic stability are analyzed simultaneously [26]. This methodology is based on the harmonic mean of the relative performance of genotypic values (HMRPGV) method, which considers the genotype mean and its variation in different environments.

The relationship between genotype and phenotype in different regions helps to anticipate more precise responses to the selection of individuals with heterogeneous habitats, whether spatial or temporal. If the genotype has phenotypic expression for a particular trait, depending on the environment, heritability measures can be altered according to variations in environmental conditions [27].

Thus, the objective of this work was to estimate the adaptability and the temporal stability, as well to select strawberry genotypes easy to propagate with lower cold requirements using a mixed linear model.

2. Materials and Methods

2.1. Location

The experiment was conducted at the University of Centro-Oeste, Guarapuava, Brazil (25°23′01″ S, 51°29′50″ W, elevation of 1025 m a.s.l.) from August 2019 to February of 2020. The soil was classified as typical dystroferric Latosol Bruno [28]. The climate was humid subtropical Cfb (temperate oceanic climate, warm summer, and without a dry season), which includes hot summers and frosty winters. The average annual temperature was 17 °C, with maximum and minimum temperatures of 23.5 °C and 12.7 °C, respectively. The average annual rainfall was 1946 mm [28].

2.2. Experimental Genotypes

In our study, 36 genotypes (Table 1) were selected from 10 populations (total of 2000 F_1 plants) from crosses among genotypes pre-selected by our strawberry breeding program and commercial cultivars with responses to neutral photoperiods ('Albion' and 'Monterey'). The 36 genotypes were multiplied to produce the nursery plants for the trials.

Table 1. Genotypes pre-selected from the strawberry breeding program obtained from intravarietal crosses.

Male Genitor Cultivars	Female Genitor Cultivar	Genotypes
RVFS 07 (Festival × Aromas)	Monterey	RVFS07M-24
RVFS 07 (Festival × Aromas)	Monterey	RVFS07M-36
RVFS 06 (Festival × Aromas)	Albion	RVFS06AL-132
RVFS 07 (Festival × Aromas)	Monterey	RVFS07M-34
RVFS 07 (Festival × Aromas)	Monterey	RVFS07M-179
RVDA 11 (Dover × Aromas)	Monterey	RVDA11M-04
RVDA 11 (Dover × Aromas)	Monterey	RVDA11M-21
RVFS 07 (Festival × Aromas)	Monterey	RVFS07M-02
RVFS 07 (Festival × Aromas)	Monterey	RVFS07M-32
RVFS 07 (Festival × Aromas)	Monterey	RVFS07M-38
RVFS 07 (Festival × Aromas)	Monterey	RVFS07M-05
RVFS 07 (Festival × Aromas)	Monterey	RVFS07M-33
RVFS 07 (Festival × Aromas)	Monterey	RVFS07M-10
RVDA 11 (Dover × Aromas)	Monterey	RVDA11M-13
RVFS 07 (Festival × Aromas)	Monterey	RVFS07M-154
RVFS 07 (Festival × Aromas)	Monterey	RVFS07M-47
RVFS 07 (Festival × Aromas)	Monterey	RVFS07M-31
RVFS 07 (Festival × Aromas)	Monterey	RVFS07M-16
RVDA 11 (Dover × Aromas)	Monterey	RVDA11M-25
RVFS 07 (Festival × Aromas)	Monterey	RVFS07M-113
RVDA 11 (Dover × Aromas)	Monterey	RVDA11M-32
RVFS 07 (Festival × Aromas)	Monterey	RVFS07M-30
RVCA 16 (Camarosa × Aromas)	Monterey	RVCA16M-01
RVDA 11 (Dover × Aromas)	Monterey	RVDA11M-28
RVFS 07 (Festival × Aromas)	Monterey	RVFS07M-151
RVDA 11 (Dover × Aromas)	Monterey	RVDA11M-10
RVFS 07 (Festival × Aromas)	Monterey	RVFS07M-124
RVFS 06 (Festival × Aromas)	Monterey	RVFS06M-29
RVDA 11 (Dover × Aromas)	Monterey	RVDA11M-29
RVFS 07 (Festival × Aromas)	Monterey	RVFS07M-42
RVFS 07 (Festival × Aromas)	Albion	RVFS07AL-28
RVFS 07 (Festival × Aromas)	Monterey	RVFS07M-48
RVFS 07 (Festival × Aromas)	Monterey	RVFS07M-88
RVFS 07 (Festival × Aromas)	Monterey	RVFS07M-80
RVDA 11 (Dover × Aromas)	Monterey	RVDA11M-03
RVFS 06 (Festival × Aromas)	Albion	RVFS06AL-36

The mother plants were kept in a greenhouse with an average temperature of 25 ± 3 °C and a humidity of $75 \pm 5\%$. The nursery plants were obtained from runners (stolons) and transplanted into polypropylene trays, with 50 cells filled with substrate. After 50 days, they were transplanted to an experimental field for evaluation in a repeat trial.

2.3. Genotype Transplantation

The nursery plants were transplanted to the field in a low tunnel cultivation system that was spaced at 30 × 40 cm. Fertilization was composed of 1650 kg ha^{-1} of simple superphosphate, 250 kg ha^{-1} of potassium chloride, and 295 kg ha^{-1} of urea and applied to the soil. Irrigation was carried out by drippers with a spacing of 30 cm using two drip lines per bed spaced at 50 cm.

The experiment was performed in a randomized block design, with three repetitions and 10 plants per plot for each genotype for a total of 1320 plants. The commercial genotypes 'Albion', 'Monterey', and 'Dover,' and the single hybrids 'RVCA 16' ('Camarosa' × 'Aromas'), 'RVFS07' ('Festival' × Aromas), 'RVFS06' (Festival × Aromas), 'RVDA 11' (Dover × Aromas), and 'RVDA44' (Camarosa × 'Sweet Charlie') were used as controls.

Seven topdressing fertilizations were carried out at 15-day intervals. Each fertilization was composed of 30 kg ha^{-1} of ammonium sulfate, 5.5 kg ha^{-1} of potassium sulfate, and 7.5 kg ha^{-1} of potassium chloride. At the beginning of flowering, boric acid and zinc sulfate were sprayed onto the leaves at 1 L 100 L^{-1} and at 1 kg 100 L^{-1}, respectively. In the fruit production stage, 0.4% of calcium chloride was applied every 15 days. Phytosanitary control was carried out with preventive spraying, according to the specific techniques recommended for the culture. The biweekly sprays were interspersed among abamectin (75 mL ha^{-1}), thiametoxan (10 mL ha^{-1}), and fipronil (250 mL ha^{-1}) products. The control of fungal diseases was carried out with alternating applications of azoxystrobin (16 g ha^{-1}), tebuconazole (75 mL ha^{-1}), and mancozeb (250 g ha^{-1}).

2.4. Statistical Analyses

The total fruit mass (TFM g/plant) was obtained for each harvest from August 2019 to February 2020. Each month was considered a collection period to perform adaptability and stability studies.

Statistical analyses were performed using a linear mixed model methodology, where the genetic parameters were estimated using the restricted maximum likelihood method (REML). The genetic values were predicted using the BLUP method [29].

Data were subjected to individual and joint analyses of variance. To analyze the individual variance, the statistical model was adopted as follows:

$$y = Xb + Zg + e$$

where:

y = data vector;
b = vector of the fixed effects of the blocks added to the general mean;
g = vector of random data effects for genotypes;
e = effect of random vector errors;
and X and Z represent the incidence matrices for vectors b and g, respectively.

Data were standardized using the correction factor obtained for cases where the coefficients of variation of heritability were verified in a broad sense according to the following expression described by Resende (2007) [25]:

$$\sqrt{h^2_{ik}}/\sqrt{h^2_{t},}$$

where:

h^2_{ik} = the broad sense of individual heritability for characteristic i in the evaluation of k;
and h^2_{t} = the broad sense of individual mean heritability to evaluate k for characteristic i.

After standardizing the data, a joint analysis of variance was performed to consider the genotypes and harvesting according to the following statistical model:

$$y = Xb + Za + Wc + e$$

where:

b = vector of the block effects (assumed as fixed) added to the general mean;
a = vector of individual genotypic effects (random);
c = vector of plot effects (randomized);

e = error vector (aleatory);

and X, Z, and W represent the incidence matrices for the said effects (b, a, and c, respectively).

Analysis of deviance (ANADEV) was performed to test the significance of the variance components according to the random effects of the model. The likelihood ratio test (LRT) was used to implement the variance components, in which the significance of the model was evaluated using the chi-square test with one degree of freedom [25].

The classification of genotypes simultaneously considering productivity and stability was performed using the harmonic mean of the genetic values ($HMGV$), which was obtained as follows:

$$HMGV_i = n/\Sigma^n_{j=1} \, 1/GV_{ij}$$

where:

n = the number of months/harvests ($n = 7$) for which genotype i was evaluated;

and GV_{ij} = the genetic value of genotype i in month/harvest j expressed by the ratio of the mean in the month/harvest.

The genotypes, considering productivity and adaptability, were selected simultaneously by the performance of their genotypic values ($RPGV$) during the months/harvest obtained by the formula:

$$RPGV_i = 1/n \left(\frac{\sum^n_{j=1} GV_{ij}}{M_j} \right)$$

where:

M_j = fruit productivity means during the month/harvest j.

Strawberry genotypes were simultaneously classified in terms of productivity, stability, and adaptability through the harmonic mean of the relative performance of genotypic values ($HMRPGV$), obtained according to the following expression:

$$HMRPGV_i = \frac{n}{\sum^n_{j=1} \frac{1}{RPGV_{ij}}}$$

The values of $RPGV\mu$ and $HMRPGV\mu$ were obtained by multiplying $RPGV$ and $HMRPGV$ by the general mean of each characteristic and then considering all months/harvests. Thus, the mean values of the genotype were provided, penalized for instability, and capitalized by adaptability. Selective precision and selection gains were obtained according to Resende [29]. Statistical model 20 was adopted for individual analyses, which refers to the evaluation of unrelated genotypes obtained from randomized blocks containing five plants per plot. In addition, model 55 was used in the conjoint analysis for genotypes in an RBD, with stability and temporal adaptability for one place and seven months/harvests using the Selegen REML/BLUP program [29]. From the $HMRPGV\mu$ values, a box plot was generated using the R software with the ggplot2 package.

3. Results

A desirable strawberry cultivar should have good productivity, post-harvest characteristics, disease and pest tolerance, adaptability, and temporal stability, distributing production uniformly throughout the cultivation period. Adapted and stable genotypes are typically identified during the final selection cycles, and only those that demonstrate superiority are tested.

The climate conditions for the experiment cultivation period are shown in Figure 1. The minimum temperature (daily average) varied from 8.26 °C in July to 17.8 °C in March 2019. The maximum temperatures varied on a daily average from 20.11 °C in July 2019 to 28.4 °C in March 2020. The lowest rainfall value was observed in July (0.63 mm as the mean per day), and the highest was in December (5.04 mm per day). The monthly mean temperatures gradually increased during the cultivation period, ranging from 14.17 °C at the time of transplantation (July 2019) to 21 °C in February 2020 at the end of the harvest period (Figure 1).

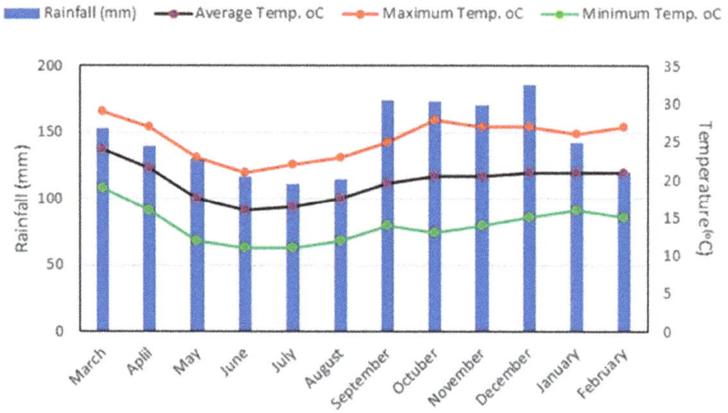

Figure 1. Rainfall and temperature data of the strawberry experimental location from March 2019 to February 2020.

For the analyses of adaptability and stability of the genotypes, the total mass of fruit was used for the analysis of the mixed models.

Significant differences were observed for total fruit mass in the sources of variation, genotypes, genotype interactions by harvest time, and permanent effects using deviance analysis based on the likelihood ratio test (Table 2). The interaction between genotype (G) and environment (E) showed variations in performance among different harvest periods (months).

Table 2. Deviance analysis (ANADEV) for total fruit mass in strawberry genotypes evaluated for seven months.

Variation Source	Deviance	LRT (X^2)
Genotypes	8512.40	63.73 **
Genotypes × Months	9019.45	570.78 **
Permanent effect $G \times E$	8492.63	43.96 **
Complete model	8448.67	-

** significant at 1% of probability by deviance analysis based on the LRT test (X^2) with 1 degree of freedom (t_{table} = 6.63).

From the estimates of the variance components obtained using REML/BLUP, for the total mass of fruit of the genotypes, heritability in the broad sense was 33%. The mean heritability of the genotypes (79%) was superior to broad-sense heritability. The data presented an accuracy of 0.89, which was considered high, and the repeatability was 36% (Table 3).

Among the 44 genotypes evaluated, 17 showed positive genetic effects, and their predicted genotypic values ranged from 128.16 (RVDA11M-25) up to 278.02 (RVFS07M-34) (Table 4). Through the analyses based on mixed models, the genotypic values were considered to evaluate the strawberry genotypes for the general performance in all seven harvests analyzed and for the individual performance for each harvest. The 11 best genotypes selected for each harvest period are presented in Table 5. As for overall performance, the mean genotypic values ranged from 158.18 g/plant (RVDA11M-13) up to 311.86 g/plant (RVFS07M-34). In general, considering every harvest, genotypes RVFS07M-34 and RVFS07M-24 were among those selected with the highest values of total fruit mass. Among the evaluated controls, Monterey, RVCA44, and RVFS07 were selected only in some specific harvests. Monterey was a unique commercial cultivar (control) ranked among the

11 most productive genotypes but only in the last harvest (February), whereas, RVCA44 was selected in harvest 1 (August) and harvest 2 (September), and RVFS07 was selected in harvest 5 (December).

Table 3. Estimation of variance components for total fruit mass in strawberry genotypes evaluated for seven months of cultivation.

Individual REML	Value
Genotypic variance (g^2)	3483.54
Genotype month variance (gm^2)	5334.64
Permanent effects variance (*perm2*)	337.04
Temporary residual variance (*e2*)	1422.48
Phenotypic variance (*f2*)	10,577.70
Heritability in the broad sense in the plot (*hp2*)	0.33
Repeatability (*r*)	0.36
Mean heritability of genotypes (*ahg2*)	0.79
Accuracy (*A*)	0.89
General mean	123.33

Table 4. Predicted genotypic values for total fruit mass obtained from 44 strawberry genotypes.

Genotype	$u + g$ [1]	CI [2]
RVFS07M-34	278.02	[222.45; 333.59]
RVFS07M-24	244.71	[189.14; 300.28]
RVDA11M-04	239.54	[183.97; 295.12]
RVFS07M-36	191.39	[135.83; 246.97]
RVFS07M-33	183.14	[127.57; 238.71]
RVFS07M-05	182.46	[126.89; 238.03]
RVFS07M-154	180.29	[124.72; 235.86]
RVFS07M-31	167.46	[111.89; 223.03]
RVFS07M-80	155.91	[100.34; 211.48]
RVFS06AL-132	153.04	[97.47; 208.61]
RVDA11M-13	151.92	[96.35; 207.50]
RVDA11M-21	151.47	[95.90; 207.04]
RVCA16M-01	151.30	[95.73; 206.87]
RVFS07M-32	148.63	[93.06; 204.20]
RVFS07M-10	147.01	[91.44; 202.58]
RVDA11M-03	139.66	[84.09; 195.23]
RVDA11M-25	128.16	[72.59; 183.73]
RVFS07M-124	122.98	[67.41; 178.55]
RVDA11M-32	119.52	[63.95; 175.09]
RVFS07M-16	116.20	[60.63; 171.77]
RVCA16	115.89	[60.32; 171.46]
RVDA11M-10	115.69	[60.12;171.26]
RVFS07M-179	112.89	[57.32; 168.46]
RVFS07M-113	102.96	[47.38; 158.53]
RVCA44	101.42	[45.85; 156.99]
RVFS07M-38	98.73	[43.16; 154.3]
RVFS07	97.12	[41.55; 152.69]
RVFS06	94.68	[39.12; 150.26]
RVFS07M-47	94.41	[38.84; 149.98]
Monterey	93.23	[37.66; 148.80]
RVDA11M-29	91.97	[36.39; 147.54]
RVFS07M-02	91.76	[36.19; 147.33]
RVFS07M-88	91.36	[35.79; 146.93]
RVDA11M-28	90.90	[35.33; 146.47]
Albion	90.58	[35.01; 146.15]
RVFS06AL-36	89.62	[34.05; 146.19]

Table 4. *Cont.*

Genotype	$u + g$ [1]	CI [2]
RVDA11	87.49	[31.92; 143.06]
RVFS07M-30	85.03	[29.46; 140.60]
RVFS07AL-28	69.84	[14.27; 125.41]
Dover	62.30	[6.73; 117.87]
RVFS07M-151	60.44	[4.87; 116.01]
RVFS06M-29	52.15	[−3.43; 107.72]
RVVFS07M-48	44.64	[−10.93; 100.21]
RVFS07M-42	38.72	[−16.85; 94.29]

[1] Predicted genotypic value; [2] confidence interval.

Table 5. Genotype selection in all harvests and in each harvest based on predicted genotypic values for the total fruit mass obtained from 44 strawberry genotypes.

General		Harvest 1 [1]		Harvest 2		Harvest 3	
Genotype	$u + g + gem$ [2]	Genotype	$u + g + ge$ [3]	Genotype	$u + g + ge$	Genotype	$u + g + ge$
RVFS07M-34	311.86	RVFS07M-47	53.38	RVFS07M-80	119.64	RVFS07M-34	698.39
RVFS07M-24	271.27	RVFS07M-34	38.59	RVFS07M-24	89.84	RVFS07M-05	414.05
RVDA11M-04	264.97	RVDA11M-04	31.38	RVFS07M-36	88.04	RVFS07M-36	379.57
RVFS07M-36	206.28	RVFS07M-24	29.96	RVFS07M-34	85.09	RVFS07M-124	350.38
RVFS07M-33	196.22	RVCA44	29.35	RVFS07M-154	84.26	RVFS07M-179	334.93
RVFS07M-05	195.40	RVFS07M-154	28.26	RVCA16M-01	75.27	RVFS07M-154	302.40
RVFS07M-154	192.75	RVFS07M-80	27.30	RVDA11M-28	65.80	RVDA11M-04	293.37
RVFS07M-31	177.11	RVFS06AL-132	25.62	RVDA11M-03	60.20	RVFS07M-10	263.49
RVFS07M-10	163.04	RVFS07M-36	24.46	RVFS07M-32	58.46	RVDA11N-03	250.05
RVFS06AL-132	159.54	RVFS07M-33	24.00	RVDA11M-21	54.19	RVFS07M-31	250.04
RVDA11M-13	158.18	RVFS07M-113	21.23	RVCA44	50.91	RVFS07M-24	247.47

Harvest 4		Harvest 5		Harvest 6		Harvest 7	
Genotype	$u + g + ge$	Genotype	$u + g + ge$	Genotype	$u + g + ge$	Genotype	$u + g + ge$
RVFS07M-34	413.91	RVFS07M-34	563.86	RVDA11M-04	462.95	RVFS07M-24	400.41
RVFS07M-31	405.00	RVFS07M-05	484.49	RVFS07M-24	340.98	RVFS07M-32	334.98
RVDA11M-04	400.64	RVFS06AL-132	442.21	RVFS07M-33	264.74	RVDA11M-04	218.63
RVFS07M-05	393.19	RVFS07M-24	416.42	RVDA11M-13	237.73	RVFS07M-80	200.98
RVFS07M-24	373.76	RVDA11M-04	403.48	RVDA11M-21	218.94	RVFS07M-33	185.94
RVFS07M-33	356.12	RVFS07M-31	397.62	RVFS07M-36	216.65	RVFS07M-34	172.44
RVFS07M-36	332.83	RVCA16M-01	373.02	RVFS07M-34	210.74	RVFS07M-16	154.77
RVFS07M-80	321.29	RVFS07M-36	349.95	RVCA16M-01	202.61	RVFS07M-154	153.83
RVFS07M-154	312.29	RVFS07M-154	341.10	RVDA11M-25	186.88	Monterey	152.35
RVDA11M-13	294.86	RVFS07M-33	331.10	RVDA11M-10	152.20	RVDA11M-21	128.65
RVDA11M-21	289.38	RVFS07	317.73	RVDA11M-03	151.93	RVDA11M-25	125.97

[1] Harvest 1 = August; harvest 2 = September; harvest 3 = October; harvest 4 = November; harvest 5 = December; harvest 6 = January; harvest 7 = February; [2] mean genotypic value in 7 harvests that capitalizes the mean interaction with all evaluated harvests; [3] predicted genotypic value in each harvest, i.e., the genotypic values capitalizing the interaction with the harvests.

Mean genotypic values penalized by instability and capitalized by adaptability were obtained. Data dispersion showed the 11 genotypes that stood out from the others (Figure 2). They had the highest yields and were the most stable and adaptable: RVFS07M-34, RVFS07M-24, RCDA11M-04, RVFS07M-154, RVFS07M-36, RVFS07M-33, RVFS07M-80, RVFS07M-10, RVDA11M-21, RVDA11M-13, and RVFS06AL-132. This demonstrated the potential of theses genotypes as commercial cultivars.

When considering the means of each harvest separately, the RVFS07M-34 (Figure 3) genotype stood out as always being among the seven best in all harvests. It had the highest predicted genotypic value (278.02 g/plant) and the highest mean genotypic value (311.86 g/plant) (Table 4 and Table 5, respectively). The RVFS07M-24 hybrid also showed high performance in almost all harvests, except in harvest 3 (October) (Table 5). In addition, RVFS07M-24 had the highest production stability in several harvesting periods, such as harvests 4 (November, 373.76 g/plant), 5 (December, 416.42 g/plant), 6 (January, 340.98 g/plant), and 7 (February, 400.41 g/plant).

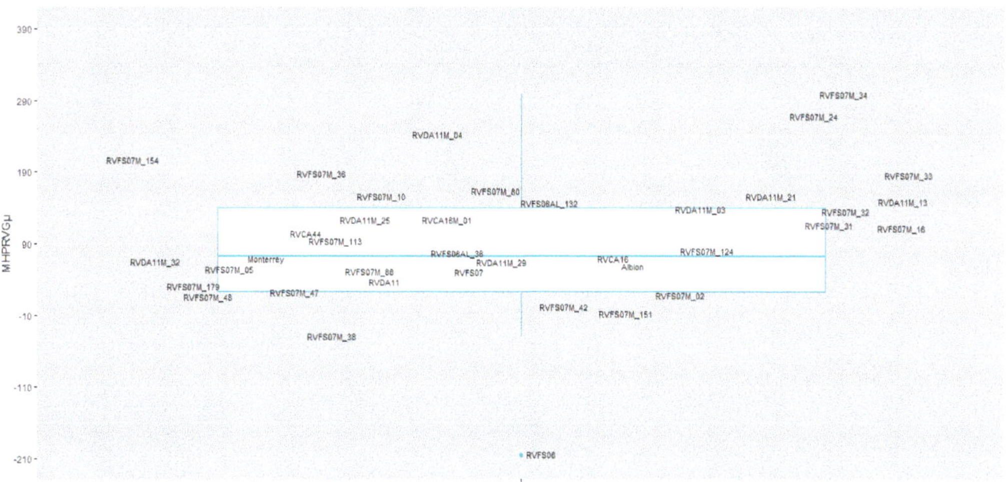

Figure 2. Box plot of adaptability and stability of the 44 evaluated genotypes of strawberry.

Figure 3. Morphoagronomic aspects of genotype RVFS07M-34 developed with high yield, stability, and temporal adaptability. (**A**) Production; (**B**) plant size, architecture, and flowering; (**C**) pseudo-fruit shape; (**D**) pseudo-fruit internal and external colors.

4. Discussion

Currently, strawberry cultivation in the southern hemisphere is based on cultivars from international breeding programs, which leads to market dependence and vulnerability [11,30], since the cultivars are not well adapted to the climate, and they do not express their maximum genetic potential. These cultivars have low productivity in tropical and subtropical conditions, due to their lack of adaptability and stability throughout the year, especially during the warmest periods [1].

To reduce dependence on imported cultivars and rescue the genetic sovereignty of the species, research has focused on the development of new strawberry genotypes classified as day-neutral and adapted to tropical and subtropical growing conditions [2].

The analysis of deviance in relation to total fruit mass in strawberry genotypes was significant for the effects of genotype, permanent effects, and the interaction between genotype and environment, demonstrating the presence of genetic variability among strawberry genotypes tested for the variable total fruit mass in response to harvest periods. Therefore, it was possible to carry out the selection of more stable genotypes for production throughout the year, which was also observed for strawberry genotypes with low chilling requirements [2], guaranteeing market fruit all year round and reducing the off-season period.

A significant interaction demonstrated the occurrence of a difference in the ranking of the genotypes in the different harvest months, indicating a complex interaction between the genotypes and the environment. According to Allard and Bradshaw [31], this type of interaction is caused by unpredictable variations in the environment, such as precipitation, temperature, relative humidity, the occurrence of pests and/or diseases, and even the production system used for cultivation. The significance in the interaction was evident when highlighting the climatic variation between the harvest months evaluated, when they occurred in three seasons of the year (winter, spring, and summer) in the southern hemisphere. The harvest seasons occurred from August (winter), September to December (spring), and February (summer), when temperatures were relatively high, and the photoperiod was long between the equator and the Tropic of Capricorn. Significant interaction results between strawberry genotypes and harvest environments have already been described in several studies [32,33]. The results underscored the importance of adaptability and stability studies in strawberry cultivars as a basic precept for recommending a cultivar developed by a breeding program [11,34,35].

The propagation of genetic material in a vegetative way in strawberry breeding programs makes it possible to take advantage of genetic variance, whether of an additive, dominant, or epistatic nature [36]. However, it is still important to consider the genotype × environment interaction. The interaction generates uncertainties in heritability estimates, correlations (genetic, phenotypic), and expected gains with selection, mainly for quantitative traits, such as total fruit mass. The effect of the complex genotype × environment interaction can be reduced by developing cultivars with greater stability and production adaptability [36], especially when the objective is to explore strawberry cultivation in a wide range of latitudes.

The broad-sense heritability of plot (hp2) of 33% was considered high ($h^2 > 0.3$ = high) [26], demonstrating good genetic control of the quantitative characteristic total fruit mass. In Table 2, the representations of additive, dominant, and epistatic h^2 are observed. The cultivated strawberry species (*Fragaria × ananassa*) was octaploid; most characteristics were under polygenic control and were highly influenced by the cultivation environment [37,38]. In a study evaluating productivity and its main components in strawberry cultivated in a subtropical climate, the authors observed heritability values in the broad sense ranging from 0.01 to 0.63 [11]. Previous studies found broad-sense heritability values above 70% for productivity in strawberry cultivars [14,15]. Therefore, the heritability results found in the present research corroborate several studies already described in literature.

Repeatability tends to be greater than broad-sense heritability, reaching the maximum value that broad-sense heritability can reach [39]. For example, one study observed heritability values of 74.81, 85.17, and 98.44% for the number of fruits per plant, mean fruit mass, and productivity, respectively [15].

Our results showed a high value for accuracy, demonstrating the experimental reliability of this research. Accuracy is the most suitable parameter for determining the proportions between variations of a genetic nature and residual quantitative traits [26].

Based on the harmonic mean of the relative performance of genotypic values (HMRPGV) method, which considers the genotype means and the variation of this mean along the environments, the 11 best genotypes were selected (Figure 2) for the variable total fruit production, as well as stability and adaptability throughout the harvest periods. The selected genotypes were RVFS07M-34, RVFS07M-24, RVDA11M-04, RVFS07M-154, RVFS07M-36, RVFS07M-33, RVFS07M-80, RVFS07M-10, RVDA11M-21, RVDA11M-13, and RVFS06AL-132 (Figure 2). Among them, only RVFS06AL-132 did not have Monterey as a female parent. Furthermore, RCDA11M-04, RVDA11M-13, and RVDA11M-21 featured Dover in the male parent's genealogy, while the others had Festival in their genealogy.

Aromas was in the genealogy of all male parents used in this study. Camarosa and Aromas cultivars have been described as the most adapted and stable commercial strawberry cultivars [36]. These cultivars probably have additive alleles already fixed for stability and adaptability characteristics, which are inherited by the progeny, since the mean h^2 of the genotypes was 79%.

Productivity is a quantitative trait of polygenic inheritance with great influence from the environment [37], which can explain the variation in the ranking of genotypes among harvest periods. The experimental genotypes were superior to the commercial cultivars used in the experiment. Monterey was a unique cultivar ranked among the top 11 genotypes and only in the last harvest. The superiority of the experimental genotypes was due to the strawberry breeding program being developed in a soil and climate condition similar to the cultivation area. These results indicate the importance of plant selection occurring as closely as possible in the cultivation environment. A previous study evaluated advanced genotypes of day-neutral strawberries, where crosses were also performed with genotypes RVFS07, RVFS11, and Monterey. Our results agree with this study, as we obtained similar results [2].

The RVFS07M-34 hybrid was the most productive genotype, with greater stability and adaptability, and was classified among the five best genotypes in harvests 1–5 (August to December). A study evaluating strawberry genotypes in two seasons with three different environments obtained significant results for the components of environmental variations and interactions [35]. According to the same authors, abiotic and biotic factors, such as temperature variation, relative humidity, pests, and diseases, can interfere with the results, and they play an important role in the environmental variation for strawberry production.

Strawberry is a microclimatic crop, and the cultivars' behaviors can vary depending on many agronomical and environmental factors (climate conditions and season) [39], highlighting the importance of our study, which analyzed several harvest periods.

5. Conclusions

The use of the mixed linear model methodology to study adaptability and stability showed the superiority of 11 genotypes that had the potential to be released as cultivars. In addition, RVFS07M-34, being the most promising, will be chosen to follow the legal procedures with the official cultivar registration agency.

Author Contributions: D.A.N. performed the experiments and wrote the manuscript. Field management activities, collection, and data analysis were carried out by D.A.N., G.C.G., L.V.B.d.O. and G.F.d.P.G.; A.R.Z., S.T.I.-S. and K.H.M. helped with the experimental design, statistical analysis, and final review of the manuscript. S.R.R. assistance in the interpretation of post-harvest data and writing of the paper, J.T.V.d.R. was responsible for the Strawberry Breeding Program for tropical and subtropical climates (project coordinator and research creator) and obtained funding, assisted in writing, and reviewed the final manuscript. All authors have read and agreed to the published version of the manuscript.

Funding: We are especially grateful to the National Research Council (CNPq), Fundação Araucária (FA), and the Superintendence of Higher Education, Science and Technology of the State of Paraná (SETI) for their financial support. J.T.V.R. and S.R.R. thank CNPq for the research grant.

Institutional Review Board Statement: Not applicable.

Informed Consent Statement: Not applicable.

Data Availability Statement: Not applicable.

Conflicts of Interest: The authors declare that they have no political, personal, religious, ideological, academic, intellectual, commercial, or any other competing interests.

References

1. Zeist, A.R.; de Resende, J.T.V. Strawberry breeding in Brazil: Current momentum and perspectives. *Hortic. Bras.* **2019**, *37*, 7–16. [CrossRef]
2. Moreira, A.F.P.; de Resende, J.T.V.; Shimizu, G.D.; Hata, F.T.; do Nascimento, D.; Oliveira, L.V.B.; Mariguele, K.H. Characterization of strawberry genotypes with low chilling requirement for cultivation in tropical regions. *Sci. Hortic.* **2022**, *292*, 110629. [CrossRef]
3. Lee, S.G.; Kim, S.K.; Lee, H.J.; Lee, H.S.; Lee, J.H. Impact of moderate and extreme climate change scenarios on growth, morphological features, photosynthesis, and fruit production of hot pepper. *Ecol. Evol.* **2018**, *8*, 197–206. [CrossRef] [PubMed]
4. Srinivasan, V.; Kumar, P.; Long, S.P. Decreasing, not increasing, leaf area will raise crop yields under global atmospheric change. *Glob. Change Biol.* **2017**, *23*, 1626–1635. [CrossRef] [PubMed]
5. Campoy, J.A.; Darbyshire, R.; Dirlewanger, E.; Quero-García, J.; Wenden, B. Yield potential definition of the chilling requirement reveals likely underestimation of the risk of climate change on winter chill accumulation. *Int. J. Biometeorol.* **2019**, *63*, 183–192. [CrossRef]
6. Chavan, S.G.; Duursma, R.A.; Tausz, M.; Ghannoum, O. Elevated CO_2 alleviates the negative impact of heat stress on wheat physiology but not on grain yield. *J. Exp. Bot.* **2019**, *70*, 6447–6459. [CrossRef]
7. Mezzetti, B.; Giampieri, F.; Zhang, Y.T.; Zhong, C.F. Status of strawberry breeding programs and cultivation systems in Europe and the rest of the world. *J. Berry Res.* **2018**, *8*, 205–221. [CrossRef]
8. Barth, E.; Resende, J.T.V.D.; Moreira, A.F.P.; Mariguele, K.H.; Zeist, A.R.; Silva, M.B.; Youssef, K. Selection of experimental hybrids of strawberry using multivariate analysis. *Agronomy* **2020**, *10*, 598. [CrossRef]
9. Barth, E.; Resende, J.T.V.; Zeist, A.R.; Mariguele, K.H.; Zeist, R.A.; Gabriel, A.; Piran, F. Yield and quality of strawberry hybrids under subtropical conditions. *Genet. Mol. Res.* **2019**, *18*, GMR18156. [CrossRef]
10. Corrêa, J.V.W.; Weber, G.G.; Zeist, A.R.; de Resende, J.T.V.; Da-Silva, P.R. ISSR analysis reveals high genetic variation in strawberry three-way hybrids developed for tropical regions. *Plant Mol. Biol. Rep.* **2021**, *39*, 566–576. [CrossRef]
11. Resende, J.T.V.; Gabriel, A.; Moreira, A.F.P.; Gonçalves, L.S.A.; Resende, N.; de Goes, C.D.M.; Zanin, D.S. Application of mixed models in the study of the adaptability and stability of short-day and neutral-day strawberry cultivars. *Res. Soc. Dev.* **2020**, *9*, e110953104. [CrossRef]
12. Ghoochani, R.; Vosough, A.; Karami, F. Heritability, genetic variability and relationship among morphological and chemical parameters of strawberry cultivars. *Biol. Forum Res. Trend* **2015**, *7*, 218.
13. Malosetti, M.; Ribaut, J.M.; van Eeuwijk, F.A. The statistical analysis of multi-environment data: Modeling genotype-by-environment interaction and its genetic basis. *Front. Physiol.* **2013**, *4*, 44. [CrossRef]
14. Singh, G.; Kachwaya, D.S.; Kumar, R.; Vikas, G.; Singh, L. Genetic variability and association analysis in strawberry (*Fragaria* × *ananassa* Duch). *Electron. J. Plant Breed.* **2018**, *9*, 169–182. [CrossRef]
15. Ramalho, M.A.P.; Dias, L.A.D.S.; Carvalho, B.L. Contributions of plant breeding in Brazil: Progress and perspectives. *Crop Breed. Appl. Biotechnol.* **2012**, *12*, 111–120. [CrossRef]
16. Cruz, C.D.; Souza Carneiro, P.C. *Modelos Biométricos Aplicados ao Melhoramento Genético (No. 575.1015195)*; Universidade Federal de Viçosa: Viçosa, Brazil, 2006.
17. Henderson, C.R. Best linear unbiased estimation and prediction under a selection model. *Biometrics* **1975**, *31*, 423–447. [CrossRef]
18. Lei, J.J.; Xue, L.; Guo, R.X.; Dai, H.P. The Fragaria species native to China and their geographical distribution. *Acta Hortic.* **2017**, *1156*, 37–46. [CrossRef]
19. Dimeglio, L.M.; Staudt, G.; Yu, H.; Davis, T.M. A phylogenetic analysis of the genus Fragaria (strawberry) using intron-containing sequence from the ADH-1 gene. *PLoS ONE* **2014**, *9*, e102237. [CrossRef]
20. Mahoney, L.L.; Sargent, D.J.; Abebe-akele, F.; Wood, D.J.; Ward, J.A.; Bassil, N.V.; Hancock, J.F.; Folta, K.M.; Davis, T.M. A high-density linkage map of the ancestral diploid strawberry, *Fragaria iinumae*, constructed with single nucleotide polymorphism markers from the IStraw90 array and genotyping by sequencing. *Plant Genome* **2016**, *9*, 1–14. [CrossRef]
21. Edger, P.P.; Poorten, J.T.; VanBuren, R.; Hardigan, M.A.; Cole, M.; McKain, M.R.; Smith, R.D.; Teresi, S.J.; Nelson, A.D.L.; Wai, C.M.; et al. Origin and evolution of the octoploid strawberry genome. *Nat. Genet.* **2019**, *51*, 541–547. [CrossRef]
22. Du, I.; Wang, J.; Wang, T.; Liu, L.; Iqbal, S.; Qiao, Y. Identification and chromosome doubling of *Fragaria mandshurica* and *F. nilgerrensis*. *Sci. Hortic.* **2021**, *289*, e110507. [CrossRef]
23. Otto, S.P. The Evolutionary Consequences of Polyploidy. *Cell* **2007**, *131*, 452–462. [CrossRef]
24. Carvalho, L.P.D.; Farias, F.J.C.; Morello, C.D.L.; Teodoro, P.E. Use of REML/BLUP methodology for selecting cotton genotypes with higher adaptability and productive stability. *Bragantia* **2016**, *75*, 314–321. [CrossRef]
25. Resende, M.D.V.d.; Duarte, J.B. Precisão e controle de qualidade em experimentos de avaliação de cultivares. *Pesq. Agr. Trop.* **2007**, *37*, 182–194.
26. Squilassi, M.G. Interação de Genótipos com Ambientes. Embrapa Tabuleiros Costeiros-Livro Técnico (INFOTECA-E), 2003. Available online: https://www.embrapa.br/busca-de-publicacoes/-/publicacao/897925/interacao-de-genotipos-com-ambientes (accessed on 18 April 2023).
27. Santos, H.G.d.; Jacomine, P.K.T.; Dos Anjos, L.H.C.; De Oliveira, V.A.; Lumbreras, J.F.; Coelho, M.R.; Cunha, T.J.F. *Sistema Brasileiro de Classificação de Solos*; Embrapa: Brasília, Brazil, 2018.
28. IDR. Instituto de Desenvolvimento Rural do Paraná, IAPAR-EMATER. Agrometeorologia. Available online: http://www.iapar.br/modules/conteudo/conteudo.php?Conte%C3%BAdo=597/ (accessed on 14 December 2019).

29. Resende, M.D.V.d. Software Selegen-REML/BLUP: A useful tool for plant breeding. *Crop Breed. Appl. Biotechnol.* **2016**, *16*, 330–339. [CrossRef]
30. Resende, J.T.V.d.; Matos, R.; Zeffa, D.M.; Constantino, L.V.; Alves, S.M.; Ventura, M.U.; Youssef, K. Relationship between salicylic acid and resistance to mite in strawberry. *Folia Hortic.* **2021**, *33*, 107–119. [CrossRef]
31. Allard, R.W.; Bradshaw, A.D. Implications of genotype-environmental interactions in applied plant breeding 1. *Crop Sci.* **1964**, *4*, 503–508. [CrossRef]
32. Pereira, W.R.; de Souza, R.J.; Yuri, J.E.; Ferreira, S. Yield of strawberry cultivars submitted to different planting dates. *Hortic. Bras.* **2013**, *31*, 500–503. [CrossRef]
33. Rosa, H.T.; Streck, N.A.; Walter, L.C.; Andriolo, J.L.; Silva, M.R.D. Crescimento vegetativo e produtivo de duas cultivares de morango sob épocas de plantio em ambiente subtropical. *Rev. Cienc. Agron.* **2013**, *44*, 604–613. [CrossRef]
34. Costa, A.F.; Teodoro, P.E.; Bhering, L.L.; Leal, N.R.; Tardin, F.D.; Daher, R.F. Biplot analysis of strawberry genotypes recommended for the State of Espírito Santo. *Genet. Mol. Res.* **2016**, *15*, 1–9. [CrossRef]
35. Gabriel, A.; Resende, J.T.V.; Zeist, A.R.; Resende, L.V.; Resende, N.C.V.; Galvão, A.G.; Camargo, C.K. Phenotypic stability of strawberry cultivars assessed in three environments. *Genet. Mol. Res.* **2018**, *17*, e18041. [CrossRef]
36. Gabriel, A.; de Resende, J.T.; Zeist, A.R.; Resende, L.V.; Resende, N.C.; Zeist, R.A. Phenotypic stability of strawberry cultivars based on physicochemical traits of fruits. *Hortic. Bras.* **2019**, *37*, 75–81. [CrossRef]
37. Gawroński, J. Evaluation of the genetic control, heritability and correlations of some quantitative characters in strawberry (*Fragaria* × *ananassa* Duch). *Acta Sci. Pol. Hortorum Cultus* **2011**, *10*, 71–76.
38. Costa, A.F.; Leal, N.R.; Ventura, J.A.; Gonçalves, L.S.A.; Amaral Júnior, A.T.D.; Costa, H. Adaptability and stability of strawberry cultivars using a mixed model. *Acta Sci. Agron.* **2015**, *37*, 435–440. [CrossRef]
39. Ariza, M.T.; Miranda, L.; Martinez-Ferri, E.; Medina, J.J.; Gómez-Mora, J.A.; Cervantes, L.; Soria, C. Consistency of organoleptic and yield related traits of strawberry cultivars over time. *J. Berry Res.* **2020**, *10*, 623–636. [CrossRef]

 *horticulturae*

Article

Effects of Different Growth Hormones on Rooting and Endogenous Hormone Content of Two *Morus alba* L. Cuttings

Hanlei Chen, Youzhen Lei, Jiajia Sun, Mingyue Ma, Peng Deng, Jin'e Quan * and Huitao Bi *

College of Forestry, Henan Agricultural University, Zhengzhou 450046, China; 18336253160@163.com (H.C.); leiyouzhen1021@163.com (Y.L.); dearperi@163.com (J.S.); 15178061569@163.com (M.M.); dengpeng@henau.edu.cn (P.D.)
* Correspondence: quanjine@163.com (J.Q.); bihuitao@126.com (H.B.)

Abstract: This study aimed to explore the effects of different concentrations of indole-3-acetic acid (IAA), indole-3-butyric acid (IBA), and indene-naphthaleneacetic acid (ABT-1) on the rooting and dynamic changes of the endogenous hormone content of Australian Mulberry (vegetable Mulberry) and Kirin mulberry (Fruit Mulberry) hardwood cuttings. As exhibited by the results, the rooting process of both vegetable mulberry and fruit mulberry could be divided into three stages, namely the initiation stage (1–18 days), the callus formation stage (18–28 days), and the adventitious root formation and elongation stage (28–48 days). The two treatments with 1000 mg·L^{-1} ABT-1 and 500 mg·L^{-1} ABT-1 achieved the highest rooting efficiencies of vegetable mulberry and fruit mulberry, significantly higher than those of other treatments ($p < 0.01$), with average rooting rates of 63.3% and 68.7%, and rooting efficiency indices of 25.3 and 34.3, respectively. During the rooting process, the contents of endogenous IAA and zeatin riboside (ZR) and the ratios of IAA/ABA and IAA/ZR presented a trend of decreasing before increasing, while the abscisic acid (ABA) and jasmonic acid (JA) contents exhibited a trend of increasing before decreasing, and the gibberellin (GA$_3$), strigolactone (SL), and IBA contents showed a continuous decreasing trend. Hence, ABT-1 was effective in inducing the synthesis of IAA, IBA, JA, and SL, reducing the contents of ABA, ZR, and GA$_3$, and promoting the rooting of vegetable mulberry and fruit mulberry cuttings. For fruit mulberry and vegetable mulberry cuttings, the optimal concentrations of ABT-1 were 500 mg·L^{-1} and 1000 mg·L^{-1}, respectively, demonstrating applicability for the efficient propagation of *Morus alba* L. cuttings.

Keywords: growth hormones; *Morus alba* L.; hardwood cuttings; endogenous hormones

Citation: Chen, H.; Lei, Y.; Sun, J.; Ma, M.; Deng, P.; Quan, J.; Bi, H. Effects of Different Growth Hormones on Rooting and Endogenous Hormone Content of Two *Morus alba* L. Cuttings. *Horticulturae* **2023**, *9*, 552. https://doi.org/10.3390/horticulturae9050552

Academic Editor: Sergio Ruffo Roberto

Received: 13 April 2023
Revised: 28 April 2023
Accepted: 29 April 2023
Published: 4 May 2023

1. Introduction

As a deciduous tree of the genus Morus in the family Moraceae, *Morus alba* L. [1] derives from northern and central China. Characterized by drought resistance, barren land tolerance,, waterlogging resistance, cold tolerance, and strong adaptability to soil, *Morus alba* L. is an excellent timber, as well as environmental protection and economic tree species [2]. In addition, as a valuable plant, *Morus alba* L. can be used both medicinally and as a food source, with its fruit, leaves, branches, and roots being highly prized. Furthermore, known as the "Chinese holy tree", *Morus alba* L. has been listed by the Ministry of Health as a dual-purpose plant for food and medicine on account of its high development and utilization value [2]. Following the promotion of scientific and technological innovation in the sericulture industry, this industry is gradually developing towards a diversified and multi-purpose direction, including fruit mulberry, vegetable mulberry, medicinal mulberry, and ecological feed mulberry [3]. In view of the large demand for *Morus alba* L. seedlings, it is essential to develop asexual propagation techniques, especially cutting propagation techniques, thereby ensuring high-quality seedlings [4].

Cutting propagation has many advantages, such as easy access to materials, simple operation, short seedling cycle, and low cost [5]. As discovered by most studies, plant growth hormones can greatly promote the rooting rate of cutting propagation [6–10]. Indole-3-acetic acid (IAA) [9], indole-3-butyric acid (IBA) [11], and indene-naphthaleneacetic acid (ABT-1) [12] (Table 1) are widely employed in the rooting of plant cuttings. Hormone treatment has been found to accelerate the rooting of cuttings of various plants such as *camptotheca acuminata* [13], *Cyclocarya paliurus* [14], *Picea abies* (L.) Karst [15] and *Malus hupehensis* [9]. As exhibited by numerous studies, the dynamic balance of endogenous hormones in cuttings is crucial for the formation of adventitious roots [16], and exogenous plant growth hormones can regulate the balance of endogenous hormones in cuttings [17]. For instance, during critical periods of adventitious root formation in *Sapindus mukorossi* [18], *Picea abies* [19] and *Tamarix taklamakanensis* [20], various endogenous hormones show dynamic change, thus enhancing rooting of cuttings. Nonetheless, there are few studies on the dynamic change of endogenous hormone content in *Morus alba* L. cuttings resulted by growth hormones, and the mechanism of the correlation between *Morus alba* L. cuttings and endogenous hormones is still unclear. Hence, to clarify the inherent mechanism of *Morus alba* L. cuttings rooting, we conduct studies on *Morus alba* L. cuttings and explore the relationship between growth hormones and endogenous hormones in rooting of cuttings.

Table 1. Introduction to plant growth hormones.

Type of Growth Hormones	Name	Molecular Formula	Molecular Structure Formula
IAA	Indole-3-acetic acid	$C_{10}H_9NO_2$	(Solaibao Technology Co., Ltd., Beijing, China)
IBA	Indole-3-butyric acid	$C_{12}H_{13}NO_2$	(Solaibao Technology Co., Ltd., Beijing, China)
ABT-1	Indene-naphthaleneacetic acid	N.A.	N.A.

Note: Indene-naphthaleneacetic acid (ABT-1) is made from a mixture of plant hormones, so the formula is not specified. Indene-naphthaleneacetic acid (ABT-1) contains 50% active ingredients, including α-Naphthalene acetic acid (NAA) 20% and Indole-3-acetic acid (IAA) 30%.

2. Materials and Methods

In this study, a new variety of Australian Mulberry (vegetable Mulberry) introduced from McLaren Vale in 2016 and a new variety of Kirin mulberry (Fruit Mulberry) introduced from China Agricultural University in 2018 were selected as study objects, and rooting responses of vegetable mulberry and fruit mulberry hardwood cuttings to different concentrations of growth hormones and endogenous hormones were explored, thereby screening for high-efficiency cutting schemes for large-scale and rapid propagation of vegetable mulberry and fruit mulberry.

2.1. Study Site and Materials

The study was conducted from 24 March to 9 May 2021 at the *Morus alba* L. cultivation base in Mawuzhai village, Laiji town, Xinmi city, Henan province (113°29′ E, 34°28′ N). Mawuzhai village is located in an area with a sub-tropical monsoon climate characterized by distinct seasons, abundant rainfall, humid air, and yellow–brown soil. The 3-year-old healthy and vigorous vegetable and fruit mulberry trees were chosen as the mother plants for cuttings. In March of the same year, disease-free, strong, and full-budded 1-year-old branches were cut from the mother plants. One week before the cuttings, an 800-fold dilution of a 50% wettable powder of carbendazim (Lanfeng Bio-Chemical Co., Ltd., Xuzhou, China) was uniformly sprayed on the substrate, followed by drying in the sun.

2.2. Study Methods

2.2.1. Study of Cuttings

In this study, a randomized complete block design was adopted, and a total of 26 treatment groups were included, with three replicates per group and 50 cuttings per replicate. A total of 3900 cuttings were required. The cuttings were uniform in thickness and length, with a diameter between 0.6–0.8 cm and a length between 12–15 cm. The upper end of the cutting was trimmed flat, and the lower end was cut at a 45° angle. In total, 50 cuttings were selected for each group, followed by soaking in an 800-fold dilution of a 50% wettable powder of carbendazim for 1 min at the base and three growth regulators, namely ABT-1 (Abiti Biotechnology Co., Ltd., Beijing, China), IAA, and IBA (Solaibao Technology Co., Ltd., Beijing, China) with 99% purity. Furthermore, 100% alcohol was used for dissolving IAA, IBA, and ABT-1 at 0.2 g, 0.5 g, 1 g and 1.5 g, respectively, and then 1 L of water was employed to dilute them to concentrations of 200, 500, 1000, and 1500 mg·L^{-1} [21], respectively, followed by soaking the base of cuttings for 4 h. The specific study design is exhibited in Tables 2 and 3. The cuttings were inserted into the substrate to a depth of approximately 7–8 cm, with a spacing of 10 cm between every two cuttings, and the cuttings were maintained slanted downwards with the buds facing upwards. Successively, the soil was compacted and watered, thereby guaranteeing close contact between the cutting base and the substrate and ensuring adequate water supply. Every 9 days after the cuttings, carbendazim was sprayed with an 800-fold dilution for disinfection, and field water management was needed, aiming at keeping the soil moisture at around 50% and removing weeds manually at proper time.

Table 2. Treatment of three plant growth hormones on vegetable mulberry cuttings.

Treatment	Type of Growth Hormones	Quality Concentration of Growth Hormones/(mg·L^{-1})
CK1	N.A.	N.A.
A1	ABT-1	200
A2		500
A3		1000
A4		1500
A5	IAA	200
A6		500
A7		1000
A8		1500
A9	IBA	200
A10		500
A11		1000
A12		1500

Note: CK1: the vegetable mulberry cuttings control group. A1: the treatment group with 200 mg·L^{-1} ABT-1, A2: the treatment group with 500 mg·L^{-1} ABT-1, A3: the treatment group with 1000 mg·L^{-1} ABT-1, A4: the treatment group with 1500 mg·L^{-1} ABT-1, A5: the treatment group with 200 mg·L^{-1} IAA, A6: the treatment group with 500 mg·L^{-1} IAA, A7: the treatment group with 1000 mg·L^{-1} IAA, A8: the treatment group with 1500 mg·L^{-1} IAA, A9: the treatment group with 200 mg·L^{-1} IBA, A10: the treatment group with 500 mg·L^{-1} IBA, A11: the treatment group with 1000 mg·L^{-1} IBA, A12: the treatment group with 1500 mg·L^{-1} IBA.

2.2.2. Observation and Statistical Analysis of Rooting Morphology

The first sampling was performed on the day after the cuttings, and subsequent samplings were conducted every 9 days. After being randomly sampled from each treatment, 5 cuttings were washed with water, and the bark was peeled off within 2 cm of the base. Afterwards, the samples were stored in a −80 °C ultra-low temperature freezer for determining the endogenous hormone content. At the sixth sampling, all cuttings were pulled out, thereby exploring number of rooted cuttings, number of roots per cutting, and root length.

Rooting rate = number of rooted cuttings/total number of cuttings × 100%

Average number of roots = sum of the number of roots/total number of cuttings

Average root length = sum of root length/total number of cuttings

Root system effectiveness index = (rooting rate × average number of roots) × average root length [22].

Table 3. Treatment of three plant growth hormones on fruit mulberry cuttings.

Treatment	Type of Growth Hormones	Quality Concentration of Growth Hormones/(mg·L^{-1})
CK2	N.A.	N.A.
B1	ABT-1	200
B2		500
B3		1000
B4		1500
B5	IAA	200
B6		500
B7		1000
B8		1500
B9	IBA	200
B10		500
B11		1000
B12		1500

Note: CK2: the fruit mulberry cuttings control group. B1: the treatment group with 200 mg·L^{-1} ABT-1, B2: the treatment group with 500 mg·L^{-1} ABT-1, B3: the treatment group with 1000 mg·L^{-1} ABT-1, B4: the treatment group with 1500 mg·L^{-1} ABT-1, B5: the treatment group with 200 mg·L^{-1} IAA, B6: the treatment group with 500 mg·L^{-1} IAA, B7: the treatment group with 1000 mg·L^{-1} IAA, B8: the treatment group with 1500 mg·L^{-1} IAA, B9: the treatment group with 200 mg·L^{-1} IBA, B10: the treatment group with 500 mg·L^{-1} IBA, B11: the treatment group with 1000 mg·L^{-1} IBA, B12: the treatment group with 1500 mg·L^{-1} IBA.

2.2.3. Determination of Endogenous Hormone Content

High-performance liquid chromatography—mass spectrometry (HPLC—MS) (Aibo CAISI Analytical Instrument Trading Co., Ltd., Shanghai, China) was employed for the determination [23] (The column, Waters ACQUITY HPLC HSS T3 C18 (100 mm × 2.1 mm i.d., 1.8 μm). Methanol and 0.1 mol/L^{-1} acetic acid were deemed as the mobile phase for gradient elution, and the contents of IAA, abscisic acid (ABA), zeatin riboside (ZR), gibberellin (GA3), jasmonic acid (JA), strigolactone (SL), and IBA were determined at 254 nm for the vegetable mulberry cuttings control group (CK1), the vegetable mulberry cuttings treatment groups with 200 mg·L^{-1} ABT-1 (A1), 500 mg·L^{-1} ABT-1 (A2) and 1000 mg·L^{-1} ABT-1 (A3), and the fruit mulberry cuttings control group (CK2), the fruit mulberry cuttings treatment groups with 200 mg·L^{-1} ABT-1 (B1), 500 mg·L^{-1} ABT-1 (B2) and 1000 mg·L^{-1} ABT-1 (B3) on days 1, 28, and 48, respectively. The study was repeated three times.

2.3. Data Processing

Statistical analysis and organization were applied to the study data using Microsoft Excel 2016, and SPSS 25.0 software was also employed for statistical analysis. Origin 2021 was adopted for drawing. The least significant difference (LSD) method was used for multiple comparisons, and two-way analysis of variance was performed for different concentrations and types of plant growth hormones.

3. Results

3.1. Observation of Rooting Progress of Vegetable Mulberry and Fruit Mulberry

As revealed by observation, for the treatment groups of both vegetable mulberry (Figure 1) and fruit mulberry (Figure 2), the incision at the base began to swell and crack to produce callus tissue at 18 days after the cuttings, and adventitious roots started to form and emerge from the bark between 18–28 days after the cuttings. From 28 to 48 days after the cuttings, numerous adventitious roots were generated and continued to elongate. Compared with the treatment group, the control group of vegetable mulberry and fruit mulberry cuttings exhibited later rooting progress. The rooting process of vegetable mulberry and fruit mulberry cuttings could be divided into three stages, namely the initiation stage (1–18 days), the callus tissue generation stage (18–28 days), and the adventitious root generation and elongation stage (28–48 days).

Figure 1. Rooting process of vegetable mulberry cuttings. (**A**): 1-day morphology of vegetable mulberry cuttings; (**B**): 18-day morphology of vegetable mulberry cuttings; (**C**): 18–28-day morphology of vegetable mulberry cuttings; (**D**): 28–48-day morphology of vegetable mulberry cuttings.

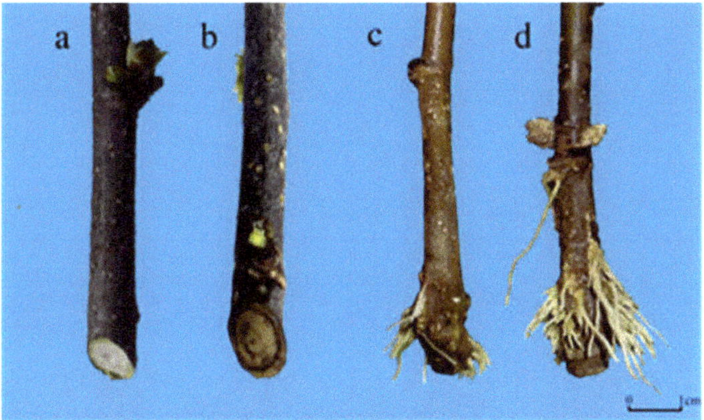

Figure 2. Rooting process of fruit mulberry. (**a**): 1-day morphology of fruit mulberry cuttings; (**b**): 18-day morphology of fruit mulberry cuttings; (**c**): 18–28-day morphology of fruit mulberry cuttings; (**d**): 28–48-day morphology of fruit mulberry cutting.

3.2. Effects of Different Treatments on Rooting of Vegetable Mulberry Hardwood Cuttings

The effects of different treatments on the rooting of vegetable mulberry hardwood cuttings are exhibited in Table 4. After treatment with plant growth hormones, the rooting rate, average number of roots, average root length, average longest root length, and root system index of vegetable mulberry were all promoted to varying degrees, increasing before decreasing with the increase in concentration of plant growth hormones. Regarding the rooting rate, the highest rooting rate was observed in ABT-1—1000 mg·L^{-1}, and the average rooting rate was 63.3%, 4.1 times higher than that of the control group (15.3%), and notably higher than that of other treatments ($p < 0.01$). For average number of roots, the average highest number of roots was noticed in ABT-1—1000 mg·L^{-1}, with 8.89 roots per cutting, followed by ABT-1—500 mg·L^{-1} and IBA—1000 mg·L^{-1} (with no remarkable difference between the two). For average root length, the optimal treatment was exhibited in IBA—1000 mg·L^{-1}, with an average root length of 4.8 cm, 3.8 cm longer than that of the control group. For average longest root length, IBA—1000 mg·L^{-1} had the highest value, with an average longest root length of 8.4 cm, 82.9% greater than that of the control group. There was no notable difference between ABT-1—1000 mg·L^{-1} and IBA—1000 mg·L^{-1}. Additionally, considering effectiveness evaluation of cuttings, the root system index is also a key indicator. ABT-1—1000 mg·L^{-1} had the greatest root system index, which was 25.3, 120.5 times greater than that of the control group, and remarkably higher than that of other treatment groups and the control group. To sum up, the best results were achieved by applying 1000 mg/L^{-1} ABT-1 in treating vegetable mulberry hardwood cuttings, thereby promoting the generation of numerous strong roots for vegetable mulberry.

Table 4. Effects of different treatments on rooting indices of vegetable mulberry hardwood cuttings.

Treatment	Rooting Rate(%)	Average Number of Roots (n)	Average Root Length (cm)	Average Longest Root Length (cm)	Root System Index
A1	43.3 ± 1.8 ^{Bc}	6.0 ± 0.2 $_{Cc}$	2.6 ± 0.2 ^{Bc}	5.1 ± 0.2 ^{Bc}	6.7 ± 0.8 ^{Bc}
A2	50.7 ± 3.5 ^{Bb}	7.2 ± 0.5 ^{Bb}	3.4 ± 0.5 ^{Bb}	6.8 ± 0.8 ^{Ab}	12.9 ± 3.4 $_{Bb}$
A3	63.3 ± 2.4 ^{Aa}	8.9 ± 0.4 ^{Aa}	4.4 ± 0.5 ^{Aa}	8.2 ± 0.5 ^{Aa}	25.3 ± 5.1 ^{Aa}
A4	21.3 ± 1.8 ^{Dd}	2.6 ± 0.5 ^{Dd}	1.3 ± 0.4 ^{Cd}	2.4 ± 0.5 ^{Cd}	0.8 ± 0.3 ^{Cd}
A5	18.7 ± 0.7 ^{Dd}	1.8 ± 0.2 ^{Dd}	1.0 ± 0.1 ^{Cd}	1.7 ± 0.1 ^{Ce}	0.3 ± 0.0 ^{Cd}
A6	43.3 ± 1.8 ^{Bc}	5.9 ± 0.2 ^{Cc}	3.0 ± 0.1 ^{Bb}	6.2 ± 0.3 ^{Bb}	7.8 ± 0.8 ^{Bc}
A7	16.7 ± 1.8 $_{Dd}$	1.4 ± 0.3 ^{De}	0.7 ± 0.1 ^{Cd}	1.3 ± 0.3 ^{De}	0.2 ± 0.1 ^{Cd}
A8	11.3 ± 2.4 ^{Ee}	0.9 ± 0.2 ^{Ee}	0.6 ± 0.1 ^{Dd}	1.2 ± 0.2 ^{De}	0.1 ± 0.0 $_{Cd}$
A9	39.3 ± 1.8 ^{Cc}	5.4 ± 0.3 ^{Cc}	3.5 ± 0.4 ^{Bb}	6.2 ± 0.5 ^{Bb}	7.6 ± 1.5 ^{Bc}
A10	44.7 ± 2.4 ^{Bc}	6.2 ± 0.4 ^{Bb}	4.3 ± 0.4 ^{Aa}	6.7 ± 0.6 ^{Ab}	12.8 ± 1.7 ^{Bb}
A11	51.3 ± 3.5 $_{Bb}$	7.2 ± 0.5 ^{Bb}	4.8 ± 0.2 ^{Aa}	8.4 ± 0.3 ^{Aa}	17.9 ± 3.4 ^{Ab}
A12	21.3 ± 1.8 ^{Dd}	2.5 ± 0.5 ^{Dd}	1.9 ± 0.5 ^{Cc}	3.2 ± 0.7 ^{Cd}	1.2 ± 0.5 ^{Cd}
CK1	15.3 ± 1.8 $_{Dd}$	1.3 ± 0.2 ^{De}	0.9 ± 0.1 ^{Cd}	1.4 ± 0.2 ^{Ce}	0.2 ± 0.1 ^{Cd}

Note: Values in the table are expressed as mean ± standard error. Different letters within the same column indicate notable differences, while same letters suggest no notable differences. In addition, capital letters denote remarkable differences between treatments at $p < 0.01$ level, while lowercase letters signify obvious differences at $p < 0.05$ level. The same applies to the following tables.

As indicated by the results of the analysis of variance for the five rooting indices (Table 5), both the type and quality concentration of plant growth hormones had extremely obvious effects on all rooting indices ($p < 0.01$), and there was a notable interaction between the type and quality concentration of plant growth hormones. Regarding the magnitude of the F value, the type of plant growth hormones had the greatest effect on vegetable mulberry hardwood cuttings, followed by the quality concentration.

Table 5. Analysis of variance for the results of vegetable mulberry cuttings.

Rooting Index	Source of Variation	Sum of Squares	Freedom	Mean Square	F Value	*p* Value
Rooting rate	Type of growth hormones (A)	3197.6	2	1598.8	107.5	0.0 **
	Quality concentration (B)	4434.2	3	1478.1	99.4	0.0 **
	A × B	1669.1	6	278.2	18.7	0.0 **
	Error	386.7	26	14.9		
	Total	55,620.0	39			
Average number of roots	Type of growth hormones (A)	87.5	2	43.7	107.4	0.0 **
	Quality concentration (B)	104.7	3	34.9	85.7	0.0 **
	A × B	43.7	6	7.3	17.9	0.0 **
	Error	10.6	26	0.4		
	Total	1037.7	39			
Average root length	Type of growth hormones (A)	31.8	2	15.9	57.0	0.0 **
	Quality concentration (B)	28.6	3	9.5	34.2	0.0 **
	A × B	12.0	6	2.0	7.1	0.0 **
	Error	7.3	26	0.3		
	Total	329.1	39			
Average longest root length	Type of growth hormones (A)	86.5	2	43.3	75.0	0.0 **
	Quality concentration (B)	99.9	3	33.3	57.7	0.0 **
	A × B	49.8	6	8.3	14.4	0.0 **
	Error	15.0	26	0.6		
	Total	1078.2	39			
Root system index	Type of growth hormones (A)	600.9	2	300.4	23.7	0.0 **
	Quality concentration (B)	1033.8	3	344.6	27.2	0.0 **
	A × B	545.8	6	91.0	7.2	0.0 **
	Error	329.9	26	12.7		
	Total	4704.3	39			

Note: ** indicates notable difference at the 0.01 level of mean difference. The same applies to the following tables.

3.3. Effects of Different Treatments on Rooting of Fruit Mulberry Hardwood Cuttings

The effects of different treatments on the rooting of fruit mulberry hardwood cuttings are exhibited in Table 6. The results indicated great differences in rooting among different treatments. The highest rooting rate was observed in ABT-1—500 mg·L^{-1}, reaching 68.7% that was 5.2 times higher than the control group, and obviously different from other treatments ($p < 0.05$). Except for IBA—1500 mg·L^{-1}, all treatments had higher values for average number of roots, average root length, and average longest root length compared with the control group. Among them, ABT-1—500 mg·L^{-1} had the average highest number of roots and average longest root length, namely 10.1 and 9.3 cm, respectively; ABT-1—200 mg·L^{-1} had the average longest root length, namely 5.2 cm. According to multiple comparisons, there was a notable difference in average number of roots between ABT-1—200 mg·L^{-1} and ABT-1—500 mg·L^{-1}, but no obvious difference in average root length and average longest root length between ABT-1—200 mg·L^{-1} and ABT-1—500 mg·L^{-1}. Regarding the root system effect index, ABT-1—500 mg·L^{-1} had the highest value of

34.3, remarkably higher than other treatment groups. Nevertheless, the rooting rates of 200 mg·L^{-1} IAA (9.3%) and 1500 mg·L^{-1} IBA (0) treatment groups were lower than that of the control group (13.3%) suggesting that high concentration of plant growth hormones does not play a role in promoting the growth of fruit mulberry even while inhibiting its growth. As illustrated by data analysis, the optimal concentration of ABT-1 for fruit mulberry cuttings was 500 mg·L^{-1}.

Table 6. Effects of different treatments on rooting index of fruit mulberry hardwood cuttings.

Treatment	Rooting Rate(%)	Average Number of Roots (n)	Average Root Length (cm)	Average Longest Root Length (cm)	Root System Index
B1	60.7 ± 2.9 [Ab]	8.5 ± 0.3 [Bb]	5.2 ± 0.6 [Aa]	9.1 ± 0.5 [Aa]	27.1 ± 5.0 [Ab]
B2	68.7 ± 2.4 [Aa]	10.1 ± 0.7 [Aa]	4.9 ± 0.1 [Aa]	9.3 ± 0.3 [Aa]	34.3 ± 4.0 [Aa]
B3	12.7 ± 1.8 [Ef]	1.2 ± 0.1 [Ff]	0.7 ± 0.1 [Ef]	1.3 ± 0.1 [Ee]	0.1 ± 0.0 [Ce]
B4	9.3 ± 1.8 [Ef]	0.8 ± 0.2 [Ff]	0.7 ± 0.1 [Ef]	1.0 ± 0.2 [Ee]	0.1 ± 0.0 [Ce]
B5	47.3 ± 1.3 [Bc]	6.7 ± 0.3 [Cc]	4.0 ± 0.1 [Bb]	7.0 ± 0.5 [Bb]	12.6 ± 1.3 [Bc]
B6	41.3 ± 1.3 [Bd]	5.8 ± 0.2 [Cc]	3.0 ± 0.2 [Cc]	5.6 ± 0.2 [Bc]	7.2 ± 1.0 [Bc]
B7	37.3 ± 2.9 [Cd]	5.2 ± 0.4 [Dd]	2.4 ± 0.3 [Cd]	4.7 ± 0.7 [Cc]	4.9 ± 1.4 [Cd]
B8	22.7 ± 1.8 [De]	3.3 ± 0.3 [Ee]	1.7 ± 0.1 [De]	3.0 ± 0.1 [Dd]	1.2 ± 0.1 [Cd]
B9	36.7 ± 1.8 [Cd]	5.0 ± 0.5 [Dd]	3.1 ± 0.1 [Bc]	5.3 ± 0.2 [Cc]	5.7 ± 1.0 [Bd]
B10	20.7 ± 2.9 [De]	2.6 ± 0.7 [Ee]	1.6 ± 0.3 [De]	2.8 ± 0.5 [Dd]	1.0 ± 0.5 [Cd]
B11	17.3 ± 1.3 [De]	1.7 ± 0.4 [Ff]	1.0 ± 0.2 [Df]	1.7 ± 0.3 [De]	0.4 ± 0.2 [Cd]
B12	0 [Fg]	0 [Gg]	0 [Eg]	0 [Ef]	0 [Ce]
CK2	13.3 ± 1.3 [Ef]	0.6 ± 0.1 [Fg]	0.6 ± 0.1 [Ef]	1.0 ± 0.2 [Ee]	0.1 ± 0.0 [Ce]

Note: Values in the table are expressed as mean ± standard error. Different letters within the same column indicate notable differences, while same letters suggest no notable differences. In addition, capital letters denote remarkable differences between treatments at $p < 0.01$ level, while lowercase letters signify obvious differences at $p < 0.05$ level.

The variance analysis of the two-factor completely randomized block design model was conducted for the observation results of fruit mulberry, with the results presented in Table 7. According to the results, the types and quality concentrations of growth hormones had notable effects on all rooting indices ($p < 0.01$), and there was an obvious interaction between growth hormone types and quality concentrations. Based on the F values, it can be inferred that the quality concentrations have the greatest effect on fruit mulberry hardwood cuttings, followed by the types of growth hormones.

Table 7. Analysis of variance for the results of fruit mulberry cuttings.

Rooting Index	Source of Variation	SUM of Squares	Freedom	Mean Square	F Value	*p* Value
Rooting rate	Type of growth hormones (A)	2840.2	2	1420.1	121.5	0.0 **
	Quality concentration (B)	8466.2	3	2822.1	241.4	0.0 **
	A × B	3314.4	6	552.4	47.2	0.0 **
	Error	304.0	26	11.7		
	Total	50,552.0	39			

Table 7. *Cont.*

Rooting Index	Source of Variation	SUM of Squares	Freedom	Mean Square	F Value	*p* Value
Average number of roots	Type of growth hormones (A)	66.6	2	33.3	76.6	0.0 **
	Quality concentration (B)	185.6	3	61.9	142.3	0.0 **
	A × B	84.8	6	14.1	32.5	0.0 **
	Error	11.3	26	0.4		
	Total	994.6	39			
Average root length	Type of growth hormones (A)	15.5	2	7.7	53.0	0.0 **
	Quality concentration (B)	62.7	3	20.9	143.1	0.0 **
	A × B	17.1	6	2.9	19.5	0.0 **
	Error	3.8	26	0.1		
	Total	296.5	39			
Average longest root length	Type of growth hormones (A)	58.4	2	29.2	80.3	0.0 **
	Quality concentration (B)	199.9	3	66.6	183.4	0.0 **
	A × B	63.1	6	10.5	28.9	0.0 **
	Error	9.4	26	0.4		
	Total	981.5	39			
Root system index	Type of growth hormones (A)	1145.4	2	572.7	53.0	0.0 **
	Quality concentration (B)	1658.0	3	552.7	51.2	0.0 **
	A × B	1492.5	6	248.8	23.0	0.0 **
	Error	280.7	26	10.8		
	Total	6807.6	39			

Note: ** indicates notable difference at the 0.01 level of mean difference. The same applies to the following tables.

3.4. Changes in Endogenous Hormone Content and Ratio during the Rooting Process of Vegetable Mulberry Cuttings

3.4.1. Changes in Endogenous IAA Content

Both the control and treatment groups exhibited a trend of first decreasing before increasing in IAA content (Figure 3A). The IAA contents of the treatment groups and the control group presented a decreasing trend during the callus formation period, possibly due to increased peroxidase activity in the cuttings after separation from the mother plant, resulting in a decrease of IAA content. During this period, callus formation occurred at the lower end of the cutting, which consumed IAA and also reduced IAA content. Simultaneously, compared with the control group, the decrease in IAA content of ABT-1—treated groups was smaller, probably because of the reversal of exogenous hormone absorption by the cuttings themselves after ABT-1 treatment. In general, the IAA content of ABT-1—1000 mg·L^{-1} was consistently higher than that of other treatments during each time period, suggesting that root growth can be promoted by high levels of IAA.

3.4.2. Changes in Endogenous ABA Content

As indicated in Figure 3B, ABA content exhibited a trend of initial increase followed by a decrease. During the initiation period, the treatment groups and the control group showed an upward trend, possibly due to the increase in ABA secretion from the cuttings in response to external stimuli for separation from the mother plant. The ABA content reached a peak during the callus formation period, followed by a decrease. In this process, the root primordia gradually developed and began to produce adventitious roots. By this time, the ABA content of ABT-1—1000 mg·L^{-1} decreased sharply, even lower than that during the initiation period, while the ABA content of the control group remained relatively high, indicating that a low level of ABA content is beneficial to the production and elongation of adventitious roots.

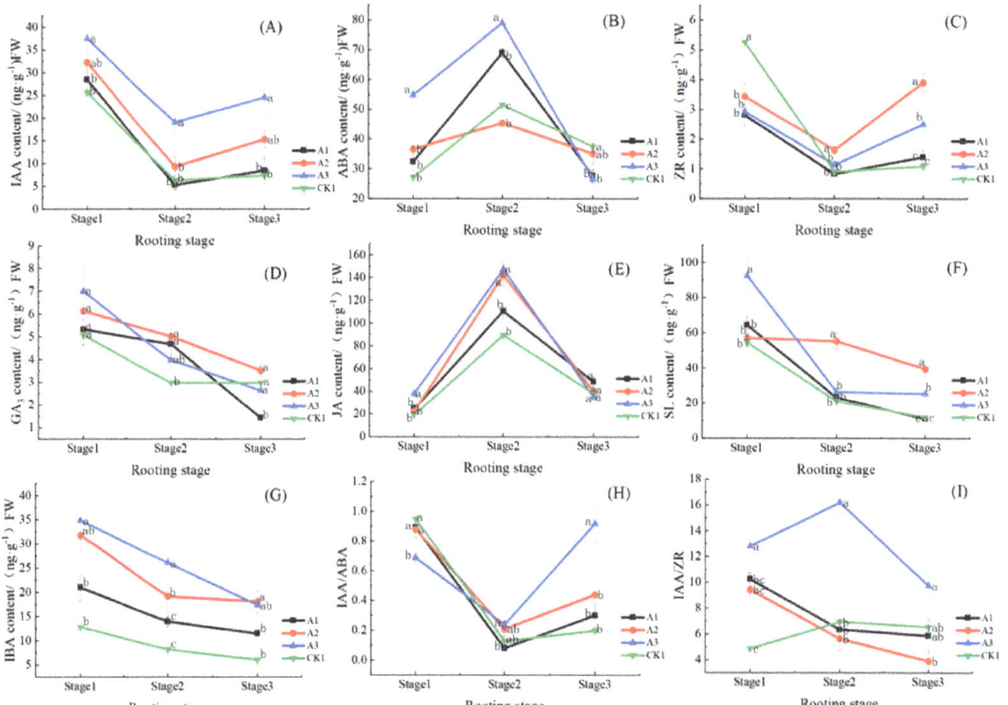

Figure 3. Effect of ABT-1 treatment on endogenous hormone content during the rooting process of vegetable mulberry cuttings. (**A–I**) represents the changes in endogenous hormone contents of IAA, ABA, ZR, GA$_3$, JA, SL, IBA, IAA/ABA, and IAA/ZR in the treatment and control groups of vegetable mulberry under the action of ABT-1 during the rooting period. Different lowercase letters indicate notable differences among the treatments (LSD multiple comparison analysis, $p < 0.05$). A1—represents 200 mg·L^{-1} ABT-1 treatment group, A2—denotes 500 mg·L^{-1} ABT-1 treatment group, A3—signifies 1000 mg·L^{-1} ABT-1 treatment group and CK1—indicates the control group. Stage 1—represents the initiation stage, Stage 2—refers to the callus formation stage and Stage 3—denotes the adventitious root formation and elongation stage.

3.4.3. Changes in Endogenous ZR

The trend of ZR content showed an initial decrease followed by an increase (Figure 3C). During the initiation period, the ZR content of the control group was higher than that of the treatment groups, suggesting that ABT-1 reduces the ZR content in cuttings. During the callus formation period, the ZR content of the control group exhibited a sharp decrease, while the treatment group only presented a slight decrease, indicating that some ZR is consumed by the formation of callus tissue in cuttings. Meanwhile, the control group demonstrated a poor ability to synthesize ZR, so ZR content exhibited a rapid decrease. During the adventitious root formation and elongation period, ZR content presented varying degrees of increase in both the treatment and control groups, thus promoting cell division and growth and benefiting growth and development of adventitious roots.

3.4.4. Changes in Endogenous GA$_3$

As indicated by Figure 3D, for the treatment groups and the control group, the contents of GA$_3$ exhibited a continuous decrease over time. During the initiation period, the GA$_3$ content of ABT-1—1000 mg·L^{-1} reached the highest level, namely 7.0 ng·g^{-1}. After callus formation, the GA$_3$ content of both treatment and control groups decreased, with that of treatment groups decreasing obviously, possibly due to the high demand for GA$_3$ in the growth of root primordia [24]. During the adventitious root formation and elongation period, the GA$_3$ contents of treatment groups were again observed as decreasing, but there was no obvious change in the GA$_3$ content of the control group. At this point, adventitious roots were formed. As the root tip emerged from the skin pore, the GA$_3$ content gradually decreased, probably because high concentrations of GA$_3$ inhibited its elongation growth.

3.4.5. Changes in Endogenous JA

Generally, vegetable mulberry JA content displayed a similar trend with ABA content, and the magnitude of change was great at each stage (Figure 3E). The JA content increased obviously over time, reaching a peak during callus formation. ABT-1—1000 mg·L^{-1} had the highest JA content at 147.6 ng·g^{-1}, 1.7 times that of the control group, laying the foundation for the formation of root primordia and adventitious roots. After callus formation, the JA content presented a sharp decrease, followed by adventitious root formation.

3.4.6. Changes in Endogenous SL

The SL content presented a continuous decreasing trend (Figure 3F). In the initiation period, ABT-1—1000 mg·L^{-1} showed the highest SL content at 92.6 ng·g^{-1}. The most notable trend of ABT-1—1000 mg·L^{-1} was observed during the callus formation, suggesting that high concentrations of ABT-1 could increase SL content. After callus formation, the SL content of ABT-1—1000 mg·L^{-1} did not change obviously, indicating that ABT-1—1000 mg·L^{-1} could stabilize the SL content earlier.

3.4.7. Changes in Endogenous IBA

After cuttings, the IBA content exhibited a gradual decreasing trend among groups (Figure 3G). ABT-1—1000 mg·L^{-1} presented the highest IBA content in the initiation period (35.0 ng·g^{-1}). During callus formation, both the treatment and control groups showed a decreasing trend, with ABT-1—500 mg·L^{-1} exhibiting the greatest decrease in IBA content. During adventitious root formation and elongation period, ABT-1—500 mg·L^{-1} did not show obvious change, while ABT-1—1000 mg·L^{-1} presented a larger magnitude of change.

3.4.8. Changes in IAA/ABA Ratio

According to studies, the rooting of plant cuttings has relations with not only the endogenous IAA and ABA contents but also the ratio between the two contents [25]. As shown in Figure 3H, the IAA/ABA ratio showed a trend of decreasing before increasing. In the initiation period, the control group exhibited the highest IAA/ABA ratio. Meanwhile, after callus formation, ABT-1—1000 mg·L^{-1} presented higher IAA/ABA ratio compared with the treatment groups with 200 and 500 mg·L^{-1}, and the control group, especially during adventitious root formation and elongation period, where the IAA/ABA ratio of ABT-1—1000 mg·L^{-1} was 4.6 times that of the control group. This indicates that a lower IAA/ABA ratio in the early stages of cuttings is conducive to root primordia differentiation and formation, while a higher IAA/ABA value after callus formation is beneficial to root system development.

3.4.9. Changes in IAA/ZR Ratio

As presented in Figure 3I, the IAA/ZR values for ABT-1—1000 mg·L^{-1} showed a trend of increasing followed by decreasing, with a slightly smaller decrease in the control group and a continuous decrease in the treatment groups with 200 and 500 mg·L^{-1} ABT-1. In the whole rooting period, the optimal treatment 1000 mg·L^{-1} ABT-1 had the higher

IAA/ZR values compared with other treatments and the control, indicating that a higher IAA/ZR value contributes to rooting of cuttings.

3.5. Changes in Endogenous Hormone Content and Ratio during the Rooting Process of Fruit Mulberry Cuttings

3.5.1. Changes in Endogenous IAA

As exhibited in Figure 4A, the changes in IAA content in the treatment groups and the control group showed a trend of decreasing before increasing. In the initiation period, the IAA contents in the treatment groups were higher than that in the control group, suggesting that ABT-1 could promote the synthesis of endogenous IAA. During callus formation, the IAA content presented a decreasing trend, possibly due to the active respiration and metabolism of internal cells of cuttings that needed a large amount of IAA to form callus tissue. After root primordia broke through the callus tissue and produced numerous new roots, IAA synthesis was promoted, and the IAA content increased. The IAA content of ABT-1—500 mg·L^{-1} (32.8 ng·g^{-1}) was greater than that of ABT-1—200 mg·L^{-1} (15.9 ng·g^{-1}) and ABT-1—1000 mg·L^{-1} (21.8 ng·g^{-1}), and the control group (20.2 ng·g^{-1}).

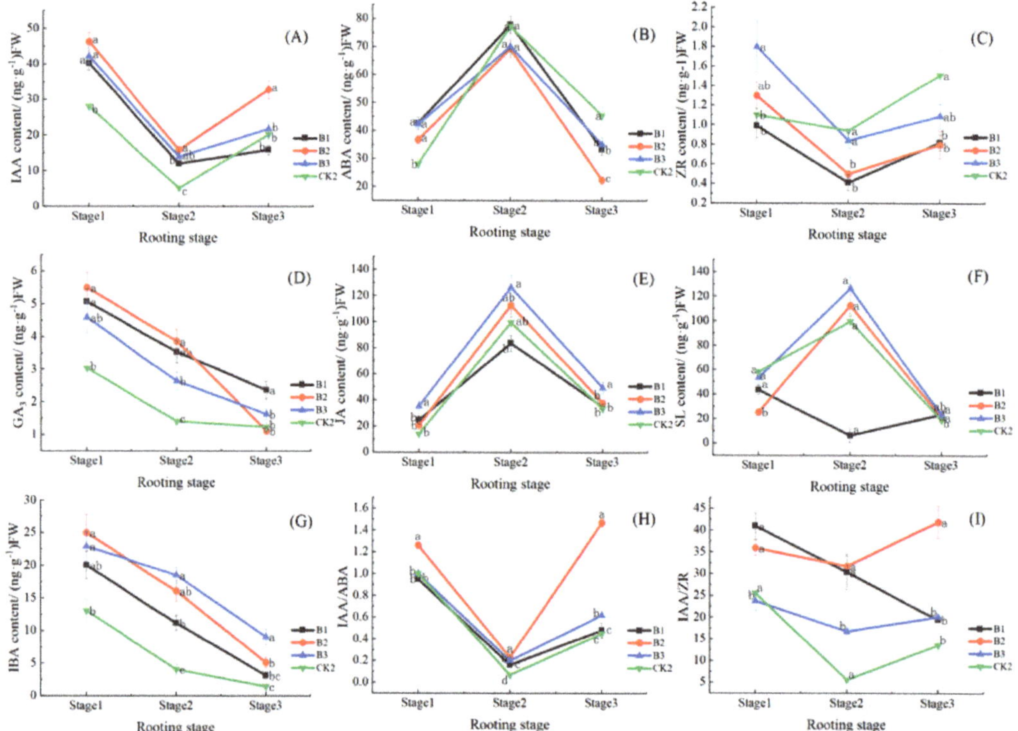

Figure 4. Effects of ABT-1 treatment on endogenous hormone level during the rooting process of fruit mulberry cuttings. Panels (**A–I**) exhibit changes of the levels of IAA, ABA, ZR, GA$_3$, JA, SL, IBA, IAA/ABA, and IAA/ZR in the ABT-1 treatment groups and control group during the rooting stages. Different lowercase letters indicate remarkable differences among the treatments (LSD multiple comparison analysis, $p < 0.05$). B1—represents 200 mg·L^{-1} ABT-1 treatment group, B2—denotes 500 mg·L^{-1} ABT-1 treatment group, B3—signifies 1000 mg·L^{-1} ABT-1 treatment group and CK2—refers to the control group. Stage 1—represents the initiation stage, Stage 2—denotes the callus formation stage and Stage 3—indicates the adventitious root formation and elongation stage.

3.5.2. Changes in Endogenous ABA

The ABA content presented a trend of first increasing and then decreasing (Figure 4B). During the initiation period, the ABA content in the treatment groups was higher than that in the control group, and the ABA content in both the treatment and control groups exhibited an increasing trend, with a peak reached during callus formation. This may be caused by the stress response of cuttings to the environment after detachment from the mother plant, resulting in an increase in ABA content and accumulation of nutrients, which is conducive to callus formation and survival of cuttings. After callus formation, the ABA content began to decrease. During the adventitious root formation and elongation period, ABT-1—500 mg·L^{-1} had the lowest ABA content, namely 22.3 ng·g^{-1}, merely 49% of that of the control group (45.4 ng·g^{-1}), suggesting that high ABA concentration inhibits adventitious root formation and elongation.

3.5.3. Changes in Endogenous ZR

The ZR content presented a trend of decreasing before increasing (Figure 4C). After the initiation period, the ZR content of the treatments and control exhibited a decreasing trend, with a more obvious decrease exhibited in the treatment group. After callus formation, the ZR content in the control group was higher than that in the treatment groups, indicating that high ZR concentration has an inhibitory impact on adventitious root formation.

3.5.4. Changes in Endogenous GA$_3$

The GA$_3$ content in the treatment groups presented a sustained decreasing trend, while that in the control group showed a slight decreasing trend (Figure 4D). In the initiation period, the GA$_3$ content in the treatment groups was higher than that in the control group, suggesting that ABT-1 treatment could effectively enhance GA$_3$ synthesis. During the callus formation period, the GA$_3$ content in the control group and the treatment groups decreased in various degrees, and the highest GA$_3$ content (3.9 ng·g^{-1}) was observed in ABT-1—500 mg·L^{-1}, while the control group had a GA$_3$ content of merely 1.4 ng·g^{-1}. During the adventitious root formation and elongation period, the GA$_3$ content in ABT-1—500 mg·L^{-1} was the lowest at 1.1 ng·g^{-1}, probably due to the consumption of large amounts of GA$_3$ for the formation of adventitious roots in cuttings at this stage.

3.5.5. Changes in Endogenous JA

The JA content exhibited a trend of increasing before decreasing (Figure 4E). In the initiation period, the JA content in the treatment groups was higher than that in the control group, and increased with time reaching a peak value (126.5 ng·g^{-1}) during callus formation, followed by decrease. As indicated by analysis, the accumulation of JA in the early stage was beneficial to callus formation, while a low level of JA in the later stage was conducive to adventitious root formation and elongation.

3.5.6. Changes in Endogenous SL

According to Figure 4F, the SL content presented a sustained decreasing trend. By viewing the overall rooting period, the largest change appeared in the control group, while the change in the optimal treatment 500 mg·L^{-1} ABT-1 was relatively small, suggesting that the effect of endogenous hormones on cuttings is the result of the combined action of multiple hormones.

3.5.7. Changes in Endogenous IBA

As indicated in Figure 4G, endogenous IBA content exhibited a similar change with the changes of GA$_3$ and SL. The highest level of endogenous IBA occurred in the initiation stage, followed by a gradual decrease till the emergence and elongation of adventitious roots. The lowest level of endogenous IBA was observed in the control group (1.4 ng·g^{-1}), while the content of ABT-1—500 mg·L^{-1} was 3.6 times higher than that of the control group, suggesting that a higher IBA content can accelerate the growth of fruit mulberry cuttings.

3.5.8. Changes in IAA/ABA Ratio

As presented in Figure 4H, the IAA/ABA ratio of both treatment and control groups exhibited a decreasing trend in the early stage, followed by an increasing trend. Compared with the treatment groups with 200 and 1000 mg·L^{-1} ABT-1 and the control group, ABT-1—500 mg·L^{-1} had a higher IAA/ABA ratio throughout the rooting process, while the control group possessed a consistently lower IAA/ABA ratio. This suggests that auxin can enhance the rooting ability of cuttings, and that IAA and ABA can act together, thereby regulating the development and growth of the root system of cuttings.

3.5.9. Changes in IAA/ZR Ratio

The IAA/ZR ratio of the treatment group with 500 and 1000 mg·L^{-1} ABT-1 and the control group showed a trend of decreasing before increasing, while that of ABT-1—200 mg·L^{-1} exhibited a continuous decreasing trend (Figure 4I). By combining with the analysis of rooting process and morphology, the IAA/ZR ratio of the optimal treatment group 500 mg·L^{-1} ABT-1 reached the lowest value during the callus formation, peaked during the emergence and elongation of adventitious roots, and was significantly higher than that of other treatments. This indicates that a higher IAA/ZR ratio is conducive to the elongation of adventitious roots.

4. Discussion

According to the results of this study, the rooting of *Morus alba* L. cuttings induced by auxins is superior to that of the control group treated with water, and the ABT-1 treatment is the most effective, followed by IAA and IBA. Auxins can accelerate the initiation of root primordia and propel the formation of adventitious roots, causing a notable improvement in rooting [26–29], which aligns with the results of *Acacia mangium* et al. [30].

IAA can stimulate cell division in the cambium and regulate the formation of callus tissue [31]. In this study, IAA promoted the rooting rate of *Morus alba* L., and the rooting rate gradually decreased with an increase in IAA concentration within the study concentration range, which was contrary to the results of Li et al. [24] and similar to the results of Raju et al. [32].This phenomenon may be caused by the fact that IAA is easily destroyed by metabolism in plants, its action is highly unstable, it easily decomposes in water, and its action time is short [31]. Moreover, the stability of IAA decreases with an increase in concentration. IBA can accelerate cell metabolism and promote the induction and formation of root primordia [33]. According to Singh et al. [34], the rooting percentage, longest root sprout length, germination rate, and root length of *Morus alba* L. cuttings treated with 2000 mg·L^{-1} IBA are all higher. These findings bear similarity to the results of the vegetable mulberry treatment, but differ from those of the fruit mulberry treatment. As indicated by the notable difference in the effect of IBA concentration required by vegetable mulberry and fruit mulberry, the plants exhibit different sensitivities to the same type of auxin, affecting the physiological and biochemical processes inside the cuttings, and further causing differences in rooting efficiency, which is consistent with the results of Hu [35]. ABT-1 can accelerate the synthesis of hormones in the rooting region and promote branch proliferation [36]. As discovered by Chen et al. [21], ABT-1 is most effective in promoting the rooting of Morus alba L., followed by IBA and IAA. As indicated by the results of this study, the optimal treatment for rooting of vegetable mulberry cuttings is 1000 mg·L^{-1} ABT-1, and 500 mg·L^{-1} ABT-1 for rooting of fruit mulberry. The different optimal treatment concentrations for vegetable mulberry and fruit mulberry may be caused by the different sensitivities of different plants to exogenous auxins. Hence, for different plant cuttings, the optimal concentration range of auxins is also different [37]. The results of this study show that the effects of IAA and IBA treatment are inferior to that of ABT-1. This is probably caused by the fact that ABT-1 is a composite of IBA and NAA, which can more effectively propel the increase in endogenous hormone content and important enzyme activity and promote plant metabolism intensity, thereby greatly improving rooting ability [38]. These findings are similar to the results of studies on *Chimonanthus praecox* [39], *Sophora japonica* [22] and *Taxus chinensis* (Pilger) Rehd F. [36].

According to most scholars, endogenous hormone expression and nutrient allocation in cuttings are regulated by exogenous hormones [40–42]. In this study, the changes in IAA content suggest that ABT-1 can increase endogenous IAA content and enhance adventitious root formation [43], similar to the results of Shang et al. [44]. This demonstrates that endogenous IAA can propel root primordium accumulation and adventitious root formation [31]. Endogenous IBA can drive root primordium formation and promote rooting [45]. As observed from the changes in IBA content in this study, IBA content remarkably increases under ABT-1 treatment, thereby enhancing adventitious root formation and elongation. ABA is an inhibitory plant hormone [46]. As discovered in various studies, the ABA content of vegetable and fruit mulberry cuttings increases, thereby regulating the adaptability of cuttings to stress and preparing them for the formation of root primordia, followed by decreases in endogenous ABA content, suggesting that low levels of ABA are conducive to root primordium differentiation and root formation [47]. In this study, changes in ABA content indicate that ABT-1 can induce a decrease in ABA content during the production of callus tissue, which is beneficial to the growth and development of adventitious roots. ZR can promote cell division and differentiation, thereby affecting adventitious root formation [22]. As shown by previous studies, low concentrations of ZR are conducive to adventitious root formation and differentiation [48,49]. In this study, the control group had a higher ZR content compared with the treatment group after the initiation period, indicating that ABT-1 has an inhibitory effect on ZR in the early rooting stage, thus promoting root primordia. Subsequently, the ZR content in the treatment group exhibited a slight increase before decreasing, suggesting that cuttings can synthesize ZR by themselves during the production and elongation of adventitious roots, thus causing an increase in ZR content. For the rooting process of vegetable and fruit mulberry cuttings, low ZR concentration facilitates the formation of callus tissue, and high ZR concentration contributes to the formation of adventitious roots, in accordance with the results of *Tilia mandshurica* [50], *Zizyphus jujuba* Mill. [51] and Hybrid Aspen [52]. GA_3 mainly enhances cell division and elongation [53]. Studies have discovered that a high concentration of GA_3 inhibits adventitious root formation, and that a low concentration of GA_3 is conducive to adventitious root formation [54]. This was confirmed by the changes in GA_3 content in this study, conforming to findings from the study of Mu et al. [50]. JA can promote plant regeneration [47]. In this study, the JA content reached its peak during the production of callus tissue, suggesting that high levels of JA can mediate the development of callus tissue and provide conditions for root primordium formation. Afterwards, the JA content gradually decreased, indicating the consumption of JA by adventitious root growth and development. Nevertheless, in comparison to the control group, the treatment group presented the higher JA content, indicating that ABT-1 can improve cell regeneration ability and stimulate callus tissue development, thereby enhancing adventitious root formation. As demonstrated in previous studies, SL can regulate root system morphology, and promote lateral roots and root hair production [55–57]. Furthermore, in this study, the optimal treatment group presented the highest SL content during the initiation period, and the SL content gradually stabilized after the production of callus tissue, suggesting that ABT-1 can propel a large amount of SL synthesis during the initiation period and provide conditions for adventitious root development.

During plant cuttings, the ratio of endogenous hormone levels bears close relation with the formation of adventitious roots [58]. According to Quan et al. [59], the higher the values of IAA/ABA and IAA/ZR are, the easier it is to form roots. As discovered in this study, the IAA/ABA and IAA/ZR values in the treatment group greatly increased after the formation of callus, indicating that *Morus alba* L. can be induced to root under the combined action of auxin and endogenous hormones.

5. Conclusions

In summary, all the three plant growth hormones of IAA, IBA and ABT-1, were able to promote the rooting of cuttings of vegetable mulberry and fruit mulberry, with ABT-1 exhibiting the best effect. Among them, the treatment with 1000 mg·L^{-1} ABT-1 presented the optimal impact on the rooting of vegetable mulberry, with a rooting rate of 63.3% and a rooting effect index of 25.3. The treatment with 500 mg·L^{-1} ABT-1 showed the optimal impact on the rooting of fruit mulberry, with a rooting rate of 68.7% and a rooting effect index of 34.3. During the rooting process, ABT-1 treatment effectively increased the contents of IAA, IBA, JA and SL, and decreased the contents of ABA, ZR, and GA$_3$, contributing to the production of adventitious roots. The rooting mechanism of plant cuttings is extremely complex, and follow-up studies will focus on molecular biology methods including transcriptomics and proteomics [60], thereby further revealing the rooting mechanism of vegetable mulberry and fruit mulberry cuttings.

Author Contributions: Conceptualization, J.Q. and H.B.; methodology, J.Q. and H.C.; software, H.C. and Y.L.; validation, Y.L., J.S., M.M. and P.D.; Data Curation, H.C., J.S., M.M. and P.D.; writing—original draft preparation, H.C. and Y.L.; writing—review and editing, J.Q. and H.B.; funding acquisition, J.Q. and H.B. All authors have read and agreed to the published version of the manuscript.

Funding: This research was funded by the National Natural Science Foundation of China (grant number 31700549) and Forestry Germplasm Resources Project of Henan Province (grant number 30802649).

Data Availability Statement: Not applicable.

Conflicts of Interest: The authors declare no conflict of interest.

References

1. Medina, M.G.; García, D.E.; Moratinos, P.; Cova, L.J. Mulberry (*Morus* spp.) as fodder resource: Research advances and considerations. *Trop. Zool.* **2009**, *27*, 343–362.
2. Nie, Y.X.; Zhang, Y.B.; Wang, Z.H.; Dong, Y.R. Main points of seedling cultivation and planting methods of mulberry. *Hubei Agric. Sci.* **2019**, *58*, 298–300. [CrossRef]
3. Ma, Y. Research on the Current Situation and Development Countermeasures of Large Scale Sericulture Base Construction in China. Master's Thesis, Jiangsu University of Science and Technology, Zhenjiang, Chian, 2021.
4. Qadri, S.F.I.; Sheikh, N.D.; Malik, M.A.; Baqual, M.F.; Mir, M.R.; Sabahat, A.; Ahamad, F. Performance of mulberry saplings in foot hills of kashmir, propagated through cuttings under polyhouse conditions. *Int. J. Agric. Stat. Sci.* **2011**, *7*, 125–130.
5. Yang, X.P.; Yu, H.; Wang, G. Studies on micro-cuttage propagation with hardwood cuttings of *Zanthoxylum bungeanum*. *J. Northwest For. Univ.* **2021**, *36*, 145–149. [CrossRef]
6. Balestri, E.; Vallerini, F.; Castelli, A.; Lardicci, C. Application of plant growth regulators, a simple technique for improving the establishment success of plant cuttings in coastal dune restoration. *Estuar. Coast. Shelf Sci.* **2012**, *99*, 74–84. [CrossRef]
7. Garay-Arroyo, A.; Sanchez, M.D.; Garcia-Ponce, B.; Azpeitia, E.; Alvarez-Buylla, E.R. Hormone symphony during root growth and development. *Dev. Dynam.* **2012**, *241*, 1867–1885. [CrossRef] [PubMed]
8. Saini, S.; Sharma, I.; Kaur, N.; Pati, P.K. Auxin: A master regulator in plant root development. *Plant Cell Rep.* **2013**, *32*, 741–757. [CrossRef] [PubMed]
9. Zhang, W.X.; Fan, J.J.; Tan, Q.Q.; Zhao, M.M.; Cao, F.L. Mechanisms underlying the regulation of root formation in *Malus hupehensis* stem cuttings by using exogenous hormones. *J. Plant Growth Regul.* **2017**, *36*, 174–185. [CrossRef]
10. Zhao, Y.Q.; Chen, Y.J.; Jiang, C.; Lu, M.Z.; Zhang, J. Exogenous hormones supplementation improve adventitious root formation in woody plants. *Front. Bioeng. Biotechnol.* **2022**, *10*, 1009531. [CrossRef]
11. Mao, J.P.; Zhang, D.; Zhang, X.; Li, K.; Liu, Z.; Meng, Y.; Lei, C.; Han, M.Y. Effect of exogenous indole-3-butanoic acid (iba) application on the morphology, hormone status, and gene expression of developing lateral roots in malus hupehensis. *Sci. Hortic.* **2018**, *232*, 112–120. [CrossRef]
12. Li, S.M.; Zheng, X.H.; Duan, H.C.; Dong, Q. Effects of ABT-1 treatment on rooting capacity and physiological and biochemical characteristic of crateva unilocalaris cuttings. *Acta Agric. Univ. Jiangxiensis* **2021**, *43*, 116–125. [CrossRef]
13. Zhou, B.; Ren, Y.P.; Mi, Y.F. Effect of iba and abt on endogenous hormones in rooted cuttings of common camptotheca fruit (camptotheca acuminata decne). *Fujian J. Agric. Sci.* **2017**, *32*, 387–393. [CrossRef]
14. Yang, W.X.; Zhuang, J.Q.; Ding, S.Y.; Zhang, M.; Tian, Y.; Wan, S.Y.; Fang, S.Z. Study on cutting cultivation technology and rooting mechanism of *Cyclocarya paliurus*. *Ecol. Chem. Eng. S* **2022**, *29*, 379–389. [CrossRef]
15. Ouyang, F.Q.; Wang, J.H.; Li, Y. Effects of cutting size and exogenous hormone treatment on rooting of shoot cuttings in norway spruce [*Picea abies* (L.) karst.]. *New For.* **2015**, *46*, 91–105. [CrossRef]

16. Haissig, B.E. Influence of Indole-3-acetic acid on adventitious root primordia of brittle willow. *Planta* **1970**, *95*, 27–35. [CrossRef]
17. Gao, J.; Liu, X.; Yuan, Y. Water cultured propagation of polygonum multiflorum and dynamic changes of physiological and biochemical characteristics during adventitious roots formation. *Chin. J. Chin. Mater. Med.* **2011**, *36*, 375–378.
18. Han, J.H.; Shao, W.H.; Liu, J.F.; Diao, S.F. Content changes of endogenous hormone and polyphenols during hardwood cuttage progress in *Sapindus mukorossi*. *Non-Wood For. Res.* **2019**, *37*, 37–43+51. [CrossRef]
19. Oyang, F.Q.; Fu, G.Z.; Wang, J.H.; Mang, J.W.; An, S.P.; Wang, M.Q.; Li, Y. Qualitative analysis of endogenesis hormone and polyphenol during rooting of cuttings in norway spruce (*Picea abies*). *Sci. Silv. Sin.* **2015**, *51*, 155–162.
20. Zhang, J.C.; Liu, Y.J.; Wang, F.L.; Ma, J.M.; Wei, L.Y.; Jiang, S.X. Physiological and biochemical characteristics of *Tamarix taklamakanensis* cuttings during rooting stages. *Acta Bot.* **2018**, *38*, 484–492. [CrossRef]
21. Chen, H.L.; Bi, S.W.; Ni, R.Y.; Lei, Y.Z.; Quan, J.E. Effects of different auxin and mass concentration on mulberry cutting rooting. *J. Henan Agric. Univ.* **2022**, *56*, 958–967. [CrossRef]
22. Wang, Y.J. Cutting Propagation Techniques and Rooting Mechanism of *Sophora japomica* L. Master's Thesis, Beijing Forestry University, Beijing, China, 2017.
23. Niu, Q.; Zong, Y.; Qian, M.; Yang, F.; Teng, Y. Simultaneous quantitative determination of major plant hormones in pear flowers and fruit by uplc/esi-ms/ms. *Anal. Methods* **2014**, *6*, 1766–1773. [CrossRef]
24. Li, C.C.; Quan, W.X.; Chen, X.J. Dynamics of endogenous hormones, anatomical structure during the cutting propagation of wild rhododendron sacbrifolium franch. *Pak. J. Bot.* **2017**, *49*, 2295–2299.
25. Hou, J.T.; Shen, C.C.; Zhang, Y.F.; Ling, N. Review on rooting mechanism of plant cuttings propagation. *J. Anhui Agric. Sci.* **2019**, *47*, 1–3, 6. [CrossRef]
26. Mishra, J.P.; Bhadrawale, D.; Rana, P.K.; Mishra, Y. Evaluation of cutting diameter and hormones for clonal propagation of *Bambusa balcooa* roxb. *J. For. Res.* **2019**, *24*, 320–324. [CrossRef]
27. Jiang, K.; Feldman, L.J. Root meristem establishment and maintenance: The role of auxin. *J. Plant Growth Regul.* **2002**, *21*, 432–440. [CrossRef]
28. Fukaki, H.; Okushima, Y.; Tasaka, M. Auxin-mediated lateral root formation in higher plants. In *International Review of Cytology—A Survey of Cell Biology*; K. W. Jeon: Nara, Japan, 2007; Volume 256, pp. 111–137.
29. Fang, T.; Motte, H.; Parizot, B.; Beeckman, T. Root branching is not induced by auxins in selaginella moellendorffii. *Front. Plant. Sci.* **2019**, *10*, 154. [CrossRef]
30. Ruan, R.G. Effect of plant growth regulators on cutting rooting of *Acacia mangium*. *J. Green Sci. Tech.* **2022**, *24*, 86–89. [CrossRef]
31. Jiang, X.; Hu, R.C.; Cai, J.B.; He, J.; Wang, Y.; Han, C.Y.; Wu, F.F.; Nie, G. Effects of different auxin on cuttage seedling of *Trifolium repens*. *J. Grass. Forage. Sci.* **2021**, *5*, 11–18. [CrossRef]
32. Raju, N.L.; Prasad, M.N.V. Influence of growth hormones on adventitious root formation in semi-hardwood cuttings of *celastrus paniculatus* willd.: A contribution for rapid multiplication and conservation management. *Agrofor. Syst.* **2010**, *79*, 249–252. [CrossRef]
33. Stefancic, M.; Stampar, F.; Osterc, G. Influence of iaa and iba on root development and quality of prunus 'gisela 5' leafy cuttings. *Hortscience* **2005**, *40*, 2052–2055. [CrossRef]
34. Singh, K.K.; Choudhary, T.; Kumar, A. Effect of various concentrations of iba and naa on the rooting of stem cuttings of mulberry (*Morus alba* L.) under mist house condition in garhwal hill region. *Indian J. Hill Farm.* **2015**, *27*, 74–77.
35. Hu, Y.J.; Omary, M.; Hu, Y.; Doron, O.; Hoermayer, L.; Chen, Q.G.; Megides, O.; Chekli, O.; Ding, Z.J.; Friml, J.; et al. Cell kinetics of auxin transport and activity in arabidopsis root growth and skewing. *Nat. Commun.* **2021**, *12*, 1657. [CrossRef] [PubMed]
36. Die, H.U.; Xinru, H.E.; Yongzheng, M.A.; Fei, Y. Effects of abt on the morphogenesis and inclusions of *Taxus chinensis* (pilger) rehd f. Baokangsis cutting rooting. *Not. Bot. Horti Agrobot.* **2021**, *49*, 12200. [CrossRef]
37. Shuting, W.; Guodong, S.; Ying, L.; Wenjun, Q.; Kai, F.; Zhaotang, D.; Jianhui, H. Role of iaa and primary metabolites in two rounds of adventitious root formation in softwood cuttings of *Camellia sinensis* (L.). *Agronomy* **2022**, *12*, 2486. [CrossRef]
38. Xie, A.D.; Jin, D.X.; Hou, W.T.; Jiang, Z.Y.; Song, C.; Wu, H.Y.; Ke, Y.F.; Ren, A. Effects of plant growth regulator abt on rooting of climbing rose cuttings and the correlations and principal components analysis. *J. Anhui Agric. Sci.* **2022**, *50*, 123–126. [CrossRef]
39. Hu, H.; Chai, N.; Zhu, H.X.; Li, R.; Huang, R.W.; Wang, X.; Liu, D.F.; Li, M.Y.; Song, X.R.; Sui, S.Z. Factors affecting vegetative propagation of wintersweet (*Chimonanthus praecox*) by softwood cuttings. *Hortscience* **2020**, *55*, 1853–1860. [CrossRef]
40. Sun, J.; Xia, J.; Zhao, X.; Su, L.; Liu, P. Effects of 1-aminobenzotriazole on the growth and physiological characteristics of tamarix chinensis cuttings under salt stress. *J. For. Res.* **2020**, *32*, 1641–1651. [CrossRef]
41. Li, S.; Huang, P.; Ding, G.; Zhou, L.; Tang, P.; Sun, M.; Zheng, Y.; Lin, S. Optimization of hormone combinations for root growth and bud germination in chinese fir (*Cunninghamia lanceolata*) clone leaf cuttings. *Sci. Rep.* **2017**, *7*, 5046. [CrossRef]
42. Xie, T.T.; Ji, J.; Chen, W.; Yue, J.Y.; Du, C.J.; Sun, J.C.; Chen, L.Z.; Jiang, Z.P.; Shi, S.Q. Gaba negatively regulates adventitious root development in poplar. *J. Exp. Bot.* **2019**, *71*, 1459–1474. [CrossRef]
43. Deng, S.C.; Tian, Y.P.; Chen, L.B.; Chen, C.L.; Xu, P.Z.; Pang, D.D.; Bao, Y.X.; Li, Y.Y. Effects of abt rooting powders on endogenous hormone levels and cuttings rooting of big-leaf tea plants. *Acta Agric. Jiangxi* **2021**, *33*, 43–48. [CrossRef]
44. Shang, W.Q.; Wang, Z.; He, S.L.; He, D.; Dong, N.L.; Guo, Y. Changes of endogenous iaa and related enzyme activities during rooting of *Paeonia suffruticosa* in vitro. *J. Northwest AF Univ. (Nat. Sci. Ed.)* **2021**, *49*, 129–136. [CrossRef]
45. Zhang, D. Study on the Interaction of Two Indoles Plant Growth Regulators with Protein and DNA. Master's Thesis, Nanchang University, Nanchang, China, 2022.

46. Heide, O.M. Stimulation of adventitious bud formation in begonia leaves by abscisic acid. *Nature* **1968**, *219*, 960–961. [CrossRef]
47. Liu, G.B.; Zhao, J.Z.; Liao, T.; Wang, Y.; Guo, L.Q.; Yao, Y.W.; Cao, J. Histological dissection of cutting-inducible adventitious rooting in *platycladus orientalis* reveals developmental endogenous hormonal homeostasis. *Ind. Crop. Prod.* **2021**, *170*, 113817. [CrossRef]
48. Uwe, D.; Philipp, F.; Hajirezaei, M.R. Plant hormone homeostasis, signaling, and function during adventitious root formation in cuttings. *Front. Plant Sci.* **2016**, *7*, 381. [CrossRef]
49. Liu, M.G.; Wang, L.; Dong, S.J.; Cai, X.D. Endogenous hormone variation in cuttings of thuja occidentalis l. In the period of adventitious root formation. *J. Shenyang Agric. Univ.* **2010**, *41*, 555–559. [CrossRef]
50. Mu, H.Z.; Jin, X.H.; Ma, X.Y.; Zhao, A.Q.; Gao, Y.T.; Lin, L. Ortet age effect, anatomy and physiology of adventitious rooting in *Tilia mandshurica* softwood cuttings. *Forests* **2022**, *13*, 1427. [CrossRef]
51. Shao, F.; Wang, S.; Huang, W.; Liu, Z. Effects of iba on the rooting of branch cuttings of chinese jujube (*zizyphus jujuba* mill.) and changes to nutrients and endogenous hormones. *J. For. Res.* **2018**, *29*, 1557–1567. [CrossRef]
52. Yan, S.P.; Yang, R.H.; Wang, F.; Sun, L.N.; Song, X.S. Effect of auxins and associated metabolic changes on cuttings of hybrid aspen. *Forests* **2017**, *8*, 117. [CrossRef]
53. Lv, G.X.; Meng, Y.D.; Qing, J.; He, F.; Liu, P.F.; Du, Q.X.; Du, H.Y.; Du, L.Y.; Wang, L. Changes of anatomical structure and physiology during softwood cutting rooting of *Eucommia ulmoides* 'huazhong no. 6'. *Sci. Silv. Sin.* **2022**, *58*, 113–124.
54. Koyama, R.; Ribeiro Junior, W.A.; Zeffa, D.M.; Faria, R.T.; Saito, H.M.; Azeredo Goncalves, L.S.; Roberto, S.R. Association of indolebutyric acid with azospirillum brasilense in the rooting of herbaceous blueberry cuttings. *Horticulturae* **2019**, *5*, 68. [CrossRef]
55. Kapulnik, Y.; Delaux, P.M.; Resnick, N.; Koltai, H. Strigolactones affect lateral root formation and root-hair elongation in *Arabidopsis*. *Planta* **2010**, *233*, 209–216. [CrossRef]
56. Ruyter-Spira, C.; Kohlen, W.; Charnikhova, T.; Zeijl, A.V.; Bezouwen, L.V.; Ruijter, N.D.; Cardoso, C.; Lopez-Raez, J.A.; Matusova, R.; Bours, R. Physiological effects of the synthetic strigolactone analog gr24 on root system architecture in arabidopsis: Another belowground role for strigolactones? *J. Plant Physiol.* **2011**, *155*, 721–743. [CrossRef] [PubMed]
57. Brewer, P.B.; Koltai, H.; Beveridge, C.A. Diverse roles of strigolactones in plant development. *Mol. Biol.* **2013**, *6*, 18–28. [CrossRef]
58. Huang, T.; Zhang, H.H.; Sheng, Q.Q.; Zhu, Z.L. Morphological, anatomical, physiological and biochemical changes during adventitious roots formation of *Bougainvillea buttiana* 'miss manila'. *Horticulturae* **2022**, *8*, 1156. [CrossRef]
59. Quan, J.E.; Ni, R.Y.; Wang, Y.G.; Sun, J.J.; Ma, M.Y.; Bi, H.T. Effects of different growth regulators on the rooting of catalpa bignonioides softwood cuttings. *Life* **2022**, *12*, 1231. [CrossRef] [PubMed]
60. Liu, L.; Jiang, P.; Huang, Y.; Ynag, Y.; Gou, J.; Zhao, D.; Wang, T.; Dai, W.; Qi, D.M.; Hu, J.Y. Preliminary study on twig cuttage and its rooting mechanism of *Rosa damascene* miller. *Seed* **2020**, *39*, 92–100. [CrossRef]

 *horticulturae*

Article

In Vitro Floral Emergence and Improved Formation of Saffron Daughter Corms

Yaser Hassan Dewir [1,2,*], Abdulla Alsadon [1], Ahmed Ali Al-Aizari [1] and Mohaidib Al-Mohidib [1]

[1] Plant Production Department, College of Food & Agriculture Sciences, King Saud University, Riyadh 11451, Saudi Arabia
[2] Horticulture Department, Faculty of Agriculture, Kafrelsheikh University, Kafr El-Sheikh 33516, Egypt
* Correspondence: ydewir@ksu.edu.sa

Abstract: In vitro cormogenesis is a potential tool for improving saffron production under controlled conditions. In this study, the effects of explant type, culture type, and medium supplements on saffron daughter corm formation in vitro were assessed. Saffron flowers emerged 30 days after culture, and the sizes of in-vitro- and ex-vitro-produced flowers and stigmas were similar. In vitro daughter corm formation and the saffron life cycle was completed after 10 and 14 weeks of culture, respectively. Using in vitro intact corms was more effective for corm production than using apical buds. Compared with apical bud explants, mother corm explants produced more corms with a higher fresh weight and diameter. Compared with solid culture, liquid cultures using bioreactors provided corms with a higher fresh weight and diameter, regardless of explant type. An ebb and flow system provided the highest cormlet fresh weight and diameter but the fewest cormlets, whereas an immersion system provided more cormlets with a smaller size. Saffron apical buds cultured with salicylic acid at 75 mg L^{-1} or glutamine at 600 mg L^{-1} exhibited the highest cormlet diameter and fresh weight. These findings will improve the process of in vitro cormogenesis and the production of saffron under controlled conditions.

Keywords: *Crocus sativus*; glutamine; Iridaceae; jasmonic acid; liquid culture; salicylic acid

Citation: Dewir, Y.H.; Alsadon, A.; Al-Aizari, A.A.; Al-Mohidib, M. In Vitro Floral Emergence and Improved Formation of Saffron Daughter Corms. *Horticulturae* **2022**, *8*, 973. https://doi.org/10.3390/horticulturae8100973

Academic Editor: Sergio Ruffo Roberto

Received: 15 September 2022
Accepted: 15 October 2022
Published: 20 October 2022

Publisher's Note: MDPI stays neutral with regard to jurisdictional claims in published maps and institutional affiliations.

1. Introduction

Saffron (*Crocus sativus*; Iridaceae) is a sterile autotriploid and stemless monocotyledonous geophyte plant [1,2]. Therefore, obtaining daughter corms from mother corms is the only available method of saffron propagation [3]. Within the life cycle of saffron, flowering occurs during autumn (October–November), and the vegetative stage, including the formation of replacement corms at the base of the shoots, occurs throughout winter. At the beginning of the dry season (April–May), the leaves senesce and wither, and the corms go into dormancy. The transition from the vegetative stage to the reproductive stage can occur shortly afterwards in the apex of the buds of underground corms [4]. Saffron is a valuable spice obtained from the stigmas of *C. sativus*, and the value of saffron is enhanced by its potential use in biomedicine [5]. Thus, the demand for saffron is expected to increase in the coming years owing to its nutraceutical and medicinal properties.

In vitro culture technologies could facilitate the sustainable indoor production of saffron through the growth of pathogen-free stock corms. Saffron microcorm production under in vitro conditions is a promising technique with respect to the rate of multiplication and the number of cormlets produced. However, few studies have investigated the production of complete plants with roots and/or corms [6–9]. Saffron flowers grown in vitro could also serve as a source of saffron spice; however, the induced in vitro flowering of saffron plants has not been reported previously.

In in vitro propagation, liquid culture/bioreactor systems enable automation and reduce the costs of plantlet production [10]. Liquid culture systems provide uniform culture conditions in which the nutrient medium is renewed without changing the container. Moreover, the container can be cleaned with ease after the culture period. Plant tissues and organs from various plant species exhibit higher levels of growth performance in a liquid medium than in solid or semi-solid media. Liquid culture/bioreactor systems and their characteristic culture conditions have several advantages over solid cultures, including the convenient handling of cultures and enhanced plant growth [11]. Although bioreactor culturing has been used to grow many plant species, to the best of our knowledge, the use of bioreactors to produce saffron microcorms has not been reported previously. The main disadvantages of saffron tissue culture are the low frequency and the small size of cormlets induced from in-vitro-derived shoots [12–14].

The aim of the present study was to compare in vitro flower induction and the formation of daughter corms in gel and liquid cultures. Additionally, different concentrations of glutamine (as a nitrogen source), salicylic acid, and jasmonic acid were tested with the aim of improving the formation of saffron daughter corms. Overall, this study demonstrates that saffron flowering can be induced in vitro. Moreover, compared with solid cultures, liquid cultures/bioreactors improved daughter corm diameter and fresh weight. Thus, the production of saffron under controlled in vitro conditions is a viable option.

2. Materials and Methods

2.1. Surface Disinfection and the Establishment of Aseptic Culture

Crocus sativus corms (3.2–3.5 cm diameters; Figure 1a,b) were obtained from Bloembollenbedrijf J.C.Koot (Vennewatersweg, The Netherlands) during two successive seasons in 2020 and 2021 and kept at room temperature. The corms were descaled, and injured or infected corms were discarded. The corms were washed three times using sterile distilled water containing Tween-20 and then rinsed with a fungicide solution (2 mL L^{-1} Ortiva; Syngentia, Switzerland) containing 200 g L^{-1} of azoxystrobin and 125 g L^{-1} of difenoconazole^{-1} for 15 min. The corms were then submerged in 10% (v/v) commercial bleach (5.2% sodium hypochlorite) for 10 min for disinfection and transferred to a 0.1% (w/v) mercuric chloride solution for 5 min. Subsequently, the corms were rinsed three more times using a sterile solution containing 2 g L^{-1} of ascorbic acid and 1 g L^{-1} of citric acid. They were then cultured in Magenta GA-7 culture vessels (77 mm × 77 mm × 97 mm; Sigma Chemical Co., St. Louis, MO, USA) containing semi-solid Murashige–Skoog (MS) medium [15] without plant growth regulators and incubated in the dark for 1 week. The cultures were checked regularly, and the percentage of corm contamination was recorded.

2.2. Saffron Flowering in In Vitro Culture and Hydroponic Culture

In vitro aseptic corms were grown in MS medium supplemented with 3% sucrose and solidified with 0.8% agar–agar. The pH of the medium was adjusted to 5.8 prior to autoclaving at 121 °C and 1.2 kg cm^{-2} pressure for 20 min. The environmental conditions of the growth chamber were adjusted to 15 °C ± 1 °C and 70 μmol m^{-2} s^{-1} photosynthetic photon flux density (PPFD) for a 16 h photoperiod using cool white fluorescent lamps. The flowering characteristics, i.e., the number of flowers per plant, stigma length, and stigma fresh and dry weights, under in vitro culture were compared with those under an aerated, volcanic-rock-based hydroponic system as described by Dewir and Alsadon [16]. The corms were obtained from the same source (Bloembollenbedrijf J.C.Koot) and were of the same size grade (3.2–3.5 cm diameters).

2.3. Effects of Explant Type and Solid/Liquid Culture on Saffron Daughter Corm Formation

Two types of liquid culture systems, temporary immersion (ebb and flow) and continuous immersion (with a net), were tested to select a suitable method for saffron cormlet production in liquid media and then compared with solid culture. Intact corms or apical buds (15 explants per bioreactor) were transferred to a 3 L balloon-type bubble bioreactor with 1.2 L of MS liquid medium supplemented with 30 g L^{-1} of sucrose and 0.5 mg L^{-1} of NAA.

The pH of the medium was adjusted to 5.8 before autoclaving at 121 °C and 1.2 kg cm^{-2} pressure for 30 min. In the immersion-type bioreactor, a supporting net was used to hold the plant material to avoid the complete submersion of explants in the liquid medium. The volume of air input was adjusted to 0.2 vvm (air volume/culture volume min^{-1}). The ebb and flow system (PLT Scientific SDN BHD, Puchong, Selangor D.E., Malaysia) was programmed to immerse the plantlets in medium four times per day for 60 min per immersion. All the bioreactors and culture vessels were maintained at 15 °C ± 1 °C and 70 μmol m^{-2} s^{-1} PPFD for a 16 h photoperiod using cool white fluorescent lamps. There were 15 explants in each treatment, and the diameter, fresh weight, and number of daughter corms were recorded after 14 weeks of culture from 9 randomly selected explants.

2.4. Effects of Salicylic Acid, Glutamine, and Jasmonic Acid on Saffron Daughter Corm Formation In Vitro

Saffron apical buds were grown in MS medium supplemented with 3% sucrose and the following additives: salicylic acid (75 or 150 mg L^{-1}), glutamine (600 or 1200 mg L^{-1}), and jasmonic acid (1 or 2 mg L^{-1}). The explants grown in MS medium without additives were used as controls. Thus, the effects of nine different treatments on daughter corm size were assessed. All media were solidified with 0.8% agar. The pH of the medium was adjusted to 5.8 prior to autoclaving at 121 °C and 1.2 kg cm^{-2} pressure for 20 min. The environmental conditions in the growth chamber were adjusted to 15 °C ± 1 °C and 70 μmol m^{-2} s^{-1} PPFD for a 16 h photoperiod using cool white fluorescent lamps. There were 15 explants in each treatment and the diameter and fresh weight of daughter corms were recorded after 14 weeks of culture from 9 randomly selected explants.

2.5. Experimental Design and Data Analysis

The experiments were conducted using a completely randomized design with three replicates per treatment. The effects of the treatments were assessed using Tukey's range test in SAS (version 6.12; SAS Institute, Inc., Cary, NC, USA).

3. Results and Discussion

3.1. In Vitro Establishment of Saffron Aseptic Culture and Flower Emergence

Preliminary trials were conducted in October 2020 to test the surface disinfection of saffron corms using sodium hypochlorite, mercuric chloride, and hydrogen peroxide for different durations and at various concentrations following the protocols described in previous studies [9,17–19]. Despite careful cleaning during corm preparation (Figure 1a–c), pathogen contamination was a serious challenge during the establishment of aseptic saffron culture (Figure 1d,e). Only 10% of the corms survived fungal and bacterial contamination, and we were unable to establish sufficient aseptic corms in this season. In the following season, we surface-disinfected saffron corms in July 2021 using the protocol described in the Materials and Methods Section, and a high percentage of aseptic corms (70%) was obtained. Teixeira da Silva et al. [20] reviewed various saffron corm disinfection methods, highlighting the numerous factors that can influence the efficiency of disinfection: cultivation conditions; the physiological state of the stock plant; the size, age, and type of the explant; the type and concentration of disinfectant; and the time and temperature of exposure. In the present study, establishing the aseptic saffron culture after the corm harvest in July, rather than in later months, increased the efficiency of corm disinfection. Bud sprouts during corm storage can increase microorganism abundance within the saffron corm, making it difficult to eradicate contaminants.

The growth of the main apical bud and the emergence of roots occurred within 1 week (Figure 1f), and all lateral buds grew within 3 weeks of culture (Figure 1g). Saffron flower buds appeared 30 days after culture (Figure 1h,i). Interestingly, the sizes of the flowers and stigmas produced in vitro and via hydroponics were similar (Figure 1j,k; Table 1). Additionally, only large saffron corms (≥2.8 cm in diameter) produced flowers in vitro; small corms did not produce flowers.

Table 1. Comparison of stigma and flower size of saffron produced under in vitro culture and via a hydroponic system (Dewir and Alsadon [16]).

	Number of Flowers/Plant	Stigma Length (mm)	Stigma Fresh Weight (mg)	Stigma Dry Weight (mg)
In vitro	1.80	41.32	42.68	5.29
Hydroponic	1.90 [NS]	42.46 [NS]	43.81 [NS]	5.68 [NS]

[NS] = non-significant according to *t*-test.

Figure 1. Surface disinfection and the establishment of in vitro saffron aseptic culture: (**a–c**) Cleaning, surface disinfection, and culture of corms on half-strength MS medium (Murashige and Skoog, 1962) supplemented with 3% sucrose; (**d,e**) Fungal contamination of in-vitro-cultured corms 1 week after incubation; (**f**) Bud break and rooting of saffron corms 7 days after incubation at 16 °C; (**g**) Shoot emergence and growth of saffron buds 3 weeks after incubation; (**h,i**) In vitro flowering of saffron corms 30 days after incubation; (**j**) Normal size and structure of in-vitro-produced saffron flowers; and (**k**) Harvested stigma of in-vitro-produced saffron.

In vitro flowering systems are considered convenient tools for studying the transition to flowering. They can be used to study specific aspects of flowering, as well as the mechanisms underlying the reproductive process [21]. Both developmental cues and environmental signals control the timing of flowering, and genetic studies have revealed the complexity of the mechanisms underlying flowering [22]. A combination of environmental, developmental, hormonal, and genetic factors determines the eventual transition to flowering [23]. However, saffron corms have no cold requirement to break dormancy or to complete flower formation, which is often the case in geophytes [24,25].

Under field conditions, floral initiation usually occurs during summer, coinciding with the revival of meristematic activity in the apical bud. Thus, corms can be cultivated directly in the field to force flowering, or they can be stored for up to 60 days at 2 °C to extend and facilitate harvesting [26,27]. Temperature, among several environmental parameters, plays a pivotal role in saffron flower induction [4,28,29]. Therefore, saffron corms are incubated at a moderately high temperature (23 °C–27 °C) to induce flowering before they are exposed to a moderately low temperature (17 °C) for floral emergence. In the present study, floral emergence occurred in vitro 30 days after the incubation of the saffron corms at 15 °C $\pm$ 1 °C. These in-vitro-produced saffron stigma can be utilized as a source of biochemicals and pharmaceuticals. They also can be explored as initial explants for callus formation and cell cultures. In-vitro-formed flowers are generally undersized compared with ex-vitro-formed flowers, as reported previously for *Phyllanthus niruri* [30] and *Spathiphyllum cannifolium* [31]. However, some in-vitro-grown plant species form normal-size flowers, e.g., *Chamomilla recutita* [32] and *Euphorbia milii* [33]. Comparing the in vitro and hydroponic flowering of saffron, the flowers formed in vitro had a similar size compared with those formed ex vitro. This may have been due to the early initiation of the saffron floral buds and the available nutrient supply from the mother corm supporting floral bud emergence and flower development.

3.2. Effects of Explant Type and Solid/Liquid Culture on Saffron Daughter Corm Formation

The use of intact mother corms or apical buds as initial explants (Figure 2a,e) influenced saffron daughter corm formation. In vitro, daughter corms were formed successfully after 10 weeks of culture (Figure 2b–g). After 14 weeks of culture, the life cycle of saffron was completed in vitro; the leaves were dried, and the saffron daughter corms were harvested (Figure 2d,g). All corms and apical bud explants formed daughter corms. However, the use of intact corms was more effective for corm production than the use of apical buds. More corms were produced, and the average fresh weight and diameter of corms were higher in mother corm explants than in apical bud explants. The number of cormlets, cormlet fresh weight, and cormlet diameter were also influenced by the culture type (i.e., continuous immersion bioreactors, ebb and flow bioreactors, and solid culture) (Figure 2h,i and Figure 3). Intact corms produced a higher number of corms than were produced by apical bud explants. The average number of cormlets per intact corm was 2.0–4.5 cormlets, owing to the presence of several lateral buds in the mother corms, whereas the apical buds produced single cormlets. For both intact corms and apical buds, liquid cultures using bioreactors resulted in corms with a higher fresh weight and diameter than those grown via solid culture. The average fresh weight and diameter of cormlets was the highest under the ebb and flow system although a low number of cormlets was obtained. Conversely, a higher number of cormlets with a smaller size was obtained using the immersion system.

Liquid medium is known to promote nutrient uptake and encourage growth and proliferation in vitro. The bioreactors designed for plant cultures provide optimal aeration and mixing with oxygen, without subjecting the propagules to shear stress [11]. Several plant species exhibit better growth in liquid/bioreactor cultures than in solid cultures, e.g., *E. milli* [34,35], *S. cannifolium* [36], *Gentiana triflora* [37], *Lessertia frutescens* [38], *Kalopanax septemlobus* [39], and *Philodendron bipinnatifidum* [40]. Piqueras et al. [41] highlighted the improved development of the cormogenic nodules and microcorms of saffron grown in liquid cultures relative to solid cultures. In the present study, the high number of cormlets and fresh mass of saffron corms grown in a bioreactor system might be attributable to enhanced water and nutrient uptake, as well as aeration, i.e., conditions that favor better growth compared with those under a nonaerated, solid culture system. Compared with continuous immersion, the ebb and flow system produced fewer cormlets, but they were larger. The higher efficacy of the ebb and flow system, in terms of fostering optimum cormlet growth, is probably due to culture ventilation and the intermittent contact between the entire tissue surface and the liquid medium, two features that are not usually combined in other liquid culture procedures [42,43].

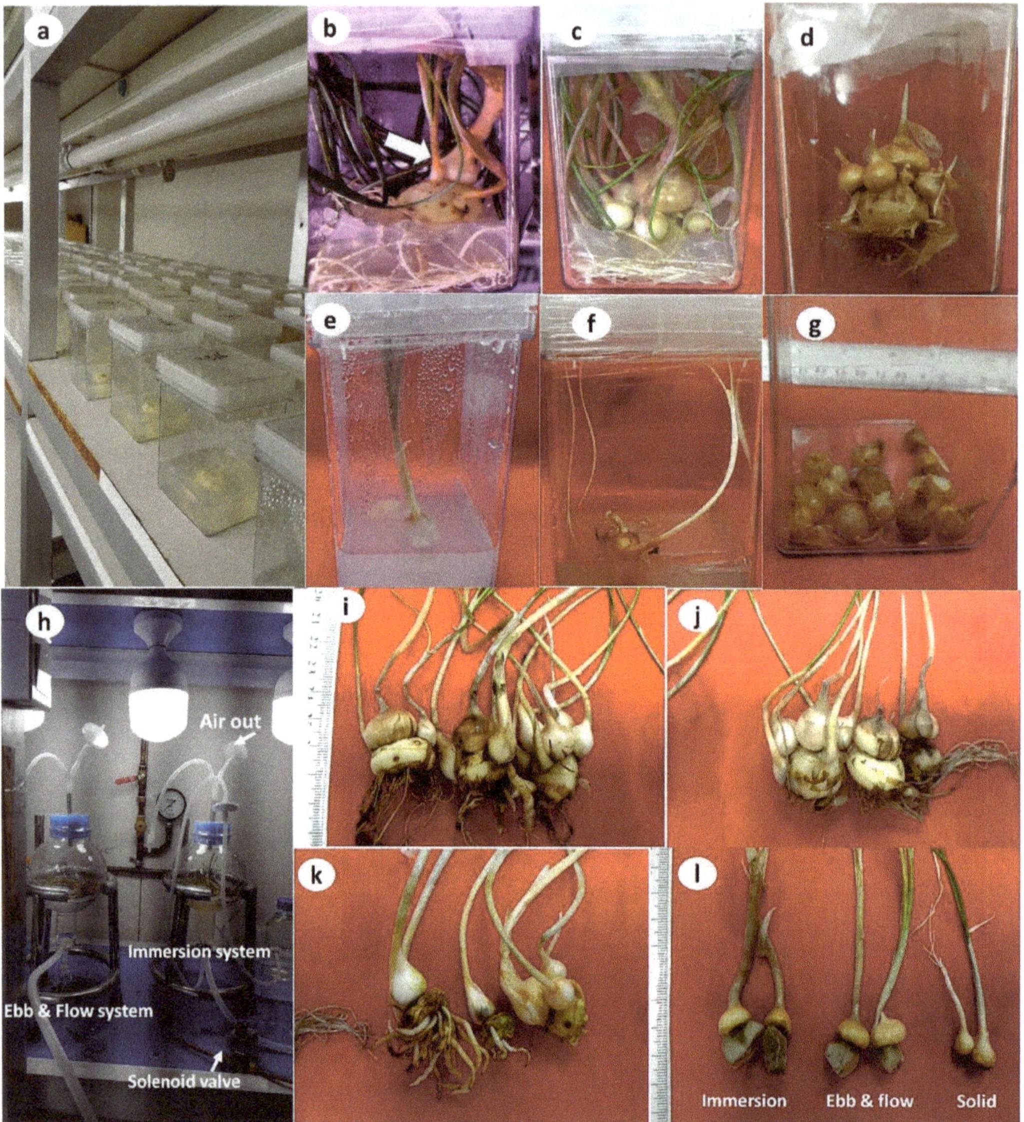

Figure 2. Effects of explant type (intact corms and apical buds) and culture system (solid culture, continuous immersion bioreactors, and ebb and flow bioreactors) on in-vitro-produced daughter saffron corm formation: (**a**) Established saffron aseptic culture; (**b–d**) Daughter corm formation following 10, 14, and 16 weeks, respectively, of solid culture using intact corm explants; (**e**) Bud break and shoot development of apical buds after 2 weeks of culture; (**f,g**) Daughter corm formation following 14 and 16 weeks, respectively, of solid culture using intact corm explants; (**h**) Layout of continuous immersion and ebb and flow bioreactors; (**i–k**) Daughter corm formation using intact corms in continuous immersion bioreactors, ebb and flow bioreactors, and solid culture, respectively, after 16 weeks of culture; (**l**) Daughter corm formation using apical buds in continuous immersion bioreactors, ebb and flow bioreactors, and solid culture, respectively, after 16 weeks of culture.

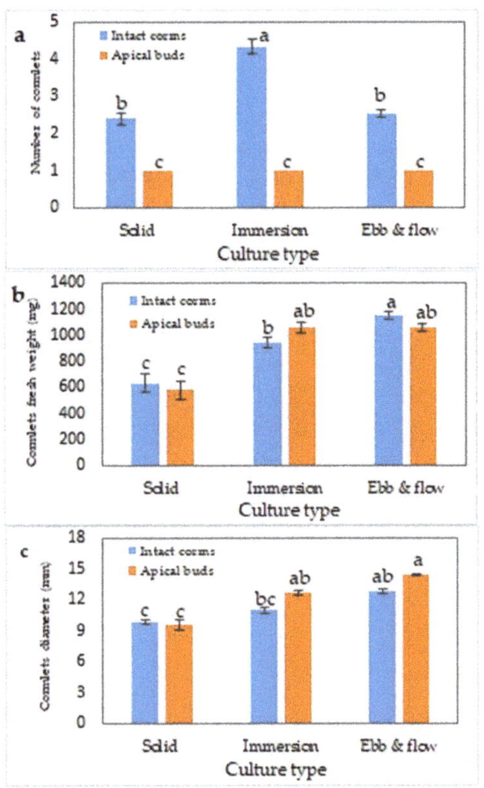

Figure 3. Effects of explant type (intact corms and apical buds) and culture type (solid culture, continuous immersion bioreactors, and ebb and flow bioreactors) on number of cormlets (**a**), cormlet fresh weight (**b**), and cormlet diameter (**c**) of saffron. Different letters show significant differences at $p \leq 0.05$.

3.3. Effects of Salicylic Acid, Glutamine, and Jasmonic Acid on Saffron Daughter Corm Formation In Vitro

The apical buds of saffron cultured on MS medium supplemented with different additives, i.e., salicylic acid, glutamine, and jasmonic acid, showed variation in cormlet diameter and fresh weight (Table 2; Figure 4). With the exception of 2 mg L^{-1} of jasmonic acid, all other treatments increased cormlet diameter and fresh weight relative to the control treatment, with salicylic acid at 75 mg L^{-1} and glutamine at 600 mg L^{-1} providing the highest and second highest increases, respectively, increasing the corm diameter by around 0.4 cm and nearly doubling the fresh weight compared with the control treatment. Salicylic acid is a phytohormone that plays important roles in many aspects of plant life, including seed germination, physiological and biochemical processes, flowering, and fruit yield [44,45]. Under field conditions, salicylic acid is effective for improving plant and bulb growth. In a plastic house experiment, soaking *Iris hollandica* bulbs in salicylic acid (200 mg L^{-1} for 4 h) improved plant vegetative growth [46]. In another study, the exogenous application of salicylic acid (250 mg L^{-1} at 30, 45, and 60 days after transplanting) increased not only the vegetative growth, but also the bulb weight, diameter, and yield of onion (*Allium cepa* 'ALR') [47]. In a study on saffron, corm dipping in a solution of salicylic acid (2 mM for 6 h) improved vegetative and reproductive characteristics, as well as yield quality [48]. Similarly, the application of salicylic acid (1–2 mM) to saffron, either during corm priming or foliar application, improved the vegetative growth and physiological and biochemical characteristics [49]. Under in vitro conditions, the exogenous application of salicylic acid is useful for the growth and development of plants. Salicylic acid has been reported to enhance microtuber formation [50] and increase starch percentage [51] in potato plants (*Solanum tuberosum*) propagated in vitro. Glutamine is an amino acid and nitrogen source that supports several metabolic processes. The addition of amino acids provides a readily available primary source of nitrogen in tissue culture systems, and uptake occurs more rapidly than the uptake of inorganic nitrogen in the same medium [52]. Glutamine has been reported to have positive effects in plant tissue cultures [53,54]. Jasmonic acid is an endogenous plant-growth regulating substance. In the present study, jasmonic acid at 1 mg L^{-1} increased corm size compared with corm size in the control, but 2 mg L^{-1} of jasmonic acid did not produce positive effects. Exogenous jasmonic acid is considered to play an important role in bulb formation, as shown in studies on garlic [55] and onion [56]. In addition, exogenous jasmonic acid supported the microtuberization of three food yam (*Dioscorea*) species [57] and promoted the enlargement of in-vitro-grown bulbs in shoot cultures of *Narcissus* plants [58] and the daughter bulb development of *Tulipa gesneriana* [59]. However, the positive or negative effects of jasmonic acid are species- and concentration-dependent.

Table 2. Effects of salicylic acid, glutamine, and jasmonic acid treatments on saffron daughter corm diameter and fresh weight after 14 weeks of culture.

Treatments	Concentration (mg L^{-1})	Average Daughter Corm Diameter (mm)	Average Daughter Corm Fresh Weight (g)
Control	0	9.93 d	0.94 c
Salicylic acid	75	14.33 a	1.99 a
	150	11.40 c	1.28 bc
Glutamine	600	13.89 ab	1.89 a
	1200	11.83 abc	1.73 ab
Jasmonic acid	1	12.49 abc	1.24 bc
	2	9.58 d	0.82 c

Different letters within a set of values denote significant differences at $p \leq 0.05$ according to Tukey's test.

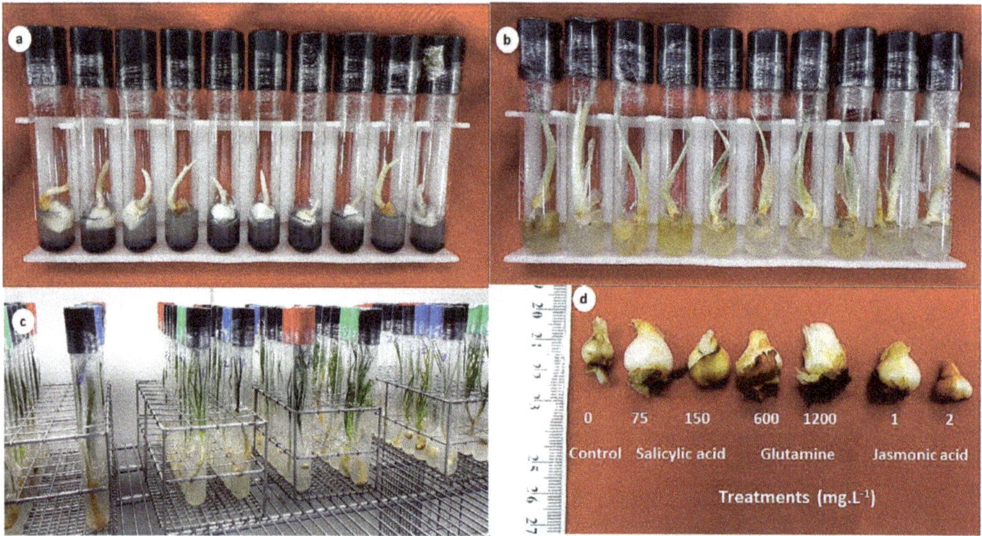

Figure 4. Influence of salicylic acid, glutamine, and jasmonic acid on saffron daughter corm formation using apical bud explants: (**a**) Surface disinfection and bud break of saffron apical buds cultured on MS medium supplemented with 3% sucrose and 1 g L^{-1} of activated charcoal 2 weeks after incubat-

ion; (**b**) Shoot development of saffron apical buds and growth 6 weeks after incubation; (**c,d**) Daughter corm formation 12 and 16 weeks, respectively, after treatment with salicylic acid (75 and 150 mg L^{-1}), glutamine (600 and 1200 mg L^{-1}), and jasmonic acid (1 and 2 mg L^{-1}).

4. Conclusions

The present results indicate that saffron flowering can be induced in vitro, and the harvested stigma of these flowers could be used as a source of spice or pharmaceuticals. Compared with solid culture, liquid cultures/bioreactors improved daughter corm diameter and fresh weight. Moreover, salicylic acid at 75 mg L^{-1} and glutamine at 600 mg L^{-1} increased corm diameter and fresh weight. These findings will help improve in vitro cormogenesis toward the production of saffron in a controlled environment. However, further investigations on optimal growth conditions, e.g., medium composition, light intensity and quality, and incubation temperature, are required to optimize in vitro cormogenesis.

Author Contributions: Conceptualization, Y.H.D. and A.A.; methodology, Y.H.D., A.A. and A.A.A.-A.; formal analysis, Y.H.D., A.A. and A.A.A.-A.; investigation and data curation, Y.H.D., A.A., A.A.A.-A. and M.A.-M.; validation, Y.H.D., A.A., A.A.A.-A. and M.A.-M.; visualization, Y.H.D., A.A., A.A.A.-A. and M.A.-M.; writing—original draft preparation, Y.H.D. and A.A.; writing—review and editing, Y.H.D., A.A., A.A.A.-A. and M.A.-M. All authors have read and agreed to the published version of the manuscript.

Funding: This project was funded by the National Plan for Science, Technology and Innovation (MAARIFAH), King Abdulaziz City for Science and Technology, Kingdom of Saudi Arabia, Award Number (15-AGR3704-02).

Institutional Review Board Statement: Not applicable.

Informed Consent Statement: Not applicable.

Data Availability Statement: All data are presented within the article.

Acknowledgments: This project was funded by the National Plan for Science, Technology and Innovation (MAARIFAH), King Abdulaziz City for Science and Technology, Kingdom of Saudi Arabia, Award Number (15-AGR3704-02).

Conflicts of Interest: The authors declare no conflict of interest.

References

1. Basker, D.; Negbi, M. The use of Saffron. *Eco. Bot.* **1983**, *37*, 228–236. [CrossRef]
2. Nemati, Z.; Harpke, D.; Gemicioglu, A.; Kerndorff, H.; Blattner, F.R. Saffron (*Crocus sativus*) is an autotriploid that evolved in Attica (Greece) from wild *Crocus cartwrightianus*. *Mol. Phylogenet. Evol.* **2019**, *136*, 14–20. [CrossRef] [PubMed]
3. Douglas, M.H.; Smallfield, B.M.; Wallace, A.R.; McGimpsey, J.A. Saffron (*Crocus sativus* L.): The effect of mother corm size on progeny multiplication, flower and stigma production. *Sci. Hortic.* **2014**, *166*, 50–58. [CrossRef]
4. Molina, R.; Valero, M.; Navarro, Y.; Guardiola, J.L.; Garcia-Luis, A. Temperature effects on flower formation in saffron (*Crocus sativus* L.). *Sci. Hortic.* **2005**, *103*, 361–379. [CrossRef]
5. Gohari, A.R.; Saeidnia, S.; Mahmoodabadi, M.K. An overview on saffron, phytochemicals, and medicinal properties. *Pharmacogn. Rev.* **2013**, *7*, 61–66. [CrossRef]
6. Dhar, A.; Sapru, R. Studies on saffron in Kashmir and in vitro production of corm and shoot-like structures. *Indian J. Genet.* **1993**, *53*, 193–196.
7. Raja, W.; Zaffer, G.; Wani, S.A. In vitro microcorm formation in saffron (*Crocus sativus* L). *Acta Hortic.* **2007**, *739*, 291–296. [CrossRef]
8. Sharifi, G.; Ebrahimzadeh, H.; Ghareyazie, B.; Karimi, M. Globular embryo-like structures and highly efficient thidiazuron-induced multiple shoot formation in saffron (*Crocus sativus* L.). *Vitr. Cell. Dev. Biol.-Plant* **2010**, *46*, 274–280. [CrossRef]
9. Zeybek, E.; Önde, S.; Kaya, Z. Improved in vitro micropropagation method with adventitious corms and roots for endangered saffron. *Cent. Eur. J. Biol.* **2012**, *7*, 138–145. [CrossRef]
10. Mehrotra, S.; Goel, M.K.; Kukreja, A.K.; Mishra, B.N. Efficiency of liquid culture systems over conventional micropropagation: A progress towards commercialization. *Afr. J. Biotechnol.* **2007**, *6*, 1684–5315.
11. Paek, K.Y.; Hahn, E.J.; Son, S.H. Application of bioreactors of large scale micropropagation systems of plants. *Vitr. Cell. Dev. Biol.-Plant* **2001**, *37*, 149–157. [CrossRef]
12. Milyaeva, E.L.; Azizbekova, N.S.; Komarova, E.N.; Akhundova, D.D. In vitro formation of regenerant corms of saffron crocus (*Crocus sativus* L.). *Russ. J. Plant Physiol.* **1995**, *42*, 112–119.

13. Chauhan, R.S.; Sharma, T.R.; Chahota, R.K.; Singh, B.M. In vitro cormlet production from micropropagated shoots of saffron (*Crocus sativus* L.). *Indian Perfum.* **1999**, *43*, 150–155.
14. Sharma, K.D.; Rathour, R.; Sharma, R.; Goel, S.; Sharma, T.R.; Singh, B.M. In vitro cormlet development in *Crocus sativus*. *Biol. Plant.* **2008**, *52*, 709–712. [CrossRef]
15. Murashige, T.; Skoog, F. A revised medium for rapid growth and bioassays with tobacco tissue cultures. *Physiol. Plant.* **1962**, *15*, 473–497. [CrossRef]
16. Dewir, Y.H.; Alsadon, A. Effects of Nutrient Solution Electrical Conductivity on the Leaf Gas Exchange, Biochemical Stress Markers, Growth, Stigma Yield, and Daughter Corm Yield of Saffron in a Plant Factory. *Horticulturae* **2022**, *8*, 673. [CrossRef]
17. Plessner, O.; Ziv, M.; Negbi, M. In vitro corm production in the saffron crocus (*Crocus sativus* L). *Plant Tissue Organ Cult.* **1990**, *20*, 89–94. [CrossRef]
18. Parray, J.A.; Kamili, A.N.; Hamid, R.; Husaini, A.M. In vitro cormlet production of saffron (*Crocus sativus* L. ashmirianus) and their flowering response under greenhouse. *GM Crops Food Biotechnol. Agric. Food Chain.* **2012**, *3*, 289–295. [CrossRef]
19. Cavusoglu, A.; Sulusoglu, M.; Erkal, S. Plant regeneration and corm formation of saffron (*Crocus sativus* L.) in vitro. *Res. J. Biotech.* **2013**, *8*, 128–133.
20. Teixeira da Silva, J.A.; Kulus, D.; Zhang, X.; Zeng, S.; Ma, G.; Piqueras, A. Disinfection of explants for saffron (*Crocus sativus*) tissue culture. *Environ. Exp. Biol.* **2016**, *14*, 183–198. [CrossRef]
21. Dewir, Y.H.; Paek, K.Y. The control of in vitro flowering and association of glutathione metabolism. In *Plant Tissue Culture and Applied Plant Biotechnology*; Kumar, A., Roy, S., Eds.; Aavishkar Publishers, Distributors: Rajasthan, India, 2011; pp. 77–108.
22. Teixeira da Silva, J.A.; Tan Nhut, D. Thin cell layers and floral morphogenesis, floral genetics and in vitro flowering. In *Thin cell Layer Culture System: Regeneration and Transformation Application*; Tan Nhut, D., Van Le, B., Tran Thanh Van, K., Thorpe, T., Eds.; Kluwer Academic Publishers: Dordrecht, The Netherlands; Boston, MA, USA; London, UK, 2003; pp. 285–342.
23. Sudhakaran, S.; Teixeira da Silva, J.A.; Sreeramanan, S. Test tube bouquets: In vitro flowering. In *Floriculture, Ornamental and Plant Biotechnology: Advances and Topical Issues*, 1st ed.; Teixeira da Silva, J.A., Ed.; Global Science Books: London, UK, 2006; Volume 2, pp. 336–346.
24. Dole, J.M. Research approaches for determining cold requirements for forcing and flowering of geophytes. *HorScience* **2003**, *38*, 341–346. [CrossRef]
25. Rees, A.R. *Ornamental Bulbs, Corms and Tubers*; CAB International: Wallingford, UK, 1992.
26. Amooaghaie, R. Low temperature storage of corms extends the flowering season of saffron (*Crocus sativus* L.). *Acta Hortic.* **2007**, *739*, 41–47. [CrossRef]
27. Mollafilabi, A.; Koocheki, A.; Moghaddam, P.R.; Mahallati, M.N. Effects of bed type, corm weight and lifting time on quantitative and qualitative criteria of saffron (*Crocus sativus* L.). *J. Agroecol.* **2017**, *9*, 607–617. [CrossRef]
28. Kafi, M.; Mohassel, M.H.R.; Koocheki, A.; Molafilabi, M. *Saffron: Production and Processing*; Ferdowsi University of Mashhad Press: Mashhad, Iran, 2002.
29. Molina, R.V.; Valero, M.; Navarro, Y.; Garcia-Luis, A.; Guardiola, J. Low temperature storage of corms extends the flowering season of saffron (*Crocus sativus* L.). *J. Hortic. Sci. Biotechnol.* **2005**, *80*, 319–326. [CrossRef]
30. Liang, O.P.; Keng, C.L. In vitro plant regeneration, flowering and fruiting of *Phyllanthus niruri* L. (Euphorbiaceae). *Int. J. Bot.* **2006**, *2*, 409–414. [CrossRef]
31. Dewir, Y.H.; Chakrabarty, D.; Ali, M.B.; Singh, N.; Hahn, E.-J.; Paek, K.-Y. Influence of GA_3, sucrose and solid medium/bioreactor culture on in vitro flowering of *Spathiphyllum* and association of glutathione metabolism. *Plant Cell Tissue Organ Cult.* **2007**, *90*, 225–235. [CrossRef]
32. Kintzios, S.; Michaelakis, A. Induction of somatic embryogenesis and in vitro flowering from inflorescences of chamomile (*Chamomilla recutita* L.). *Plant Cell Rep.* **1999**, *18*, 684–690. [CrossRef]
33. Dewir, Y.H.; Chakrabarty, D.; Hahn, E.J.; Paek, K.Y. The effects of paclobutrazol, light emitting diodes (LEDs) and sucrose on flowering of *Euphorbia millii* plantlets in vitro. *Eur. J. Hortic. Sci.* **2006**, *71*, 240–244.
34. Dewir, Y.H.; Chakrabarty, D.; Hahn, E.J.; Paek, K.Y. Reversion of inflorescence development in *Euphorbia millii* and its application to large-scale micropropagation in an air-lift bioreactor. *J. Hortic. Sci. Biotechnol.* **2005**, *80*, 581–587. [CrossRef]
35. Dewir, Y.H. *Mass Propagation of the Christ Plant Using Bioreactor and Hydroponics*; Scholars' Press: Newcastle upon Tyne, UK, 2016; p. 108. ISBN 978-3-659-84521-5.
36. Dewir, Y.H.; Chakrabarty, D.; Hahn, E.J.; Paek, K.Y. A simple method for mass propagation of *Spathiphyllum cannifolium* using an airlift bioreactor. *Vitr. Cell. Dev. Biol.-Plant* **2006**, *42*, 291–297. [CrossRef]
37. Dewir, Y.H.; Singh, N.; Govender, S.; Pillay, P. In vitro flowering and shoot multiplication of *Gentiana triflora* in air-lift bioreactor cultures. *J. Appl. Hortic.* **2010**, *12*, 30–34. [CrossRef]
38. Shaik, S.; Dewir, Y.H.; Singh, N.; Nicholas, A. Micropropagation and bioreactor studies of the medicinally important plant Lessertia (*Sutherlandia*) *frutescens* L. *South Afr. J. Bot.* **2010**, *76*, 180–186. [CrossRef]
39. Kim, S.J.; Dewir, Y.H.; Moon, H.K. Large-scale plantlets conversion from cotyledonary somatic embryos of *Kalopanax septemlobus* tree using bioreactor cultures. *J. Plant Biochem. Biotechnol.* **2011**, *20*, 241–248. [CrossRef]
40. Alawaadh, A.A.; Dewir, Y.H.; Alwihibi, M.S.; Aldubai, A.A.; El-Hendawy, S.; Naidoo, Y. Micropropagation of lacy tree Philodendron (*Philodendron bipinnatifidum* Schott ex Endl.). *HortScience* **2020**, *55*, 294–299. [CrossRef]

41. Piqueras, A.; Han, B.H.; Escribano, J.; Rubio, C.; Hellín, E.; Fernández, J.A. Development of cormogenic nodules and microcorms by tissue culture, a new tool for the multiplication and genetic improvement of saffron. *Agronomie* **1999**, *19*, 603–610. [CrossRef]
42. Etienne, H.; Berthouly, M. Temporary immersion systems in plant micropropagation. *Plant Cell Tissue Organ Cult.* **2002**, *69*, 215–231. [CrossRef]
43. Hwang, H.-D.; Kwon, S.-H.; Murthy, H.N.; Yun, S.-W.; Pyo, S.-S.; Park, S.-Y. Temporary Immersion Bioreactor System as an Efficient Method for Mass Production of In vitro Plants in Horticulture and Medicinal Plants. *Agronomy* **2022**, *12*, 346. [CrossRef]
44. Arif, Y.; Sami, F.; Siddiqui, H.; Bajguz, A.; Hayat, S. Salicylic acid in relation to other phytohormones in plant: A study towards physiology and signal transduction under challenging environment. *Environ. Exp. Bot.* **2020**, *175*, 104040. [CrossRef]
45. Kang, G.; Wang, C.; Sun, G.; Wang, Z. Salicylic acid changes activities of H_2O_2-metabolizing enzymes and increases the chilling tolerance of banana seed-lings. *Environ. Exp. Bot.* **2003**, *50*, 9–15. [CrossRef]
46. Alwan, N.M.; Aziz, N.K.; Sadiq, S.M. Effect of the period of soaking and concentrations of salicylic acid in the growth and production of iris. *IOSR J. Agric. Vet. Sci. (IOSR-JAVS)* **2018**, *11*, 34–41.
47. Tripathy, P. Effect of salicylic acid on growth and bulb yield of onion (*Allium cepa* L.). *Int. J. Bio-Resour. Stress Manag.* **2016**, *7*, 960–963.
48. Khayyat, M.; Jabbari, M.; Fallahi, H.R.; Samadzadeh, A. Effects of corm dipping in salicylic acid or potassium nitrate on growth, flowering, and quality of saffron. *J. Hortic. Res.* **2018**, *26*, 13–21. [CrossRef]
49. Ansarian Mahabadi, S.; Alahdadi, I.; Ghorbani Javid, M.; Soltani, E. Effects of Salicylic Acid and Application Methods on Daughter Corm Yield and Physiological Characteristics of Saffron (*Crocus sativus* L). *J. Saffron Res.* **2019**, *6*, 203–218.
50. Koda, Y.; Takahashi, K.; Kikuta, I. Potato tuber inducing activities of salicylic acid and related compounds. *J. Plant Growth Regul.* **1992**, *11*, 215–221. [CrossRef]
51. Alutbi SDAl-Saadi, S.A.A.M.; Madhi, Z.J. The effect of salicylic acid on the growth and Microtuberization of potato (*Solanum tuberosum* L.) cv. Arizona propagated in vitro. *Am. J. Potato Res.* **2017**, *95*, 395–412.
52. George, E.F.; Hall, M.A.; Klerk, G.-J.D. *Plant Propagation by Tissue Culture*; The Background; Springer: Dordrecht, The Netherlands, 2008; Volume 1.
53. Vasudevan, A.; Selvaraj, N.; Ganapathi, A.; Kasthurirengan, S.; Ramesh Anbazhagan, V.; Manickavasagam, M. Glutamine: A suitable nitrogen source for enhanced shoot multiplication in *Cucumis sativus* L. *Biol. Plant.* **2004**, *48*, 125–128. [CrossRef]
54. Greenwell, Z.L.; Ruter, J.M. Effect of glutamine and arginine on growth of *Hibiscus moscheutos* "in vitro". *Ornam. Hortic.* **2018**, *24*, 393–399. [CrossRef]
55. Ravnikar, M.; Zel, J.; Plaper, I.; Spacapan, A. Jasmonic acid stimulates shoot and bulb formation of garlic in vitro. *J. Plant Growth Regul.* **1993**, *12*, 73–77. [CrossRef]
56. Koda, Y. Possible involvement of jasmonates in various morphogenic events. *Physiol. Plant.* **1997**, *100*, 639–946. [CrossRef]
57. Jasik, J.; Mantell, S.H. Effects of jasmonic acid and its methylester on in vitro microtuberisation of three food yam (*Dioscorea*) species. *Plant Cell Rep.* **2000**, *19*, 863–867. [CrossRef]
58. Santos, I.; Salema, R. Promotion by jasmonic acid of bulb formation in shoot cultures of *Narcissus triandrus* L. *Plant Growth Regul.* **2000**, *30*, 133–138. [CrossRef]
59. Sun, Q.; Zhang, B.; Yang, C.; Wang, W.; Xiang, L.; Wang, Y.; Chan, Z. Jasmonic acid biosynthetic genes TgLOX4 and TgLOX5 are involved in daughter bulb development in tulip (*Tulipa gesneriana*). *Hortic. Res.* **2022**, *9*, uhac006. [CrossRef]

 *horticulturae*

Article

Phytochemicals and Antioxidant Activities of Conventionally Propagated Nodal Segment and In Vitro-Induced Callus of *Bougainvillea glabra* Choisy Using Different Solvents

Mohammad Nasim Nasrat [1,2], **Siti Zaharah Sakimin** [1] **and Mansor Hakiman** [1,3,*]

1   Department of Crop Science, Faculty of Agriculture, Universiti Putra Malaysia (UPM),
     Serdang 43400, Selangor, Malaysia
2   Department of Horticulture, Faculty of Agriculture, Herat University, Herat 3001, Afghanistan
3   Laboratory of Sustainable Resources Management, Institute of Tropical Forestry and Forest Products,
     Universiti Putra Malaysia (UPM), Serdang 43400, Selangor, Malaysia
*   Correspondence: mhakiman@upm.edu.my; Tel.: +60-162-221-070

Abstract: *Bougainvillea*, popularly known as 'Bunga kertas' in Malaysia, is thoroughly explored for nutritional and medicinal purposes. *Bougainvillea* has been shown to possess alkaloids and flavonoids which are widely used in folk medicine to treat different illnesses such as inflammatory, diarrheal, ulcer, and diabetic. Despite its major conventional therapeutic importance, only limited attempts have been made to investigate this species' chemical and pharmacological properties in relation to its medicinal uses. Therefore, this study was conducted to determine the effect of in vitro-induced callus under different light conditions and plant growth regulators on phytochemical and antioxidant activities using different extraction solvents. Based on the results, the maximum days (17.67) to callus initiation were recorded when nodal was cultured on woody plant medium (WPM) supplemented with 7.5 µM 2,4-D + 0.5 µM BAP under light condition. On the contrary, the minimum days (7) to callus initiation were obtained when nodal was treated with 2.5 and 5 µM 2,4-D + 1 and 1.5 µM BAP under dark conditions. However, higher fresh and dry weight of callus was obtained when nodal was cultured on woody plant medium fortified with 7.5 µM 2,4-D + 1.5 µM BAP under dark and light conditions. In the analysis of the phenolics content and antioxidant activities, aqueous extract of conventionally propagated nodal part exhibited the highest phenolic content and antioxidant activities. However, the highest iron (II) chelating activity was produced from the aqueous extract of the calli induced under a dark condition. Hence, it can be concluded that the callus culture of *Bougainvillea* produced plant secondary metabolites and antioxidant activities comparable to the mother plants.

Keywords: ornamental plant; cytokinin; auxin; incubation condition; callus induction; antioxidant; phenolic

Citation: Nasrat, M.N.; Sakimin, S.Z.; Hakiman, M. Phytochemicals and Antioxidant Activities of Conventionally Propagated Nodal Segment and In Vitro-Induced Callus of *Bougainvillea glabra* Choisy Using Different Solvents. *Horticulturae* **2022**, *8*, 712. https://doi.org/10.3390/horticulturae8080712

Academic Editor: Luigi De Bellis

Received: 24 July 2022
Accepted: 1 August 2022
Published: 8 August 2022

1. Introduction

Medicinal plants or herbs are enriched in various phytochemicals, which have been shown to have various biological effects and should be researched further [1,2]. The World Health Organization (WHO) recognized the importance of medicinal plants in millions of people's primary care. It is estimated that more than 80% of the world's population relies on these resources as their primary source of health-related problems [3]. One of the medicinal plants that has currently gain much attention because of its therapeutic constituents is *Bougainvillea glabra* [4]. The genus of *Bougainvillea* belongs to the family Nyctaginaceae, which is one of the utmost valuable ornamental and medicinal plants native to South America (Brazil, Peru, and northern Argentina). The name derives from the French navigator Louis Antoine de Bougainville, who was the first to discover this plant in Brazil in 1786 [5,6]. *B. glabra* is a perennial and evergreen shrub that is widely growing in warm climates like Indonesia, Ethiopia, Philippines, Thailand, Malaysia, Vietnam, Taiwan, India, Australia, Mexico, South Africa, United States, Central America, Caribbean, and the Mediterranean [7].

Species of the *Bougainvillea* genus are frequently explored for their nutritional and therapeutic properties. The plants of the *Bougainvillea* species are considered to be utilized in traditional medicine for a variety of ailments, as phytopharmacological studies have reported, including anti-inflammatory and antipyretic [8], antidiarrheal, pain ailments and antiulcer [9], antimicrobial [10], antidiabetic [11], immunomodulatory [12], hypoglycaemic [13], antihyperlipidemic [14], anti-cough, sore throat, blood vessel troubles, leucorrhoea, hepatitis [15], anti-skin problems (tyrosinase) [16], analgesic [17], antiviral [18], antifungal [19], neuroprotective [4], and anthelmintic [20]. Various parts of the *Bougainvillea* plants are used to treat various ailments. In Panama, the flowers are used for treating hypotension. In India, leaves, flowers, and stem barks treat many illnesses, including stomach acidity, diarrhea, cough, blood vessel problems, sore throat, and hepatitis. In Thailand, the flowers treat stomachache, nausea, and diarrhea [1].

According to the report by Abarca-Vargas and Petricevich [15] on *Bougainvillea glabra*, 35 volatile compounds, 4 phenolic compounds, and 21 flavonoids compounds were extracted from leaves, bracts, and branched. Thus, the demand for raw materials is increased in tropical areas with temperate and cool weather. *B. glabra* is vegetatively propagated by stem cuttings. However, in the traditional propagation method, the production of secondary metabolites is known to be unsuitable due to external factors in the environment such as climate, plant pests and diseases, and fertilizer application. Therefore, the plant tissue culture technique is the most suitable technique for the consistent production of secondary metabolites under a controlled condition. A massive amount of raw materials is needed to extract the high amount of secondary metabolites which is not possible to be produced through conventional methods due to land availability. This problem can be overcome by producing the secondary metabolites in the laboratory by plant tissue culture technique [21]. Plant tissue culture is the most effective way to propagate rare, endangered, and valuable medicinal and commercial plant species on a wide scale in a short time while also protecting them [22]. Furthermore, callus formation investigations are helpful in understanding the metabolic pathways of secondary metabolites. On the other hand, the selection of plant growth regulators (PGRs) and environmental (cultural) conditions have an impact on callus growth in culture. Specific PGRs at appropriate concentrations can play an important role during callogenesis. Auxin and cytokinin type and concentration are essential determinants of in vitro callogenesis and regeneration. The ratio of auxin to cytokinin is the most crucial factor in this case [23]. Because the extraction of secondary metabolites requires a considerable amount of biomass, tissue culture using multiple shoots and callus induction is one of the methods used.

Extraction is the first and vital step in analyzing secondary metabolite constituents from plant materials [24]. Due to the presence of diverse molecules with varying chemical features, the polarity of the solvent used for extraction and the method of extraction play critical roles in both the efficiency and efficacy of plant secondary metabolites [24–28]. Moreover, extraction from plant products is complicated and difficult due to the vast range of structures and polarity of chemical compounds. Solvent, time, solid-to-solvent ratio, number of extractions, temperature, and partial size of the sample material are all important extraction parameters [29].

As far as we know, there are not enough studies on the influence of white light and dark on callus induction and phytochemical/antioxidant activity in *B. glabra*. Hence, we studied the effect of light quality on morphological and biochemical components of in vitro grown node-derived callus cultures of *B. glabra*. This research will help understand the effect of light on the production of commercially essential secondary metabolites and their optimization in the in vitro cultures of *B. glabra*.

2. Materials and Methods

2.1. In Vitro Callus Induction

2.1.1. Plant Materials and Sterilization

The nodal segments of conventionally propagated plants of *Bougainvillea glabra* Choisy were collected from the campus, Faculty of Agriculture, Universiti Putra Malaysia, Serdang, Selangor, Malaysia. The samples of *B. glabra* plants used in the study were deposited at Biodiversity Unit, Institute of BioScience, Universiti Putra Malaysia with a plant confirmation voucher number MFI 0197/21.

For the sterilization process, after removing the leaves from cut branches, nodal segments were cut into pieces (0.5–1 cm) and the explants were washed under running tap water containing a few drops of detergent for 30 min to remove dust particles. After that, the explants were washed with sterile distilled water once and put in a glass jar containing 500 mg/250 mL (*w/v*) streptomycin + 500 mg/250 mL (*w/v*) bavistin for pre-treatment of the explants for one hour [30]. Then, the explants were disinfected by immersion in 70% ethanol for 25 s and washed two times with autoclaved distilled water. Then, the explants' surface was sterilized with a bleaching agent with a concentration of 30% of Clorox® (5.25% (*w/v*) of sodium hypochlorite and a few drops of tween-20 as an emulsifier per 100 mL solution for 15 min by using a shaker with 300 rpm [31]. After that, the explants were washed three times with autoclaved distilled water and prepared for culturing.

Chemicals and Reagents

Folin-Ciocalteu reagent, potassium persulphate, iron (II) chloride, sodium hydroxide, ammonium chloride, gallic acid, sodium nitrite, sodium carbonate, dimethyl sulfoxide (DMSO), hexane, methanol, acetone, and ethyl alcohol were purchased from (R&M Chemical, Selangor, Malaysia). 2,2′-azino-bis (3-ethylbenzothiazoline-6-sulfonic acid) ABTS was purchased from (Alfa Aesar, Mumbai, India). 2,2-diphenyl-1-picrylhydrazyl DPPH (Sigma-Aldrich, Baden-Württemberg, Germany). Ferrozine reagent (Acros Organics, Burgenland, Austria), rutin hydrate (Sigma, China), Trolox (6-hydroxy-2,5,7,8-tetramethylchromane-2-carboxylic acid) (Sigma-Aldrich, Shanghai, China). N6-benzylaminopurine (R&M Chemical, Cardiff, UK), 2,4-dichlorophenoxyaceticacid (BDH, Poole, UK), streptomycin sulfate (Sigma, St. Louis, MO, USA), gelrite (Duchefa Biochemie, Haarlem, The Netherlands), Clorox® (Clorox Sdn. Bhd., Kuala Lumpur, Malaysia), Polysorbate 20 or Tween 20 (YKL Multi Sdn. Bhd., Bukit Mertajam, Malaysia) were also used in the study. All of the chemicals and reagents were of analytical grade.

2.1.2. Preparation of Basal Medium, Aseptic Condition and Glassware

A basal medium was prepared using a formulation described by Lloyd and McCown (WPM) (1981). The basal medium was fortified with 30 g/L sucrose, 3.0 g/L gelrite as a gelling agent, and pH was adjusted to 5.7 ± 0.5 using 1 M sodium hydroxide (NaOH) or 1 M hydrochloric acid (HCl). Then, the basal medium was autoclaved at a temperature of 121 °C for 20 min at a pressure of 1.05 kg/cm^2. The laminar airflow chamber was exposed to ultraviolet (UV) light for 30 min to sterilize the surface of the working area.

The glassware used for the culture comprised of 35 mL culture tubes for surface sterilization with 10 mL of solid WPM medium and 150 to 250 mL conical flasks, 300 mL modified jars with transparent polypropylene caps with 30 mL of solid WPM medium for the callus induction experiment. Before using the glassware, they were thoroughly washed under running tap water with liquid detergent and then rinsed with distilled water. Next, the clean glassware was sterilized by autoclaving at 121 °C and 104 kPa pressure for 20 min.

2.1.3. Callus Induction as Affected by Cytokinin and Auxin

The survived nodal explants were selected after four weeks for callus induction under different culture conditions. The prepared basal medium was supplemented with the combination of auxin; 2,4-dichlorophenoxyacetic acid (2,4-D) at a concentration of (2.5, 5, and 7.5 µM) and cytokinin; 6-benzylaminopurine (BAP) at a concentration of (0.5, 1, and 1.5 µM). The WPM basal medium without PGRs served as control to find out and compare the influence of PGRs on callus induction of *B. glabra* under different cultural conditions. Each treatment of this experiment was replicated three times, with 9 explants per replication and every three explants cultured in a 250 mL conical flask (Figures 1 and 2A) for callus initiation, but after the callus initiated, only two calluses derived from nodal segment cultured in a 250 mL conical flask (Figures 1 and 2B–E) in order to have better access to the media and space. The data on days to callus initiation were taken every day but the data such as callus frequency, callus morphology, fresh and dry weight of the callus were taken after every four weeks of incubation four times for, in total, 16 weeks. The following Equation (1) was used for determining the callus induction frequency as given below [32]:

$$\text{Callus induction frequency (\%)} = \frac{\textit{Number of explants induced callus}}{\textit{Number of explants cultured}} \times 100 \qquad (1)$$

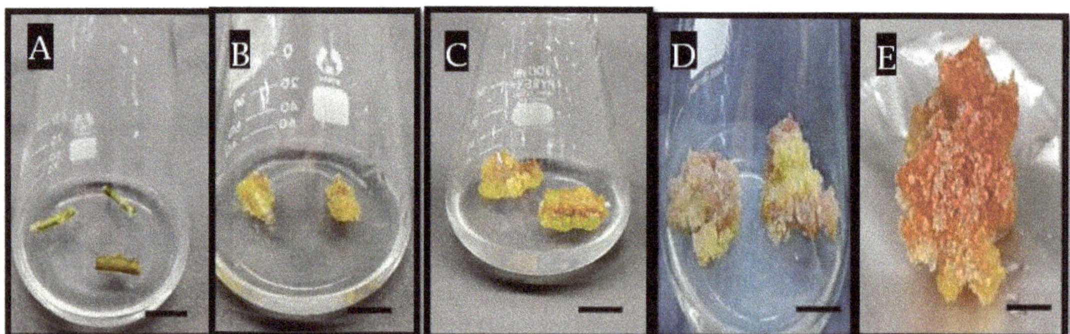

Figure 1. In vitro callus induction of *Bougainvillea glabra* Choisy under light incubation condition. (**A**) Sterilized nodal segment on WPM basal medium; (**B**) callus induction from a node section on WPM medium supplemented with 5 µM 2,4-D + 1 µM BAP after 4 weeks; (**C**) callus after 6 weeks in the same medium and PGRs condition; (**D**) callus after 8 weeks; (**E**) callus induction from a node section on WPM medium fortified with 7.5 µM 2,4-D + 1.5 µM BAP after 10 weeks with a red spot.

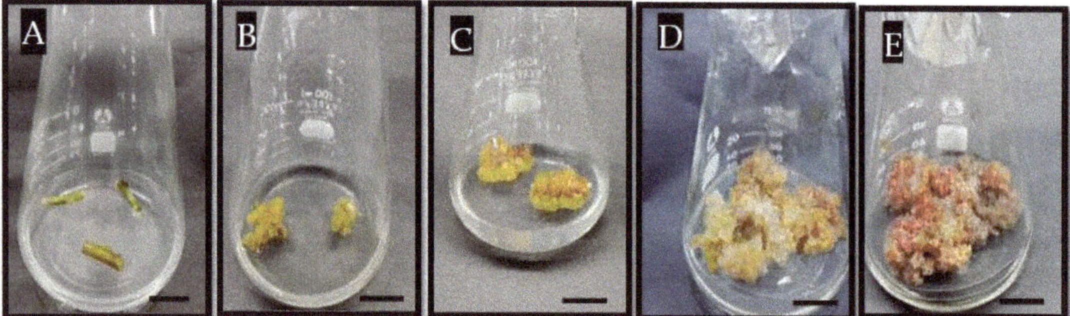

Figure 2. In vitro callus induction of *Bougainvillea glabra* Choisy under dark incubation condition. (**A**) Sterilized nodal segment on WPM basal medium; (**B**) callus induction from a node section on WPM medium supplemented with 2.5 µM 2,4-D + 1.5 µM BAP after 3 weeks; (**C**) callus after 4 weeks in the same medium and PGRs condition; (**D**) callus after 8 weeks; (**E**) callus induction from a node section on WPM medium fortified with 7.5 µM 2,4-D + 1.5 µM BAP after 10 weeks with a red spot.

2.1.4. Culture Maintenance

All the cultures were incubated in a culture room maintained at a temperature of 25 ± 2 °C under 16 h light and 8 h dark using white fluorescence light irradiation of 45 µmol/m^2/s or complete dark. Subculture of experimental materials (developing culture) was done every four weeks after each culture. Similar basal medium and PGRs composition from previous culture was used.

2.1.5. Fresh and Dry Weight of Callus

Growth parameters were made in terms of an increase in the fresh and dry weights of cultured tissue. The cultured tissue was carefully taken from the culture vessel and cleaned of agar particles that had adhered at the point of contact. After that, the tissue was placed on pre-weighed aluminum foil and the weight was calculated using a single pan digital balance of fresh weight of callus which presented as gram (g). The tissues were oven-dried at 55 °C to a consistent weight on the same foils for estimation of their dry weight as milligrams (mg) after recording their fresh weight.

2.2. Quantification of Phenolic Content and Antioxidant Activities of B. glabra Nodal Segments and In Vitro-Induced Calli

2.2.1. Planting Materials and Preparation of Extract

The nodal segments of conventionally propagated plants of *B. glabra* were collected from the Faculty of Agriculture, Universiti Putra Malaysia. In addition, the treatment of 7.5 µM 2,4-D + 1.5 µM BAP under dark and light incubation conditions which exhibited the highest biomass accumulation in vitro were obtained after 16 weeks in the Tissue Culture Laboratory, Faculty of Agriculture, Universiti Putra Malaysia were selected as a suitable treatment and quantified for phenolic content and antioxidant activities. The collected samples were oven-dried at a temperature of 55 °C for 48 h or until the weight remained constant. The dried nodes and in vitro-induced calluses were used to determine phenolic content and antioxidant properties. Furthermore, the extraction was conducted following the method employed by Hakiman and Maziah [33] with minor modifications. Dried nodes and in vitro-induced callus samples of *B. glabra* were ground using a commercial blender (Brand: Panasonic). An amount of 1 g dry weight of each sample was weighed and placed in a 150 mL conical flask. Each solvent, such as distilled water, ethanol, acetone, and hexane, received a total volume of 50 mL, and the flasks were covered with aluminum foil. The samples were placed in conical flasks on an orbital shaker at room temperature for 1 h in the dark. The samples were filtered with Whatman No. 1 filter paper and the extract was used for further analysis.

2.2.2. Total Phenolic Acids Content

Total phenolic acid content was evaluated following a method proposed by Singleton and Rossi [34]. To begin, test tubes were filled with 0.5 mL of extracts and 4.5 mL of distilled water. Then, 0.5 mL of Folin–Ciocalteu phenol reagent was added and thoroughly mixed using a vortex machine. After 5 min, 5 mL sodium carbonate (7%) was added. By adding 2 mL of distilled water, the final volume was adjusted to 12.5 mL. At room temperature, the reaction mixtures were incubated for 90 min. At 750 nm, the absorbance was measured. By establishing a standard curve of absorbance against various amounts of gallic acid, the total phenolic acids content was measured. The total phenolic acids content of the extract was measured in milligrams of gallic acid equivalents per gram of dry weight (mg GAE/g DW).

2.2.3. Total Flavonoids Content

Total flavonoids content was generated using the method established by Marinova et al. [35]. First, 0.5 mL of extracts was added to 2 mL of distilled water in test tubes. After that, 150 µL of 5% sodium nitrite was added to the mixture, which was then incubated for 5 min after 150 µL of 10% aluminum chloride was added. After that,

1 mL of 1 M sodium hydroxide and 1.2 mL distilled water were added at the sixth minute. A spectrophotometer was used to measure absorbance at 510 nm after the mixture was properly mixed. To assess the level of the total flavonoid content of the extract, a standard curve of absorbance against various concentrations of rutin was established. The extract's total flavonoids concentration was calculated as mg rutin equivalents per gram dry weight of the sample (mg RE/g DW).

2.2.4. DPPH Free Radical Scavenging Activity

DPPH (2,2-diphenyl-1-picrylhydrazyl) free radical scavenging assay was determined to quantify the ability of the DPPH molecule to neutralize free radicals by donating a hydrogen atom. Thus, the color shift from purple to violet is visible. DPPH free radical scavenging assay was performed according to the procedure explained by Wong et al. [36]. The DPPH was first generated in methanol at a concentration of 0.1 mM, and the initial absorbance of methanolic DPPH was measured with a spectrophotometer at 515 nm. The extracts were then combined in 1.5 mL of 0.1 mM methanolic DPPH solution. After shaking the mixture and incubating it at room temperature for 30 min, the absorbance was measured at 515 nm. The control was Trolox (6-hydroxy-2,5,7,8-tetramethylchromane-2-carboxylic acid), and the DPPH value was measured in mg Trolox equivalent per gram dry weight of the sample (mg TE/g DW)

2.2.5. ABTS Scavenging Activity

The ABTS (2,2′-azino-bis (3-ethylbenzothiazoline-6-sulfonic acid)) scavenging activity was determined following Re et al. [37]. Before the experiment, 7 mM ABTS stock solution was mixed in a ratio of 1:1 with 2.45 mM potassium persulfate and incubated at room temperature for 16 h in the dark. The ABTS$^+$ solution was diluted with ethanol after incubation until the absorbance was detected at 0.700 ± 0.05 at 734 nm. The extract was then combined with 0.9 mL of diluted ABTS$^+$ solution. After 15 min, the absorbance of the reaction mixture was measured at 734 nm. As a control, Trolox was utilized. The ABTS scavenging activity was measured in milligrams of Trolox equivalent per gram of dry weight (mg TE/g DW).

2.2.6. Iron (II) Chelating Activity

Iron (II) chelating activity was conducted by the procedure described by Dinis et al. [38] with minor modifications. An amount of 400 μL of the extract was mixed with 50 μL of 2 mM ferrous chloride. The reaction was started by adding 200 μL of 5 mM ferrozine to the reaction liquid and incubating it at room temperature for 10 min. The mixture's absorbance was measured at 562 nm in comparison to a blank sample. The decrease in absorbance was due to increased iron chelating activity. The following Equation (2) was used to calculate the iron (II) chelating activity:

$$\text{Chelating (\%)} = (1 - \text{A562 Sample}/\text{A562 Control}) \times 100 \tag{2}$$

2.3. Experimental Design and Statistical Analysis

The data were analyzed using the Statistical Analysis System (SAS) ver. 9.4. All of the experiments were performed in triplicates in a two or three-factorial Completely Randomized Design (CRD) and the results are expressed as mean $\pm$ SE. Duncan's Multiple Range Test (DMRT) was used to compare means at a $p = 0.05$ level. Pearson's correlation analysis was used to determine the correlation between variables with indicator $r < 0.25$ indicating a weak correlation, $r < 0.75$ indicating an intermediate correlation, $r < 1$ indicating a strong correlation, $r = 1$ indicating perfect correlation.

3. Results

3.1. In Vitro Callus Induction of Bougainvillea glabra via Nodal Segment

3.1.1. The Main Effect of 2,4-D, BAP and Light Regimes on Callus Induction of *B. glabra*

Results from the present study showed that from the different concentrations of 2,4-D, the minimum days for callus initiation was obtained in the treatment of 5 μM 2,4-D but it was not significantly different from the treatment of 2.5 μM 2,4-D. Meanwhile, the maximum days for callus initiation were recorded from the treatment of 7.5 μM 2,4-D. However, for the treatments of BAP, the minimum days for callus initiation was exhibited from the 1.5 μM BAP but it was not significantly different from 0.5 and 1 μM BAP, respectively. The type of explants to induce callus is also an important factor that needs to be considered in callus induction experiments due to various factors affecting the development of cell culture systems, such as genotype, explant type, plant growth regulators (PGR), culture medium, and culture condition. Based on the results in (Table 1), with WPM medium supplemented with various concentrations of 2,4-D, an increment of the callus frequency was observed as 2,4-D concentration increased until 5 μM. The highest callus frequency was recorded using 5 μM 2,4-D, which is significantly different from 7.5 μM 2,4-D, but it was not significantly different from 2.5 μM 2,4-D. On the other hand, there was no callus formation on the control media without plant growth regulators, indicating that growth regulators had a significant effect on callus induction on the nodal explant. Following that, there was a decrease in percent callus induction growth, followed by an increase of the concentrations of 2, 4-D, and BAP.

Table 1. Main effects of 2,4-D and BAP on callus induction in the nodal segment of *Bougainvillea glabra* under different light conditions.

Treatment	Days to Callus Initiation	Callus Frequency (%)	FW of Callus (g)	DW of Callus (mg)
Control	—	—	—	—
2,4-D (μM)				
2.5	12.09 ± 0.72 b	88.54 ± 1.89 a	2.98 ± 0.17 c	122.12 ± 3.76 c
5	11.72 ± 0.76 b	90.08 ± 1.57 a	3.85 ± 0.15 ab	164.58 ± 8.90 a
7.5	14.26 ± 0.82 a	76.04 ± 0.77 b	4.06 ± 0.26 a	150.09 ± 6.77 b
BAP (μM)				
0.5	13.02 ± 0.85 a	86.39 ± 2.56 a	3.15 ± 0.19 b	132.89 ± 8.13 b
1	12.79 ± 0.79 a	85.67 ± 1.69 a	3.51 ± 0.12 b	140.39 ± 19.10 b
1.5	12.73 ± 0.86 a	81.77 ± 2.56 b	4.23 ± 0.24 a	167.50 ± 12.48 a
Culture Condition				
Light	14.73 ± 0.73 a	75.69 ± 3.81 b	3.52 ± 0.23 a	141.56 ± 17.44 a
Dark	8.45 ± 0.73 b	81.03 ± 4.05 a	3.43 ± 0.20 b	132.77 ± 8.63 b
F-value				
2,4-D	744.57 ***	3020.89 ***	121.66 ***	143.7 ***
BAP	6.72 ***	17.97 ***	16.42 ***	15.05 ***
Culture Condition	1836.00 ***	117.57 ***	5.91 **	4.35 *
2,4-D*BAP	22.19 ***	83.51 ***	31.82 ***	42.68 ***
2,4-D*Condition	53.02 ***	4.19 **	0.40 ns	0.81 ns
BAP*Condition	27.05 ***	1.38 **	2.48 ns	1.56 ns
2,4-D*BAP*Condition	7.5 ***	6.35 ***	0.96 ns	0.53 ns
CV (%)	5.58	2.77	12.56	12.15

Values are means ± SE. the same letter within the same column for each factor indicates no significant difference ($p < 0.05$). The means separation is done by using Duncan's multiple range test (DMRT). F value represented * = <0.05, ** = $p < 0.01$, *** = $p < 0.001$ and ns = no significant differences. 2,4-dichlorophenoxyacetic acid (2,4-D), 6-benzyl amino purine (BAP), FW (Fresh Weight), and DW (Dry Weight).

In terms of biomass accumulation, maximum callus biomass was recorded when the nodal explants were cultured on WPM medium supplemented with 5 μM 2,4-D. However, for the treatments of BAP, 1.5 μM BAP exhibited significantly higher callus biomass, respectively. Based on the results, it was seen that there was an increase in the fresh and dry weight of callus as the concentration of BAP was increased up to 1.5 μM, but for the treatments of 2,4-D, the callus biomass increased as the 2,4-D increased until 5 μM and significantly decreased with 7.5 μM. Primary metabolism, particularly carbon and nitrogen metabolism, is closely linked to biomass accumulation. Carbon metabolism meets the energy requirements resulting from carbohydrate synthesis, and hence, contributes to cell growth and structural components [39]. The products of nitrogen metabolism, on the other hand, are primarily amino acids and proteins, which are then used to regulate cellular processes [40]. Over all, both cellular development and division result in increased fresh and dry weight.

The callus induction of *B. glabra* was also evaluated under two growth conditions, dark and light. The results showed that the dark condition was more favorable for earlier callus initiation and higher callus frequency than the light condition. However, the calluses induced under light incubation conditions produced significantly higher callus fresh and dry weight than the calluses induced under complete dark incubation conditions. Based on the observation in this experiment, the nodal segments incubated under complete dark conditions grew faster than in a light-containing photoperiod. In the dark condition, the minimum days for callus initiation was (8.45), which is significantly different from the light condition with (14.73) days. In addition, the higher callus frequency was also obtained from the cultured incubated under the dark condition, which is significantly different from the cultured incubated under the light condition, respectively. However, callus grown under a set photoperiod presented the higher callus fresh and dry weight than callus incubated under the complete dark condition.

Based on the observation in this experiment, the interaction effects of various concentrations of 2,4-D and BAP were significant ($p < 0.05$) for all parameters, but the interaction effect between 2,4-D *Condition and BAP *Condition were only significant for the parameters such as days to callus initiation and callus frequency; also, the statistical analysis indicated that the triple interaction of 2,4-D, BAP, and culture condition was significantly different for days to callus initiation and callus induction rate, but was not significantly different for the biomass production (Table 1). A similar trend also was observed in the WPM medium supplemented with different concentrations of BAP. The maximum callus frequency was obtained from 0.5 μM and 1 μM BAP, respectively. However, by increasing the BAP concentration, the callus frequency decreased.

3.1.2. Synergistic Effect of Cytokinin, Auxins, and Light Regime on Callus Induction

The results in Table 2 exhibit that the minimum days of callus initiation were recorded from the treatments of 2.5 μM 2,4-D + 1.5 μM BAP and 5 μM 2,4-D + 1 μM BAP under the dark incubation condition, respectively. However, the maximum days for callus initiation were recorded from the treatment of 7.5 μM 2,4-D + 0.5 μM BAP under light incubation conditions. As shown, the period of callus induction and growth of callus varied. They depend on the type and concentration of growth regulators and lighting conditions. Meanwhile, the highest callus induction rate was produced from the combination of 2.5 μM 2,4-D + 1.5 μM BAP under a dark incubation condition and 5 μM 2,4-D + 0.5 μM BAP under both light and dark incubation conditions, respectively. However, the lowest callus induction rate was obtained from the treatments of 7.5 μM 2,4-D + 1.5 μM BAP under both light and dark conditions.

In this experiment, for the WPM medium fortified with various concentrations of 2,4-D and BAP, a different trend of callus fresh and dry weight was observed. The callus fresh and dry weight was drastically increased by three-fold in the treatment of 7.5 μM 2,4-D + 1.5 μM BAP under dark incubation condition but it was not significantly different from the same treatment under the light condition. As the concentration decreased to 2.5 μM 2,4-D and 0.5 μM BAP, the callus fresh weight was decreased under dark and light incubation conditions, respectively. Overall, no significant biomass differences were observed for light regimes.

Table 2. Interaction effects of 2,4-D and BAP on callus induction in the nodal segment of *Bougainvillea glabra* under different light conditions.

Condition	2,4-D (µM)	BAP (µM)	Days to Callus Initiation	Callus Frequency %	FW of Callus (g)	DW of Callus (mg)	Callus Morphology
Light	Control	0	—	—	—	—	—
	2.5	0.5	15.17 ± 0.44 fg	78.33 ± 1.67 de	2.53 ± 0.15 hi	120.00 ± 11.54 fg	Y, B, & C
		1	15.33 ± 0.44 ef	76.67 ± 1.67 def	3.21 ± 0.24 fgh	141.67 ± 20.48 de	Y, B, & C
		1.5	14.12 ± 0.56 g	93.33 ± 1.67 b	3.03 ± 0.03 gh	93.33 ± 3.33 i	Y, B, & C
	5	0.5	16.01 ± 0.18 def	100.00 ± 0 a	3.67 ± 0.17 defgh	165.00 ± 7.63 bc	Y, B, & C
		1	13.00 ± 0.58 h	86.67 ± 1.67 c	3.7 ± 0.53 defg	113.33 ± 8.82 ghi	Y, B, & C
		1.5	16.00 ± 0.58 def	80.67 ± 1.2 d	3.77 ± 0.34 defg	156.00 ± 7.81 bcd	Y, B, & C
	7.5	0.5	17.67 ± 0.88 ab	73.33 ± 1.67 f	3.00 ± 0.06 gh	116.67 ± 8.81 fgh	R, Y, & C
		1	16.73 ± 0.50 bcd	73.33 ± 1.67 f	3.55 ± 0.24 efgh	146.67 ± 9.28 de	R, Y, & C
		1.5	17.27 ± 0.43 bcd	73.33 ± 1.67 f	5.23 ± 0.16 a	221.67 ± 13.01 a	R, Y, & C
Dark	Control	0	—	—	—	—	—
	2.5	0.5	10.17 ± 0.30 j	86.67 ± 1.67 c	2.11 ± 0.39 i	100.00 ± 5.7 hi	Y, B, & F
		1	10.58 ± 0.33 ij	80.00 ± 1.53 d	3.68 ± 0.16 defgh	159.00 ± 2.3 bc	Y, B, & F
		1.5	7.00 ± 0 k	100.00 ± 0 a	3.30 ± 0.3 fgh	118.00 ± 18.9 fgh	Y, B, & F
	5	0.5	8.00 ± 0.12 k	100.00 ± 0 a	4.36 ± 0.46 bcd	182.00 ± 9.6 bc	W, B, & F
		1	7.00 ± 0 k	93.33 ± 1.2 b	2.93 ± 0.03 h	126.67 ± 3.38 ef	R, Y, & F
		1.5	10.28 ± 0.17 j	93.33 ± 0.88 b	4.71 ± 0.54 abc	185.33 ± 12.81 b	R, Y, & F
	7.5	0.5	11.11 ± 0.11 ij	80.00 ± 1.53 d	3.25 ± 0.05 fgh	113.67 ± 3.18 fgh	Y, B, & F
		1	11.08 ± 0.22 ij	80.00 ± 1.53 d	3.98 ± 0.02 cdef	154.67 ± 3.18 cd	R, Y, & F
		1.5	11.72 ± 0.03 i	73.33 ± 0.67 f	5.33 ± 0.16 a	230.67 ± 5.20 a	W, B, & F

Values are means ± SE. the same letter within the same column for each factor indicates no significant difference ($p < 0.05$). The means separation is done by using Duncan's multiple range test (DMRT). 2,4-dichlorophenoxyacetic acid (2,4-D), and 6-benzylaminopurine (BAP). W white, Y yellow, R red, B brown, F friable, and C compact.

Based on the results, the morphogenesis responses based on the intensity of callus formation, texture, types, and concentration of PGRs, cultural condition, and in vitro morphogenesis were different according to the treatment used and are presented in (Table 2) and (Figures 1 and 2A–E). The cultural condition was capable of inducing a callus that could be classified into two types. The first type, which was induced under light incubation conditions, was compact and yellow to brown and red (Figure 1A–E). The second type, which was induced under dark incubation conditions, was friable and yellow to white and brown (Figure 2A–E).

3.2. Quantification of Phenolics Contents and Antioxidant Activities of In Vitro-Induced Calli and Conventionally Propagated Plant of Bougainvillea glabra

3.2.1. Total Phenolic Acid and Total Flavonoid Content

The results showed that the content of total phenolic acid, total flavonoid, and antioxidant properties except for iron (II) chelating activity of conventionally propagated nodal segment of *B. glabra* were significantly different ($p \leq 0.05$) from the in vitro-induced calli under light and dark incubation conditions (Tables 3–5). The aqueous extract of the node recorded the highest total phenolic acid followed by aqueous extract of the calli induced under photoperiod. In addition, there were no significant differences between the aqueous extract of calli induced under the dark condition and the ethanol extract of node and calli induced under photoperiod and dark conditions. The results also showed no significant differences between acetone extract of the node and calli induced under photoperiod and dark conditions. Meanwhile, the lowest amount of TPC was obtained from hexane extract of the node, but there were no significant differences between hexane extract of node and calli induced under photoperiod and dark conditions.

Table 3. Total phenolic acids and flavonoids contents of node and in vitro-induced calli of *Bougainvillea glabra*.

Source of Sample	Type of Solvent									
	Phenolic Acids (mg GAE/g DW)					Flavonoids (mg RE/g DW)				
	Aqueous	Ethanol	Acetone	Hexane	Mean	Aqueous	Ethanol	Acetone	Hexane	Mean
Node	21.88 ± 0.57 a	3.33 ± 0.38 cd	1.84 ± 0.22 ef	0.25 ± 0.05 g	6.82 ± 1.87 A	42.05 ± 0.18 a	21.46 ± 0.31 b	8.92 ± 0.04 d	1.00 ± 0.04 i	18.36 ± 4.67 A
Callus induced in light	6.43 ± 0.26 b	2.73 ± 0.26 d	1.52 ± 0.03 ef	0.20 ± 0.02 g	2.72 ± 0.7 B	10.30 ± 0.11 c	8.46 ± 0.22 e	7.34 ± 0.14 f	0.92 ± 0.11 i	6.75 ± 1.07 B
Callus induced in dark	3.90 ± 0.17 c	2.63 ± 0.08 d	1.00 ± 0.27 f	0.25 ± 0.07 g	1.95 ± 0.43 C	6.67 ± 0.11 g	6.67 ± 0.15 g	4.46 ± 0.07 h	0.88 ± 0.17 i	4.67 ± 0.72 C
Mean	10.74 ± 1.83 A	2.90 ± 0.17 B	1.45 ± 0.16 C	0.24 ± 0.03 D		19.67 ± 5.62 A	12.20 ± 2.33 B	6.91 ± 0.65 C	0.93 ± 0.06 D	

Values are means ± SE, the same letter within the same column for each factor indicates no significant difference ($p < 0.05$). The means separation is done using Duncan's multiple range test (DMRT).

In total flavonoid content analysis, the highest value was recorded from aqueous extract of the node, followed by ethanol extract of the node. On the other hand, the lowest TFC was obtained from hexane extract of the node and calli induced under photoperiod, and dark conditions ranged between 0.88 and 1 mg RE/g DW, respectively.

3.2.2. Antioxidant Activities

The antioxidant activities of *B. glabra* nodal extracts, propagated via conventional method and in vitro-induced calli via tissue culture techniques, were measured using three in vitro antioxidant assays including 2,2-diphenyl-1-picryl-hydrazyl-hydrate (DPPH) free radical scavenging activity, 2,2-azino-bis (3-ethylbenzothiazoline-6-sulfonic acid) (ABTS), and iron (II) chelating activity. All antioxidant assays were measured using a UV-Vis spectrophotometer at a specific absorbance. The results showed that DPPH free radical scavenging activity and ABTS scavenging activity of the conventionally propagated nodal segment were significantly different at $p \leq 0.05$ than in vitro-induced calli under dark and light incubation condition. However, the highest iron (II) chelating activity was obtained from in vitro-induced calli under a dark incubation condition. The DPPH free radical scavenging activity recorded from sources of samples and extraction solvents ranged between 0.14 to 7.64 mg TE/g DW. The highest DPPH free radical scavenging activity was obtained from aqueous extract of the node, followed by ethanol extract of the node, and aqueous extract of the calli induced under photoperiod. In addition, there were no significant differences between hexane extract with all of the samples. The lowest DPPH free radical scavenging activity was recorded from hexane extract of calli induced under photoperiod and dark conditions.

Table 4. 2,2-diphenyl-1-picryl-hydrazyl-hydrate (DPPH) free radical scavenging activity and 2,2-azino- bis (3-ethylbenzothiazoline-6-sulfonic acid) (ABTS) scavenging activity of node and in vitro-induced calli of *Bougainvillea glabra*.

Source of Sample	Type of Solvent									
	DPPH (mg TE/g DW)					ABTS (mg TE/g DW)				
	Aqueous	Ethanol	Acetone	Hexane	Mean	Aqueous	Ethanol	Acetone	Hexane	Mean
Node	7.64 ± 0.01 a	2.98 ± 0.11 b	1.01 ± 0.01 e	0.24 ± 0.08 gh	2.97 ± 0.68 A	1.51 ± 0 a	0.72 ± 0.01 d	0.58 ± 0.01 ef	0.46 ± 0.02 gh	0.82 ± 0.12 A
Callus induced in light	2.12 ± 0.03 c	1.37 ± 0.06 d	0.89 ± 0.05 f	0.19 ± 0.08 hi	1.14 ± 0.56 B	1.12 ± 0 b	0.68 ± 0.01 de	0.54 ± 0.03 fg	0.40 ± 0.01 j	0.69 ± 0.08 B
Callus induced in dark	0.80 ± 0.18 f	0.59 ± 0.07 g	0.40 ± 0.06 gh	0.14 ± 0.02 ij	0.48 ± 0.44 C	0.98 ± 0 c	0.40 ± 0.01 hi	0.30 ± 0.01 ij	0.16 ± 0.01 jk	0.46 ± 0.09 C
Mean	3.52 ± 0.64 A	1.64 ± 0.46 B	0.76 ± 0.31 C	0.19 ± 0.33 D		1.20 ± 0.08 A	0.60 ± 0.05 B	0.48 ± 0.04 C	0.34 ± 0.05 D	

Values are means ± SE, the same letter within the same column for each factor indicates no significant difference ($p < 0.05$). The means separation is done using Duncan's multiple range test (DMRT).

Table 5. Iron (II) chelating activity of node and in vitro-derived calli of *Bougainvillea glabra*.

Source of Sample	Type of Solvent				
	Iron (II) Chelating Activity (%)				
	Aqueous	Ethanol	Acetone	Hexane	Mean
Node	29.64 ± 0.77 b	20.08 ± 1.31 de	6.64 ± 2.18 h	17.02 ± 1.24 ef	18.35 ± 2.55 B
Callus induced in light	26.87 ± 2.45 bc	20.78 ± 0.73 de	5.39 ± 1.4 h	15.25 ± 1.12 fg	17.07 ± 2.47 B
Callus induced in dark	43.30 ± 0.13 a	22.78 ± 0.53 cd	11.96 ± 0.56 g	22.92 ± 1.7 cd	25.24 ± 3.44 A
Mean	33.27 ± 2.64 A	21.21 ± 0.61 B	8.00 ± 1.27 C	18.40 ± 1.35 D	

Values are means ± SE, the same letter within the same column for each factor indicates no significant difference ($p < 0.05$). The means separation is done using Duncan's multiple range test (DMRT).

In the analysis of ABTS scavenging activity, there was a significant interaction between the sources of samples and extraction solvents on ABTS scavenging activity. The highest ABTS scavenging activity was significantly obtained by aqueous extract of the node followed by aqueous extract of the calli induced under the photoperiod, and aqueous extract of the calli induced under the dark condition Meanwhile, there were no significant differences between ethanol extract of the node and calli induced under photoperiod conditions. In addition, there were no significant differences between ethanol, acetone, and hexane extract of the calli induced under dark conditions. On the other hand, the lowest ABTS scavenging activity was recorded from hexane extract of calli induced under dark conditions. Thus, all of these extracts had some ABTS radical scavenging abilities in their phytochemical components, and the contents may have some ABTS radical scavenging ability equivalence.

In the present study on iron (II) chelating activity, a significant interaction was found between extraction solvents and sources of samples. In contrast to other antioxidant activities conducted in which aqueous and other node extracts produced higher antioxidant activities, the iron (II) chelating activity was produced from the aqueous extract of the calli induced under dark conditions. The highest iron (II) chelating activity was produced from an aqueous extract of the calli induced under the dark condition. Meanwhile, there were no significant differences between the aqueous extract of the node and calli induced under photoperiod and hexane extract of the calli induced under dark conditions. In addition, there were no significant differences between ethanol extract of the node, calli induced under photoperiod, and dark conditions. Furthermore, the lowest iron (II) chelating activity was produced by the acetone extract of calli induced under the photoperiod

3.2.3. Correlation Analysis between Variables

Pearson's correlation analysis was used to examine the relationship between phenolic contents (phenolic acids and flavonoids) and antioxidant activities, such as DPPH free radical scavenging activity, ABTS scavenging activity, and iron (II) chelating activity (Table 6).

Table 6. Pearson's correlation analysis between variables.

Variable	TPC	TFC	ABTS	DPPH	Fe^{2+}
TPC	1				
TFC	0.92 **	1			
ABTS	0.90 **	0.80 **	1		
DPPH	0.79 **	0.87 **	0.84 **	1	
Fe^{2+}	0.46 **	0.30 ns	0.54 **	0.24 ns	1

Notes: **: Significant correlation at $p < 0.05$; ns: Non-significant correlation; TPC: Total phenolic acids content; TFC: Total flavonoids content; DPPH: DPPH free radical scavenging activity; ABTS: ABTS scavenging activity; Fe2+: Iron (II) chelating activity, respectively.

The correlation showed that all variables except iron (II) chelating activity were positively correlated ranged from intermediate positive to a high positive correlation. The variables of total phenolic acids, flavonoids, DPPH free radical scavenging activity, and ABTS scavenging activity produced a highly significant strong correlation (r > 0.75). Meanwhile, iron (II) chelating activity showed a low positive correlation against TPC and ABTS (r < 0.75) and a non-significant correlation against TFC and DPPH free radical scavenging activity. The Pearson's correlation analysis showed that the phenolic compounds present in *B. glabra* nodes and calli induced under different light condition extract are strong as scavenging and chelating agents.

4. Discussion

4.1. In Vitro Callus Induction of B. glabra

Bougainvillea glabra is an important medicinal plant with high medicinal values. Plants propagated via traditional methods are vulnerable to a variety of diseases and pests, as well as weather and land availability, all of which have an adverse effect on the medicinal qualities of the harvested plants [41]. The overall objective of this study was to induce calli under different PGRs and cultural conditions and evaluate the secondary metabolites content of the in vitro-induced calli and conventionally propagated plant. In the in vitro callus induction, plant growth regulators are one of the most important factors affecting the growth of explants. Synthetic auxins, such as 2,4-D, are important PGRs that are applied in a variety of embryogenic cell and tissue culture methods, as well as callus production and cell suspension culture [42]. Meanwhile, some research has confirmed the positive effect of 2,4-D on callus formation during the physiological and molecular process in many circumstances, and those studies have shown that 2,4-D regulates the endogenous IAA metabolism, promotes specific proteins, and controls DNA methylation [43]. On the other hand, cytokinin plays an important part in inducing callus by promoting cell division and differentiation [44].

In the present study, the minimum days of callus initiation (7) was recorded from the treatments of 5 µM 2,4-D + 1 µM BAP under a dark incubation condition; however, the maximum days for callus initiation (17.67) was recorded from the treatment of 7.5 µM 2,4-D + 0.5 µM BAP under light incubation conditions. The study results also showed that the highest callus induction rate (100%) was produced from the combination of 2.5 µM 2,4-D + 1.5 µM BAP under a dark incubation condition; however, the lowest callus induction rate (73.33%) was obtained from the treatments of 7.5 µM 2,4-D + 1.5 µM BAP under both light and dark conditions. Similarly, the synergistic effect of plant growth regulators and cultural condition has been studied by Renu et al. [45] in the nodal segment of *Catharanthus roseus*, revealing the minimum days for callus initiation and higher callus frequency obtained when WPM basal medium supplemented with BAP 0.5 mg + 2,4-D 2 mg under dark condition.

Meanwhile, a study conducted by Behbahani et al. [46] showed that the earlier callus initiation and higher callus frequency obtained when the nodal segment was cultured on a WPM basal medium fortified with 1 mg 2,4-D under dark incubation condition. Moreover, previous studies by Azad et al. [47] of *Phellodendron amurense*, Hoque et al. [48] of *Trapa* sp., Thammina et al. [49] of *Euonymus alatus*, Hesami and Daneshvar. [42] of *Ficus religiosa*, Pandey et al. [50] of *Boerhaavia diffusa* reported that the explants cultured on basal medium supplemented with 2.2–8.8 µM 2,4-D + 1.1–4.4 µM BAP significantly increased the callus induction rate and minimized the callus initiation under a dark condition. Meanwhile, they have reported that a higher concentration of 2,4-D with a higher concentration of BAP significantly reduced the callus frequency and increased the days for callus initiation. It was noted that a combination of 2,4-D with lower BAP concentrations was found to increase the callus induction frequency. This improved effect could be related to BAP, which helps 2,4-D influence rapid cell division and the synthesis of zeatin-like hormones in the plant for effective callus induction [51]. Furthermore, the frequency of callus induction decreased with the increase in the concentration of 2,4-D. The increase in the concentration of 2,4-D causing a significantly delayed response. Bhojwani and Dantu [52] reported that 2,4-D may have herbicidal properties at high concentrations, which could reduce callus formation.

Other than PGRs, light is one of the most important elicitors, as it affects several physiological processes such as photosynthesis; hence, light can influence the growth, development, and morphogenesis of a variety of plant species in vitro [53,54]. Light, on the other hand, is known to affect cell division rates and ethylene evolution, which can affect callogenesis and rhizogenesis. Therefore, the duration and timing of light exposure are critical in explant morphogenesis [55,56]. In our experiments, the explants exposed to light in the early stages of culture developed more severe browning than those exposed to darkness, implying that darkness is preferable for callus induction in the early stages. According to Habibah et al. [56], in a dark environment, it is thought that the release of phenolic substances can be inhibited, and the rate of callus induction can be increased. However, dark incubation conditions may be associated with negative effects in some parameters such as callus biomass. Overall, callus initiation was faster in the dark than in the light, but the callus deteriorated when sub-cultured in the dark continuously, independent of media.

Based on the callus biomass accumulation, a different trend of callus fresh and dry weight was observed. The callus fresh weight was drastically increased by three-fold in the treatment of 7.5 µM 2,4-D + 1.5 µM BAP with 5.33 g. As the concentration decreased to 2.5 µM 2,4-D and 0.5 µM BAP, the callus fresh weight was decreased to 2.11 g and 2.53 g under dark and light incubation conditions, respectively. The yield of biomass and secondary metabolites can be used to determine the success of callus culture. These can be achieved with the proper balance of plant growth regulators, nutritional media, and growing conditions. The primary metabolism, particularly carbon and nitrogen metabolism, is closely linked to biomass accumulation. Carbon metabolism meets the energy requirements resulting from carbohydrate synthesis, and hence, contributes to cell growth and structural components [39]. The products of nitrogen metabolism, on the other hand, are primarily amino acids and proteins, which are then used to regulate cellular processes [40].

4.2. Quantification of Secondary Metabolites and Antioxidant Activity of In Vitro-Induced Calli and Conventionally Propagated Nodal Segment of B. glabra

The results of phytochemicals and antioxidant activities analysis revealed that total phenolic acid, total flavonoid content, DPPH free radical scavenging activity, and ABTS scavenging activity of aqueous extract of the conventionally propagated nodal segment were significantly different from other extraction solvents of in vitro-induced calli of *B. glabra*. The interrelating effect of the extraction solvent and source of the sample has been studied by Mahendra et al. [57] in conventionally propagated plant parts and calli induced under the light regime of *Decalepis arayalpathra*. They revealed maximum TPC and TFC from aqueous extract of the conventionally propagated nodal segment and calli induced under photoperiod, respectively. Similarly, the current study results confirm the findings of Esmaeili et al. [58] that high polar solvents such as water of conventionally propagated plant parts produced higher TPC, and TFC compared with in vitro induced calli. In addition, node cultures of *Eucalyptus camaldulensis* produced under a 16-h light photoperiod revealed an increase in the level of phenolic compounds [59]. On the other hand, a study conducted by Zahid et al. [60] indicated the hexane extract of micropropagation and conventional propagated 'Bentong' ginger *Zingiber officinale* Roscoe produced the lowest amount of TPC and TFC compared to other extraction solvents, which are in agreement with our findings.

In the antioxidant activities, the aqueous extract of conventionally propagated nodal segment produced higher DPPH free radical scavenging activity and ABTS scavenging activity compared to in vitro-induced calli. This is in agreement with the findings of Esmaeili et al. [58] that the highest antioxidant properties was obtained from the aqueous extract of field-grown plant parts compared to the in vitro induced calli. Moreover, previous studies by Islam et al. [61] on *B. glabra*, Murali and Prabakaran [62] on *Ociumum basilicum* L., and Mahendra et al. [57] on *Salacia macrosperma* showed that the aqueous and ethanol extracts of the conventionally propagated plant parts exhibited the higher DPPH free radical scavenging and ABTS scavenging activity compared to the in vitro induced calli of less polar solvents such as acetone and hexane. Furthermore, studies conducted by López-Laredo et al. [63] on *Tecoma stans*, Shah et al. [64] on *Silybum marianum*, Mohammad et al. [65] on *Olea europaea* L., and Rameshkumar et al. [51] on *Nilgirianthus ciliate* indicated that the calli induced under the photoperiod produced a higher antioxidant activity than the calli induced under dark incubation conditions which are in line with our findings. On the other hand, a study conducted by Zahid et al. [60] indicated that the hexane extract of micropropagated and conventional propagated 'Bentong' ginger *Zingiber officinale* Roscoe produced the lowest inhibition of DPPH than other extraction solvents, which are also in agreement with our findings.

In contrast to other antioxidant activities conducted in which aqueous and other node extracts produced higher antioxidant activities, the iron (II) chelating activity was produced from the aqueous extract of the calli induced under dark conditions. It is clear that the chelating powers of water extracts of the calli induced under dark conditions were higher than the other three extracts and sources of samples. Similarly, a study conducted by Hakkim et al. [66] reported that the in vitro induced callus had a higher iron (II) chelation than that of the field-grown stem, leaves, and inflorescence of *Ocimum sanctum* L. Furthermore, according to Costa et al. [67], the synergistic effect of solvents and the source of the sample exhibited that the aqueous extract of the in vitro-induced callus showed higher iron (II) chelation than conventionally propagated plant parts and other extraction solvents of *Thymus lotocephalus*. In addition, study conducted by Song et al. [68] confirmed that shoot of the field-grown plant of *Mertensia maritima* L. produced higher TPC, TFC, DPPH free radical scavenging activity, and ABTS scavenging activity. However, the in vitro-induced calli chelate had more iron (II) than field-grown shoots, supporting our findings. So, the in vitro-induced callus can be a good source of iron (II) chelation, and this indicates that polyphenols are not primary chelating particles [69].

5. Conclusions

The results of this study demonstrated that the growth of *B. glabra* calli is greatly influenced by the combination of plant growth regulators and cultural conditions. No callus was induced on a PGR-free medium. The media composed of WPM supplemented with 2.5 µM 2,4-D + 1.5 µM BAP and 5 µM 2,4-D + 1 µM BAP significantly reduced the days for callus initiation (7 days) under a dark incubation condition, while the maximum days (17.67) was recorded when WPM fortified with 7.5 µM 2,4-D + 0.5 µM BAP under light incubation condition. Meanwhile, 5 µM 2,4-D + 0.5 µM BAP significantly increased the callus frequency (100%) under both light and dark incubation condition. By increasing the PGRs concentration and combination, the callus frequency significantly decreased.

Moreover, the light regime did not have any influence on biomass accumulation, while the PGRs significantly influenced and the maximum biomass was recorded when the WPM basal medium was supplemented with 7.5 µM 2,4-D + 1.5 µM BAP under both dark and light conditions with FW: 5.33 g per callus and DW: 230.67 mg per callus, respectively. Furthermore, the morphology of the callus was influenced by PGRs and cultural condition; the calli that induced under a light condition had good texture and were compact in nature with yellow to brown and red colors, while the calli induced under a dark condition were friable with yellow to brown and white colors.

The study on phenolics content found that the aqueous extract of the conventionally propagated node exhibited the highest phenolic acids (21.88 mg GAE/g DW) and flavonoids content (42.05 mg RE/g DW). Furthermore, in antioxidant activities conducted, the highest DPPH free radical scavenging activity (7.64 mg TE/g DW), and ABTS scavenging activity (1.51 mg TE/g DW), were also recorded from aqueous extract of the conventionally propagated node. On the other hand, the highest iron (II) chelating activity was exhibited from the aqueous extract of the calli induced under the dark condition with 43.3% inhibition, respectively. In conclusion, higher phenolics content and antioxidant properties recorded from the conventionally propagated nodal part showed that the plant propagated through conventional propagation technique has the potential to produce high secondary metabolites. Meanwhile, both conventionally propagated parts and callus samples suggest that in addition to the plant material, the callus may also be used as a supplement raw material to obtain secondary metabolites for the pharmaceutical industry. Thus, the in vitro-derived *B. glabra* callus could plausibly act as a novel source of metabolites for food and pharmaceutical industries in their preparation of food preservatives and medicines, respectively.

Author Contributions: Conceptualization, M.N.N. and M.H.; methodology, M.N.N. and M.H.; resources, S.Z.S. and M.H.; data curation, M.N.N.; writing—original draft preparation, M.N.N.; writing—review and editing, M.N.N., S.Z.S. and M.H.; supervision, S.Z.S. and M.H.; funding acquisition, S.Z.S. and M.H. All authors have read and agreed to the published version of the manuscript.

Funding: This research was funded by the Ministry of Higher Education Malaysia (Translational Research Program (TR@M) and Universiti Putra Malaysia (Putra Grant-Young Putra Initiative), grant number 5526700 and 9629000, respectively, and the APC was funded by the Universiti Putra Malaysia.

Institutional Review Board Statement: Not applicable.

Informed Consent Statement: Not applicable.

Data Availability Statement: The data presented in this study are available in the article.

Acknowledgments: The authors would like to thank the Higher Education Development Project (HEDP) for Afghanistan and Herat University for the scholarship.

Conflicts of Interest: The authors declare no conflict of interest.

References

1. Saleem, H.; Thet, H.; Naidu, R.; Sirajudeen, A.; Ahmed, N. HPLC–PDA Polyphenolic Quantification, UHPLC–MS Secondary Metabolite Composition, and in Vitro Enzyme Inhibition Potential of *Bougainvillea glabra*. *Plants* **2020**, *9*, 388. [CrossRef] [PubMed]
2. Asuk, A.A.; Agiang, M.A.; Dasofunjo, K.; Willie, A.J. The Biomedical Significance of the Phytochemical, Proximate and Mineral Compositions of the Leaf, Stem Bark and Root of *Jatropha Curcas*. *Asian Pac. J. Trop. Biomed.* **2015**, *5*, 650–657. [CrossRef]
3. Arteaga Figueroa, L.; Abarca-Vargas, R.; García Alanis, C.; Petricevich, V.L. Comparison between Peritoneal Macrophage Activation by *Bougainvillea X buttiana* Extract and LPS and/or Interleukins. *Biomed Res. Int.* **2017**, *2017*, 4602952. [CrossRef] [PubMed]
4. Soares, J.J.; Rodrigues, D.T.; Gonçalves, M.B.; Lemos, M.C.; Gallarreta, M.S.; Bianchini, M.C.; Gayer, M.C.; Puntel, R.L.; Roehrs, R.; Denardin, E.L.G. Paraquat Exposure-Induced Parkinson's Disease-like Symptoms and Oxidative Stress in Drosophila Melanogaster: Neuroprotective Effect of *Bougainvillea Glabra* Choisy. *Biomed. Pharmacother.* **2017**, *95*, 245–251. [CrossRef] [PubMed]
5. Naito, F.Y.B.; de Nazaré Almeida dos Reis, L.; Batista, J.G.; Nery, F.M.B.; Rossato, M.; Melo, F.L.; de Cássia Pereira-Carvalho, R. Complete Genome Sequence of *Bougainvillea* Chlorotic Vein Banding Virus in *Bougainvillea Spectabilis* from Brazil. *Trop. Plant Pathol.* **2020**, *45*, 159–162. [CrossRef]

6. Pavan Kumar, P.; Janakiram, T.; Bhat, K.V. Microsatellite Based DNA Fingerprinting and Assessment of Genetic Diversity in *Bougainvillea* Cultivars. *Gene* **2020**, *753*, 144794. [CrossRef]

7. Marasini, P.; Khanal, A. Assessing Rooting Media and Hormone on Rooting Potential of Stem Cuttings of *Bougainvillea*. *J. Inst. Agric. Anim. Sci.* **2018**, *35*, 197–201. [CrossRef]

8. Abarca-Vargas, R.; Petricevich, V.L. Extract from *Bougainvillea Xbuttiana* (Variety Orange) Inhibits Production of Lps-Induced Inflammatory Mediators in Macrophages and Exerts a Protective Effect in Vivo. *Biomed Res. Int.* **2019**, *2019*, 2034247. [CrossRef]

9. Ogunwande, I.A.; Avoseh, O.N.; Olasunkanmi, K.N.; Lawal, O.A.; Ascrizzi, R.; Flamini, G. Chemical Composition, Anti-Nociceptive and Anti-Inflammatory Activities of Essential Oil of *Bougainvillea gabra*. *J. Ethnopharmacol.* **2019**, *232*, 188–192. [CrossRef]

10. Kumara Swamy, M.; Sudipta, K.M.; Lokesh, P.; Neeki, A.M.; Rashmi, W.; Bhaumik, H.S.; Darshil, H.S.; Vijay, R.; Kashyap, S.S.N. Phytochemical Screening and in vitro Antimicrobial Activity of *Bougainvillea Spectabilis* Flower Extracts. *Int. J. Phytomed.* **2012**, *4*, 375–379.

11. Saleem, H.; Htar, T.T.; Naidu, R.; Zengin, G.; Ahmad, I.; Ahemad, N. Phytochemical Profiling, Antioxidant, Enzyme Inhibition and Cytotoxic Potential of *Bougainvillea glabra* Flowers. *Nat. Prod. Res.* **2020**, *34*, 2602–2606. [CrossRef] [PubMed]

12. Kalsum, N. Preliminary Studies of the Immunomodulator Effect of the Propolis Trigona Spp. Extract in a Mouse Model. *IOSR J. Agric. Vet. Sci.* **2017**, *10*, 75–80. [CrossRef]

13. Ghogar, A.; Jiraungkoorskul, W. Antifertility Effect of *Bougainvillea spectabilis* or Paper Flower. *Pharmacogn. Rev.* **2017**, *11*, 19–22. [CrossRef] [PubMed]

14. Adebayo, G.I.; Alabi, O.T.; Owoyele, B.V.; Soladoye, A.O. Anti-Diabetic Properties of the Aqueous Leaf Extract of *Bougainvillea glabra* (Glory of the Garden) on Alloxan-Induced Diabetic Rats. *Rec. Nat. Prod.* **2009**, *3*, 187–192.

15. Abarca-Vargas, R.; Petricevich, V.L. *Bougainvillea* Genus: A Review on Phytochemistry, Pharmacology, and Toxicology. *Evid. Based Complement. Altern. Med.* **2018**, *2018*, 9070927. [CrossRef]

16. Choudhary, N.; Kapoor, H.C.; Lodha, M.L. Cloning and Expression of Antiviral/Ribosome-Inactivating Protein from *Bougainvillea Xbuttiana*. *J. Biosci.* **2008**, *33*, 91–101. [CrossRef]

17. Abarca-Vargas, R.; Peña Malacara, C.F.; Petricevich, V.L. Characterization of Chemical Compounds with Antioxidant and Cytotoxic Activities in *Bougainvillea x buttiana* Holttum and Standl, (Var. Rose) Extracts. *Antioxidants* **2016**, *5*, 45. [CrossRef]

18. Bortolotti, M.; Bolognesi, A.; Polito, L. Bouganin, an Attractive Weapon for Immunotoxins. *Toxins* **2018**, *10*, 323. [CrossRef]

19. Jehani, M.; Pathak, D.M. In Vitro Antifungal Activity of Plant Extracts (Sterilized and Unsterilized) against *Macrophomina Phaseolina* (Tassi) Goid. Cause Stem Canker of Pigeonpea [*Cajanus Cajan* (L.) Millsp.]. *Int. J. Plant Prot.* **2019**, *12*, 105–109. [CrossRef]

20. Vaquero, L.R.; Estrada, M.E.V.; Aparicio, A.R.J.; Arellano, S.L.E.; Lozano, S.E. Evaluation of Methanolic Extract of *Bougainvillea glabra* Choisy "Variegata" against Spodoptera Frugiperda1 under Laboratory Conditions. *Southwest. Entomol.* **2016**, *41*, 983–990. [CrossRef]

21. Haida, Z.; Nakasha, J.J.; Hakiman, M. In Vitro Responses of Plant Growth Factors on Growth, Yield, Phenolics Content and Antioxidant Activities of *Clinacanthus nutans* (Sabah Snake Grass). *Plants* **2020**, *9*, 1030. [CrossRef] [PubMed]

22. Singh, V.; Vipul, A. Phytochemical Analysis and in Vitro Antioxidant Activities of Leaves, Stems, Flowers, and Roots Extracts of *Bougainvillea spectabilis* Willd. *Int. J. Green Pharm.* **2018**, *12*, 277–284. [CrossRef]

23. Ahmad, A.; ul Qamar, M.T.; Shoukat, A.; Aslam, M.M.; Tariq, M.; Hakiman, M.; Joyia, F.A. The Effects of Genotypes and Media Composition on Callogenesis, Regeneration and Cell Suspension Culture of Chamomile (*Matricaria chamomilla* L.). *PeerJ* **2021**, *9*, e11464. [CrossRef] [PubMed]

24. Sasidharan, S.; Chen, Y.; Saravanan, D.; Sundram, K.; Yoga, L. Extraction, Isolation and Characterization of Bioactive Compounds from Plants Extracts. *Afr. J. Tradit. Complementary Altern. Med.* **2011**, *8*, 1–10. [CrossRef]

25. Bucić-Kojić, A.; Planinić, M.; Tomas, S.; Bilić, M.; Velić, D. Study of Solid-Liquid Extraction Kinetics of Total Polyphenols from Grape Seeds. *J. Food Eng.* **2007**, *81*, 236–242. [CrossRef]

26. Fatima, H.; Khan, K.; Zia, M.; Ur-Rehman, T.; Mirza, B.; Haq, I.U. Extraction Optimization of Medicinally Important Metabolites from *Datura innoxia* Mill.: An in Vitro Biological and Phytochemical Investigation. *BMC Complement. Altern. Med.* **2015**, *15*, 376. [CrossRef]

27. Ishida, T.; Rossky, P.J. Solvent Effects on Solute Electronic Structure and Properties: Theoretical Study of a Betaine Dye Molecule in Polar Solvents. *J. Phys. Chem. A* **2001**, *105*, 558–565. [CrossRef]

28. Waszkowiak, K.; Gliszczyńska-Świgło, A.; Barthet, V.; Skręty, J. Effect of Extraction Method on the Phenolic and Cyanogenic Glucoside Profile of Flaxseed Extracts and Their Antioxidant Capacity. *J. Am. Oil Chem. Soc.* **2015**, *92*, 1609–1619. [CrossRef]

29. Cos, P.; Vlietinck, A.J.; Vanden Berghe, D.; Maes, L. Anti-Infective Potential of Natural Products: How to Develop a Stronger in Vitro "Proof-of-Concept." *J. Ethnopharmacol.* **2006**, *106*, 290–302. [CrossRef]

30. Mahmoud, S.N.; Al-ani, N.K. Effect of Different Sterilization Methods on Contamination and Viability of Nodal Segments of *Cestrum nocturnum* L. *Int. J. Res. Stud. Biosci.* **2016**, *4*, 4–9. [CrossRef]

31. Ahmad, I.; Zamir, R.; Shah, S.T.; Wali, S. In Vitro Surface Sterilization of the Shoot Tips of *Bougainvillea Spectabilis* Willd. *Pure Appl. Biol.* **2016**, *5*, 1171–1175. [CrossRef]

32. Mostafiz, S.; Wagiran, A. Efficient Callus Induction and Regeneration in Selected *Indica* rice. *Agronomy* **2018**, *8*, 77. [CrossRef]

33. Hakiman, M.; Maziah, M. Non Enzymatic and Enzymatic Antioxidant Activities in Aqueous Extract of Different *Ficus deltoidea* Accessions. *J. Med. Plants Res.* **2009**, *3*, 120–131. [CrossRef]

34. Singleton, V.L.; Rossi, J.A.J. Colorimetry to Total Phenolics with Phosphomolybdic Acid Reagents. *Am. J. Enol. Vinic.* **1965**, *16*, 144–158.

35. Marinova, D.; Ribarova, F.; Atanassova, M. Total Phenolics and Total Flavonoids in Bulgarian Fruits and Vegtables. *J. Univ. Chem. Technol. Metal* **2005**, *40*, 255–260.

36. Wong, S.P.; Leong, L.P.; William Koh, J.H. Antioxidant Activities of Aqueous Extracts of Selected Plants. *Food Chem.* **2006**, *99*, 775–783. [CrossRef]

37. Re, R.; Brühwiler, P.; Mourad, S.; Verdejo, R.; Shaffer, M. Development and Characterisation of Carbon Nanotube-Reinforced Polyurethane Foams. *Free Radic. Biol. Med.* **1999**, *26*, 1231–1237. [CrossRef]

38. Dinis, T.C.P.; Madeira, V.M.C.; Almeida, L.M. Action of Phenolic Derivatives (Acetaminophen) as Inhibitors of Membrane Lipid Peroxidation and as Peroxyl Radical Scavengers. *Arch. Biochem. Biophys.* **1994**, *315*, 161–169. [CrossRef]

39. Sturm, A.; Tang, G.Q. The Sucrose-Cleaving Enzymes of Plants Are Crucial for Development, Growth and Carbon Partitioning. *Trends Plant Sci.* **1999**, *4*, 401–407. [CrossRef]

40. Barneix, A.J.; Causin, H.F. The Central Role of Amino Acids on Nitrogen Utilization and Plant Growth. *J. Plant Physiol.* **1996**, *149*, 358–362. [CrossRef]

41. Tariq, U.; Ali, M.; Abbasi, B.H. Morphogenic and Biochemical Variations under Different Spectral Lights in Callus Cultures of *Artemisia absinthium* L. *J. Photochem. Photobiol. B Biol.* **2014**, *130*, 264–271. [CrossRef] [PubMed]

42. Hesami, M.; Daneshvar, M.H. In Vitro Adventitious Shoot Regeneration through Direct and Indirect Organogenesis from Seedling-Derived Hypocotyl Segments of *Ficus religiosa* L.: An Important Medicinal Plant. *HortScience* **2018**, *53*, 55–61. [CrossRef]

43. Pan, Z.; Zhu, S.; Guan, R.; Deng, X. Identification of 2,4-D-Responsive Proteins in Embryogenic Callus of Valencia Sweet Orange (*Citrus sinensis* Osbeck) Following Osmotic Stress. *Plant Cell. Tissue Organ Cult.* **2010**, *103*, 145–153. [CrossRef]

44. Mok, D.W.S.; Mok, M.C. Cytokinin Metabolism and Action. *Annu. Rev. Plant Physiol. Plant Mol. Biol* **2001**, *52*, 89–118. [CrossRef] [PubMed]

45. Renu, S.; Kharb, P.; Rani, K. Rapid Micropropagation and Callus Induction of Catharanthus Roseus in Vitro Using Different Explants. *World J. Agric. Sci.* **2011**, *7*, 699–704.

46. Behbahani, M.; Shanehsazzadeh, M.; Hessami, M.J. Optimization of Callus and Cell Suspension Cultures of *Barringtonia racemosa* (Lecythidaceae Family) for Lycopene Production. *Sci. Agric.* **2011**, *68*, 69–76. [CrossRef]

47. Azad, M.A.K.; Yokota, S.; Ohkubo, T.; Andoh, Y.; Yahara, S.; Yoshizawa, N. In Vitro Regeneration of the Medicinal Woody Plant Phellodendron Amurense Rupr. through Excised Leaves. *Plant Cell. Tissue Organ Cult.* **2005**, *80*, 43–50. [CrossRef]

48. Hoque, A.; Razvy, M.; Biswas, M.; Kbir, A. Micropropagation of Water Chestnut (*Trapa* Sp.) through Local Varieties of Rajshahi Division. *Asian J. Plant Sci.* **2006**, *5*, 409–413.

49. Thammina, C.; He, M.; Lu, L.; Cao, K.; Yu, H.; Chen, Y.; Tian, L.; Chen, J.; Mcavoy, R.; Ellis, D.; et al. In Vitro Regeneration of Triploid Plants of *Euonymus alatus* "compactus" (Burning Bush) from Endosperm Tissues. *HortScience* **2011**, *46*, 1141–1147. [CrossRef]

50. Pandey, A.; Verma, O.; Chand, S. In Vitro Propagation of *Boerhaavia diffusa* L.: An Important Medicinal Plant of Family Nyctagimaceae. *Indian J. Genet. Plant Breed.* **2019**, *79*, 89–95. [CrossRef]

51. Rameshkumar, R.; Satish, L.; Pandian, S.; Rathinapriya, P.; Rency, A.S.; Shanmugaraj, G.; Pandian, S.K.; Leung, D.W.M.; Ramesh, M. Production of Squalene with Promising Antioxidant Properties in Callus Cultures of *Nilgirianthus ciliatus*. *Ind. Crops Prod.* **2018**, *126*, 357–367. [CrossRef]

52. Bhojwani, S.S.; Dantu, P.K. *Plant Tissue Culture: An Introductory Text*; Springer India: New Delihi, India, 2013; ISBN 9788132210252.

53. Younas, M.; Drouet, S.; Nadeem, M.; Giglioli-Guivarc'h, N.; Hano, C.; Abbasi, B.H. Differential Accumulation of Silymarin Induced by Exposure of *Silybum Marianum* L. Callus Cultures to Several Spectres of Monochromatic Lights. *J. Photochem. Photobiol. B Biol.* **2018**, *184*, 61–70. [CrossRef]

54. Gao, J.; Li, J.; Luo, C.; Yin, L.; Li, S.; Yang, G.; He, G. Callus Induction and Plant Regeneration in *Alternanthera philoxeroides*. *Mol. Biol. Rep.* **2011**, *38*, 1413–1417. [CrossRef] [PubMed]

55. Jaramillo, J.; Summers, W.L. Dark–Light Treatments Influence Induction of Tomato Anther Callus. *HortScience* **2019**, *26*, 915–916. [CrossRef]

56. Habibah, N.A.; Moeljopawiro, S.; Dewi, K.; Indrianto, A. Callus Induction and Flavonoid Production on the Immature Seed of *Stelechocarpus burahol*. *J. Phys. Conf. Ser.* **2018**, *983*, 012186. [CrossRef]

57. Mahendra, C.; Murali, M.; Manasa, G.; Sudarshana, M.S. Biopotentiality of Leaf and Leaf Derived Callus Extracts of *Salacia macrosperma* Wight—An Endangered Medicinal Plant of Western Ghats. *Ind. Crops Prod.* **2020**, *143*, 111921. [CrossRef]

58. Esmaeili, A.K.; Taha, R.M.; Mohajer, S.; Banisalam, B. Antioxidant Activity and Total Phenolic and Flavonoid Content of Various Solvent Extracts from in Vivo and in Vitro Grown *Trifolium pratense* L. (Red Clover). *Biomed Res. Int.* **2015**, *2015*, 643285. [CrossRef]

59. Arezki, O.; Boxus, P.; Kevers, C.; Gaspar, T. Changes in Peroxidase Activity, and Level of Phenolic Compounds during Light-Induced Plantlet Regeneration from *Eucalyptus camaldulensis* Dehn. Nodes in Vitro. *Plant Growth Regul.* **2001**, *33*, 215–219. [CrossRef]

60. Zahid, N.A.; Jaafar, H.Z.E.; Hakiman, M. Micropropagation of Ginger (*Zingiber Officinale* Roscoe) 'Bentong' and Evaluation of Its Secondary Metabolites and Antioxidant Activities Compared with the Conventionally Propagated Plant. *Plants* **2021**, *10*, 630. [CrossRef]

61. Islam, M.Z.; Hossain, M.T.; Hossen, F.; Akter, M.S.; Mokammel, M.A. In-Vitro Antioxidant and Antimicrobial Activity of *Bougainvillea glabra* Flower. *Res. J. Med. Plant* **2016**, *10*, 228–236. [CrossRef]

62. Murali, M.; Prabakaran, G. Effect of Different Solvents System on Antioxidant Activity and Phytochemical Screening in Various Habitats of *Ocimum basilicum* L. (Sweet Basil) Leaves. *Int. J. Zool. Appl. Biosci.* **2018**, *3*, 375–381. [CrossRef]

63. López-Laredo, A.R.; Ramírez-Flores, F.D.; Sepúlveda-Jiménez, G.; Trejo-Tapia, G. Comparison of Metabolite Levels in Callus of *Tecoma stans* (L.) Juss. Ex Kunth. Cultured in Photoperiod and Darkness. *Vitr. Cell. Dev. Biol. Plant* **2009**, *45*, 550–558. [CrossRef]

64. Shah, M.; Ullah, M.A.; Drouet, S.; Younas, M.; Tungmunnithum, D.; Giglioli-Guivarc'h, N.; Hano, C.; Abbasi, B.H. Interactive Effects of Light and Melatonin on Biosynthesis of Silymarin and Anti-Inflammatory Potential in Callus Cultures of *Silybum marianum* (L.) Gaertn. *Molecules* **2019**, *24*, 1207. [CrossRef]

65. Mohammad, S.; Khan, M.A.; Ali, A.; Khan, L.; Khan, M.S.; Mashwani, Z.u.R. Feasible Production of Biomass and Natural Antioxidants through Callus Cultures in Response to Varying Light Intensities in Olive (*Olea europaea* L.) Cult. Arbosana. *J. Photochem. Photobiol. B Biol.* **2019**, *193*, 140–147. [CrossRef]

66. Hakkim, F.; Shankar, C.; Girija, S. Chemical Composition and Antioxidant Property of Holy Basil (*Ocimum sanctum* L.) Leaves, Stems, and Inflorescence and Their in Vitro Callus Cultures. *J. Agric. Food Chem.* **2007**, *55*, 9109–9117. [CrossRef] [PubMed]

67. Costa, P.; Gonçalves, S.; Valentão, P.; Andrade, P.B.; Coelho, N.; Romano, A. *Thymus lotocephalus* Wild Plants and in Vitro Cultures Produce Different Profiles of Phenolic Compounds with Antioxidant Activity. *Food Chem.* **2012**, *135*, 1253–1260. [CrossRef]

68. Song, K.; Sivanesan, I.; Ak, G.; Zengin, G.; Cziáky, Z.; Jekő, J.; Rengasamy, K.R.R.; Lee, O.N.; Kim, D.H. Screening of Bioactive Metabolites and Biological Activities of Calli, Shoots, and Seedlings of *Mertensia maritima* (L.) Gray. *Plants* **2020**, *9*, 1551. [CrossRef]

69. Zengin, G.; Aktumsek, A. Investigation of Antioxidant Potentials of Solvent Extracts from Different Anatomical Parts of *Asphodeline anatolica* E. *Afr. J. Tradit. Complement. Altern. Med.* **2014**, *11*, 481–488. [CrossRef]

 *horticulturae*

Article

Establishment of an Efficient In Vitro Propagation Method for a Sustainable Supply of *Plectranthus amboinicus* (Lour.) and Genetic Homogeneity Using Flow Cytometry and SPAR Markers

Mohammad Faisal * and Abdulrahman A. Alatar

Department of Botany and Microbiology, College of Science, King Saud University, P.O. Box 2455, Riyadh 11451, Saudi Arabia; aalatar@ksu.edu.sa
* Correspondence: faisalm15@yahoo.com

Abstract: *Plectranthus amboinicus* (Lour.) Spreng is a medicinally important aromatic perennial herb used for the treatment of skin diseases, constipation, asthma, flu, fever, cough, and headache as well as a flavoring ingredient in traditional drinks, food, and meat stuffing. In this study, a high-performance in vitro propagation system of *P. amboinicus* through direct shoot organogenesis was developed using axillary node explants cultured on MS (Murashige and Skoog) medium augmented with 0.5, 2.5, 5.0, 7.5, and 10.0 µM of 6-benzyladenine (BA) or kinetin (Kin), alone or with 0.1, 0.5, 2.5, and 5.0 µM of indole-3-acetic acid (IAA) or α-naphthalene acetic acid (NAA). To optimize the regeneration potential of node explants, the effects of basal media strength and pH were also investigated. After 8 weeks of culture, explants cultured in full strength MS basal medium (pH 5.7) with 5.0 µM BA and 2.5 µM NAA exhibited the highest percentage (97.1%) of regeneration and the maximum number (19.3) of shoots per explant. Individual elongated shoots were rooted on half strength MS basal medium containing 0.25 µM indole 3-butyric acid (IBA) after 4 weeks of culture, producing 5.3 roots/shootlets with a root induction frequency of 93.7%. First time genetic stability of in vitro raised *P. amboinicus* plants was determined using SPAR markers, such as DAMD and ISSR, as well as flow cytometric tests, assuring the availability of authenticated raw materials for commercial production of the plant and its bioactive components.

Keywords: acclimatization; aromatic plant; node segments; micropropagation; tissue culture

Citation: Faisal, M.; Alatar, A.A. Establishment of an Efficient In Vitro Propagation Method for a Sustainable Supply of *Plectranthus amboinicus* (Lour.) and Genetic Homogeneity Using Flow Cytometry and SPAR Markers. *Horticulturae* **2022**, *8*, 693. https://doi.org/10.3390/horticulturae8080693

Academic Editor: Sergio Ruffo Roberto

Received: 23 June 2022
Accepted: 27 July 2022
Published: 1 August 2022

Publisher's Note: MDPI stays neutral with regard to jurisdictional claims in published maps and institutional affiliations.

1. Introduction

The *Plectranthus* genus is among the most prominent members of the Lamiaceae family. *Plectranthus amboinicus* (Lour.) Spreng, native to eastern and southern Africa, is an important medicinal succulent perennial herb which tends to be creeping or climbing; it has highly aromatic leaves with short erect hairs [1]. In folk medicine of different countries, it is used for the treatment of skin diseases, constipation, asthma, flu, fever, cough, and headache [2–4]. The high concentration of active compounds that are found in *P. amboinicus* is predominantly responsible for its varied therapeutic potential as well as its culinary uses. The plant's raw leaves are eaten or added as a flavoring ingredient in traditional drinks, food, and meat stuffing [1,5]. Biochemical studies revealed that the essential oil extracted from this plant has high amounts of thymol [6], carvacrol [7], α-humulene, α-terpineol, β-caryophyllene, β-selinene, γ-terpinene, and p-cymene [4,8]. Previous literature revealed that this plant has several pharmacological effects, including antimicrobial [9], antibacterial [10,11], antifungal [8,12], and antiviral [13–16] activities. In addition to its antiepileptic [17], antitumorigenic [18,19], anti-inflammatory [20,21], and antioxidant [22,23] effects, *P. amboinicus* shows antagonistic activities against respiratory [24], cardiovascular [25], oral [26], digestive [27], and genitourinary [28] disorders.

Furthermore, the leaf paste of this plant is effective in the treatment of wounds [29] and skin diseases [30,31].

Since 1960, tissue culture has been used as an alternative method for conventional propagation in many plant species, where a large number of plants can be produced in a relatively short time and small space from a small piece of plant tissue [32]. Culturing of plant cells and tissues in vitro has become a reliable and necessary technique for propagation, improvement, and mass-multiplication of several plant species [33]. Micropropagation has been used to propagate many species of *Plectranthus*, such as *P. bourneae* [34–36], *P. edulis* [37,38], *P. amboinicus* [39,40], *P. barbatus* [41], *P. zeylanicus* [42], and *P. esculentus* [43].

One of the major problems facing the micropropagation system is the occurrence of morphological, physiological, and molecular changes in in vitro regenerated plants [44]. Therefore, it is necessary to evaluate the genetic stability of the micropropagated plants to ensure their conformity with the donor plants. Polymerase chain reaction (PCR) based on SPAR (single primer amplification reaction) markers is one of the techniques used to analyze the genetic stability of micropropagated plants. Among SPAR markers, inter-simple sequence repeat (ISSR) and directed amplification of minisatellite-region DNA (DAMD) are successfully used to reveal the genetic homogeneity between micropropagated and donor plants in many plant species [33,45–51]. ISSR and DAMD techniques are cost-effective, quick, simple, avoid the use of radioactivity and DNA blotting, and are susceptible to automation [33,51,52]. In addition, genetic changes in regenerated plants can be detected by flow cytometry, which has been shown to be an efficient and reliable tool for estimation of ploidy level and genome size by calculating the nuclear DNA content [53]. The flow cytometry system has been used to evaluate the genetic fidelity of regenerated plants in many medicinal plants, including *Bacopa monnieri* [44], *Ficus carica* [54], *Salix lapponum* [55], and *Brassica juncea* [53].

The objective of this research was to develop an efficient and reproducible method for in vitro propagation of *P. amboinicus* from nodal explants for a sustainable large-scale production. Genetic homogeneity of micropropagated plantlets was assessed for the first time through DNA-based SPAR markers and flow cytometry to ensure the propagation and supply of true-to-type plantlets.

2. Materials and Methods

2.1. Shoot Materials and Explants Preparation

Shoots of the *Plectranthus amboinicus* were harvested during March–April 2021 from a plant maintained in the growth chamber at the Botany & Microbiology Department, King Saud University, which was full of healthy growth. The plant samples were identified with the help of a taxonomist in the department and verified at the department's herbarium, where vouchers were deposited. The excised shoots were thoroughly cleaned in water with Tween-20 (2-3 drops) for a period of twenty minutes in order to eliminate any dust and dirt. After that, they were cut into pieces that were 2–3 cm in size and placed under running tap water for a period of 30 min. All the materials were moved into a biosafety laminar-air flow hood, where they were sterilized for three minutes with a freshly prepared 0.1 percent (w/v) mercury chloride ($HgCl_2$) solution (Riedel-de Haan AG, Seelze, Germany) while being gently stirred. The sterilized shoots (nodal segments) were given a thorough washing in deionized ultra-pure water to eliminate any residues of the sterilant before being cut into pieces measuring 0.5–0.7 cm in length.

2.2. Culture Media and Growth Condition

The sterile shoot segments were grown in nutritional medium that included various concentrations and combinations of growth regulators. Murashige and Skoog [56] plant cell culture basal medium (MS; M5519, Sigma-Aldrich, Inc., St. Louis, MO, USA) was used in all assays; it was composed of macro and micro-salts, vitamins, 3 percent (w/v) sucrose (AppliChem GmbH, Darmstadt, Germany), and 0.8 percent (w/v) agar-agar (Avonchem Ltd., Wellington House, Cheshire, UK) and had a pH of 5.7. The pH of the MS basal was adjusted with 1 N aqueous solution of NaOH or HCl before steam sterilization in an ALP-autoclave (CLG-32L, ALP Co., Tokyo, Japan) for 20 min at 121 °C (15 psi). In a growth chamber (Conviron Adaptis-CMP6010, USA), all the cultured vials were incubated under 50 µmol m^{-2} s^{-1} 1 light illuminance provided by Philips 39-Watt T5 linear fluorescent tubes (F39T5/841/HO/ALTO, Philips, Amsterdam, The Netherlands) with a day–night photoperiod of 16/8 h and a temperature of 24 ± 2 °C.

2.3. Growth Regulators, Shoot Induction, and Proliferation

For shoot induction and proliferation, the sterilized nodal sections of *P. amboinicus* were cultivated on MS medium supplemented with 0, 0.5, 2.5, 5.0, 7.5, and 10.0 µM of 6-benzyladenine (BA; Duchefa Biochemie B.V., Haarlem, The Netherlands) or kinetin (Kin; Sigma-Aldrich Chemicals Co., MO, USA) at different concentrations, either individually or in combination with 0, 0.1, 0.5, 2.5., and 5.0 µM of auxins such as indole-3-acetic acid (IAA; Duchefa Biochemie B.V., Haarlem, The Netherlands) or α-naphthalene acetic acid (NAA; Duchefa Biochemie B.V., Haarlem, The Netherlands). The effects of the strength of the MS basal medium (one-quarter; one-third; half; and full strength) and various pH levels (4.7, 5.2, 5.7, and 6.2) on the in vitro morphogenic response of *P. amboinicus* nodal sections were also evaluated using optimum phytohormonal combinations and concentrations of BA (5.0 µM) and NAA (2.5 µM). In order to achieve a higher number of shoots from each nodal section, the responding plant materials were sub-cultured onto the fresh media every three weeks. After 8 weeks of in vitro growth and proliferation, data on the percentage of the explants' regeneration as well as the number of axillary shoots per explant were recorded.

2.4. Rooting of Shootlets and Acclimation

Individually harvested, the in vitro proliferating shootlets of *P. amboinicus* with 2–3 pairs of substantially developed leaves were transplanted to a half strength MS basal medium supplemented with 0, 0.25, 0.5, 1.0, and 2.0 µM of indole 3-butyric acid (IBA) solidified with 0.8% (*w/v*) agar-agar and having a pH of 5.7. Data on the frequency of shootlets producing roots and the number of roots per shootlets were recorded after 4 weeks of transfer in rooting media.

For ex vitro acclimation, well-rooted plantlets were gently washed under the laboratory running water to remove the medium and agar residues. Then, the plantlets were transferred to pots containing sterile planting soil substrates (Planta-Guard^{TM}, Germany). Maintaining the plantlets at high humidity for the first few days following transplanting is a critical factor for the survival of micropropagated plants. To retain humidity, the ex vitro transplanted plants were covered with clear polybags, which were removed after 4 weeks and irrigated with half strength MS basal medium containing macro and micro-salts devoid of vitamins. After 1 month, the percent survival of acclimated plants was calculated as follows:

$$\frac{\text{Total number of plants survived}}{\text{Total number of plants trasnferred}} \times 100$$

2.5. Flow Cytometric Profile of Plants

Nuclei isolated from leaf samples (approximately 100 mg) of regenerated plants and donor plants of *P. amboinicus* were used to generate flow cytometric profiles. As previously reported by Galbraith et al. [57], the nuclei were isolated by cutting leaf samples with a new surgical blade in micro-Petri dishes (35 × 15 mm^2) containing 2.0 mL of precooled isolation buffer (20 mM MOPS, 45 mM MgCl$_2$, 30 mM sodium citrate, 0.1% (*v/v*) Triton X-100; pH 7.0). Using a double-layered nylon membrane of 30-micron thickness, the homogenates were filtered with the help of a micro syringe to eliminate any remaining leaf debris from the mixtures. The filtered suspensions were transferred to labelled Eppendorf tubes containing 2.5 µL of 10 mg/mL DNAse-free RNAse and incubated for 10 min. After 30 min of staining with 50 µgml^{-1} of propidium iodide (Sigma-Aldrich, Inc., St. Louis, MO, USA), the nuclei samples were passed and analyzed using a flow cytometry machine (Muse™ Cell Analyzer, Merck KGaA, Darmstadt, Germany).

2.6. DNA Extraction and Molecular Characterization

Genomic DNA was isolated from young leaves of the donor plant as well as micropropagated plants of *P. amboinicus* using the CTAB (cetyltrimethylammonium bromide) method [58]. Quantification and DNA purity was determined by a NanoDrop 200c Spectrophotometer (Thermo Fisher Scientific, Waltham, MA, USA). The DNA samples were diluted in Ultrapure Milli-Q water to a final concertation of 25 ng/μL and stored in a laboratory refrigerator at 4 °C until further use. Five DAMD (Directed amplifications of minisatellite-region DNA; Table 1) primers (Metabion International AG, Planegg, Germany) and nine ISSR (Inter-simple sequence repeat; Table 2) primers (GeneLink, Inc., Orlando, FL, USA) were used for PCR.

Table 1. DAMD primers screening for the genetic stability of *Plectranthus amboinicus*.

Primers	Sequence $(5' \rightarrow 3')$	Annealing Temperature (°C)	Size Range (bp)	Number of Bands
HBV3	GCTCCTCCCCTCCT	53	2500–500	9
HBV5	GGTGAAGCACAGGTG	56	2000–300	11
HVR	GGTGTAGAGAGGGGT	50	20,000–400	17
M13	GAGGGTGGCGGTTCT	57	1500	1
33.6	GGAGGTGGGCA	47	3500–400	10
			Total =	48
			Average band per primer =	9.6

Table 2. ISSR primers screening for the genetic stability of *Plectranthus amboinicus*.

Primers	Sequence $(5' \rightarrow 3')$	Annealing Temperature (°C)	Size Range (bp)	Number of Bands
UBC-811	(GA)8C	49	3000–400	11
UBC-825	(AC)8T	46	3000–500	14
UBC-827	(AC)8G	50	3000–500	9
UBC-834	(AG)8YT	50	2000–300	10
UBC-841	(GA)8YC	50	3000–400	11
UBC-855	(AC)8YT	50	20,000–500	12
UBC-866	(CTC)6GT	55	20,000–500	13
UBC-868	(GAA)6	46	5000–700	12
UBC-880	(GGAGA)3	50	5000–800	11
			Total =	150
			Average band per primer =	11.4

PCR amplification was carried out in 20 μL volumes containing 2.0 μL of 10× PCR buffer, 1.2 μL MgCl$_2$ (25 mM), 0.4 μL dNTPs (10 mM), 1 μL (10 pmole) primers, 0.2 μL Taq polymerase (3 Unit), and 1.2 μL Template DNA (25/μL ng). Both DAMD and ISSR reactions were carried out with a Bio-Rad thermal cycler (T100; Bio-Rad Laboratories, Inc., Hercules, CL, USA) with initial denaturation for 2 min (1 cycle) at 94 °C, denaturation for 5 min (35 cycles) at 94 °C, annealing for 2 min at 47–57 °C, extension for 1 min at 72 °C, and a final extension for 7 min at 72 °C. The amplified DNA fragments were separated and resolved on 1.5% (w/v) agarose gel (Sigma Aldrich, Inc., St. Louis, MO, USA) containing 4 μL ethidium bromide (Sigma Aldrich, USA) in a horizontal electrophoresis system (Biometra, Göttingen, Germany) with 1× TBE (Tris-Boric acid-EDTA). The electrophoresis was run for 2 h and the DNA bands were visualized on a UV Gel-Documentation System (G:BOX F3, Syngene, Cambridge, UK). Each sample were run thrice and only intense and reproducible bands were scored.

2.7. Statistical Analysis

The experiments in this study were repeated thrice using a completely randomized design (CRD) with 20 explants per treatment, and the results were expressed as mean value and standard error (SE). One-way analysis of variance (ANOVA, San Francisco, CA, USA) with Tukey's honestly significant difference test (Tukey's HSD) analysis was performed for comparisons of significant differences between treatments at a *p*-value of less than 0.05. The data were processed and analyzed using IBM SPSS version 24 (SPSS Inc., Chicago, IL, USA) and the graphs were created using the Microsoft Excel by Microsoft for Windows for Mac.

3. Results

3.1. Effects of Cytokinin on Shoot Induction

Only 1–2 shoots were formed by the sterile nodal segments of *P. amboinicus* cultured on cytokinin-free medium, which were used as a control. Adding BA or Kin to the basal medium, on the other hand, significantly increased the number of shoots that sprouted from the cultivated shoot segments. Nodal segments of *P. amboinicus* respond quite differently in vitro depending on the cytokinin type and concentration used (Table 3). The cultures were carefully investigated at periodic intervals (sub-culture) to see if any morphogenic alterations in the nodal sections had occurred. It was determined that all tested concentrations of both BA and Kin cytokinins, i.e., 0.5, 2.5, 5.0, or 10 µM, enhanced the induction and proliferation of shoots. Within ten days of culture, the latent axillary bud swelled, and after four weeks, it differentiated into numerous microshoots (Figure 1A,B). In general, of the different concentrations of BA and Kin tested, the number of shootlets increased with increasing the concentration from 0.5 to 5.0 µM. In contrast, concentrations of BA and Kin in the media more than 5.0 µM resulted in fewer shoots with non-degenerative callusing (slight) from the cut ends of each explant. When compared to Kin, BA was shown to be more effective for the induction and proliferation of shootlets. The MS basal medium combined with 5.0 µM of BA exhibited the greatest number of 7.5 shootlets per explant in 79.3 percent of cultures.

Table 3. Effect of 6-benzyladenine (BA) and kinetin (Kin) on shoot regeneration from axillary node explants of *Plectranthus amboinicus*.

Plant Growth Regulators (µM)		Regeneration %	Shoot Number per Explant
BA	**Kin**		
0.0	0.0	23.01 ± 1.15 f	1.26 ± 0.15 g
0.5		46.67 ± 2.02 e	2.13 ± 0.20 f
2.5		61.31 ± 2.31 cd	4.73 ± 0.45 c
5.0		79.37 ± 2.30 a	7.57 ± 0.30 a
7.5		70.66 ± 1.88 b	5.06 ± 0.35 c
10		62.10 ± 1.51 cd	3.23 ± 0.15 e
	0.5	41.10 ± 1.73 e	2.03 ± 0.14 f
	2.5	55.33 ± 1.20 d	4.13 ± 0.21 cde
	5.0	69.43 ± 2.61 bc	6.36 ± 0.26 b
	7.5	62.00 ± 1.73 cd	4.27 ± 0.15 cd
	10	42.67 ± 1.45 e	3.37 ± 0.23 de

Data are expressed as mean ± SE recorded after 8 weeks of culture on shoot induction medium. Different letters in each column indicate significant differences among treatments according to Tukey's honestly significant difference test (Tukey's HSD) at a *p*-value of less than 0.05.

Figure 1. (**A**) Initiation of shoots from axillary node explants of *Plectranthus amboinicus*. (**B**) Multiple shootlets of *P. amboinicus* after 4 weeks of culture. (**C**) Proliferated multiple shoots of *P. amboinicus* after 8 weeks of culture. (**D**) Rooted *P. amboinicus* shoot after 4 weeks of transfer on the rooting media.

3.2. Effects of Auxin with Optimal Concentration of BA

In another set of experiments, the effectiveness of the optimum concentration of BA in combination with concentrations of 0, 0.1, 0.5, 2.5, or 5.0 µM of auxins (i.e., IAA or NAA) for the induction and proliferation of shootlets was assessed (Table 4). The better multiplication response was obtained from a shoot section of *P. amboinicus* grown on medium with 5.0 µM BA and 2.5 µM NAA (Figure 1C). On this media, 97.1 percent of the explants generated 19.3 shootlets (Table 4). However, NAA concentrations over 2.5 µM, along with optimal BA concentration, reduced the frequency of shoots with non-degenerative callusing from the explants. When IAA was coupled with BA, it improved the morphogenic response but produced fewer shootlets than BA with NAA in the growth medium. The maximum number of shootlets, 15.3 and 12.2, was observed from each shoot section on 5.0 µM BA or Kin supplied medium containing 2.5 µM of IAA, respectively (Table 4).

Table 4. Combined effect of 6-benzyladenine (BA) or kinetin (Kin) with α-naphthalene acetic acid (NAA) or indole-3-acetic acid (IAA) on shoot regeneration from axillary node explants of *Plectranthus amboinicus*.

Plant Growth Regulators (μM)				Regeneration %	Shoots Number per Explant
BA	Kin	NAA	IAA		
5.0				79.37 ± 2.30 cde	7.53 ± 0.50 gh
	5.0			69.43 ± 2.61 ghi	6.27 ± 0.46 h
5.0		0.1		81.01 ± 2.70 cde	12.43 ± 1.14 cd
5.0		0.5		85.04 ± 2.01 bcd	14.70 ± 1.15 bc
5.0		2.5		97.10 ± 2.16 a	19.37 ± 1.21 a
5.0		5.0		72.76 ± 1.45 efgh	7.97 ± 0.53 fgh
5.0			0.1	80.33 ± 2.53 cde	11.43 ± 0.97 de
5.0			0.5	80.71 ± 2.04 cde	10.63 ± 0.87 def
5.0			2.5	92.67 ± 3.10 ab	15.36 ± 1.32 b
5.0			5.0	67.31 ± 1.65 hi	6.98 ± 0.42 gh
	5.0	0.1		77.67 ± 1.45 defg	11.70 ± 0.75 de
	5.0	0.5		85.10 ± 1.53 bcd	12.20 ± 1.05 cd
	5.0	2.5		90.43 ± 1.45 ab	16.27 ± 1.16 b
	5.0	5.0		70.31 ± 1.98 fgh	7.01 ± 0.93 gh
	5.0		0.1	73.44 ± 1.83 efgh	9.30 ± 0.77 efg
	5.0		0.5	80.11 ± 1.45 cde	8.53 ± 0.50 fgh
	5.0		2.5	88.33 ± 1.54 abc	12.27 ± 0.67 cd
	5.0		5.0	61.01 ± 1.73 i	5.97 ± 0.56 h

Data are expressed as mean ± SE recorded after 8 weeks of culture on shoot induction medium. Different letters in each column indicate significant differences among treatments according to Tukey's honestly significant difference test (Tukey's HSD) at a *p*-value of less than 0.05.

3.3. Effects of Basal Medium Strength and pH

In addition, varying strengths (one-quarter; one-third; half; and full MS) of basal medium were evaluated when combined with 5 μM BA and 2.5 μM NAA in order to get the best response for in vitro propagation and mass multiplication of *P. amboinicus* from nodal sections (Figure 2). Substantial differences were observed between the strengths of the medium used in this set of experiments. From the results, it was evidenced that the explants responded better in the MS basal medium supplied with the full strength of macro and micro-salts and vitamins. Shoot sections grown on MS basal media containing one-quarter macro and micro-salts as well as vitamins generated fewer shootlets. In subsequent studies, different pH values of MS basal medium (4.7, 5.2, 5.7, and 6.2) were employed to examine their influence on in vitro morphogenesis (Figure 3).

The pH 5.7 basal medium was shown to be optimal for the induction and proliferation of shoots, but the highly and severely acidic basal medium had a detrimental effect on in vitro shoot proliferation. The MS basal medium, which was supplied with full strength macro and micro-salts, vitamins, 5 μM BA, 2.5 μM NAA, and pH 5.7, provided the best response in terms of percent shoot induction (97.1) and number of shoots (19.3), according to the data collected throughout the investigation (Figure 3). The clusters of shootlets that formed were either sub-cultured onto new medium for further proliferation or individually transplanted into auxin-containing media for rhizogenesis.

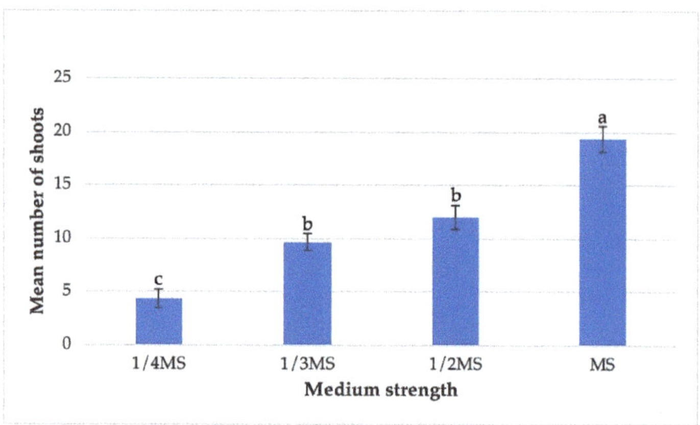

Figure 2. Effect of MS medium strength with optimized concentration of 6-benzyladenine (BA) on shoot regeneration from axillary node explants of *Plectranthus amboinicus*. Data are expressed as mean ± SE recorded after 8 weeks of culture on shoot induction medium. Bars denoted with different letters indicate significant differences among treatments according to Tukey's honestly significant difference test (Tukey's HSD) at a *p*-value of less than 0.05.

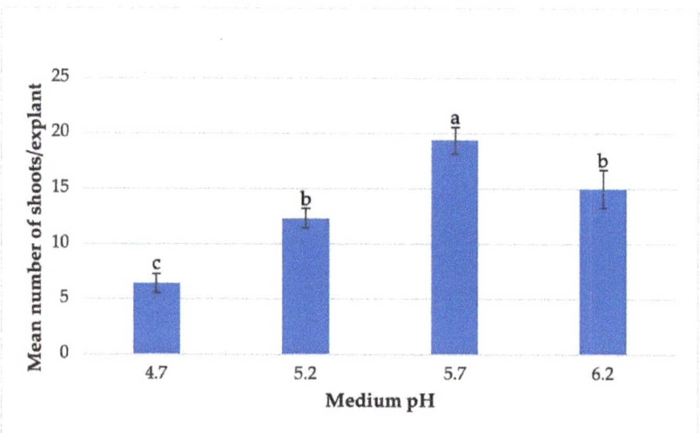

Figure 3. Effect of different pH of the MS medium with optimized concentration of 6-benzyladenine (BA) on shoot regeneration from axillary node explants of *Plectranthus amboinicus*. Data are expressed as mean ± SE recorded after 8 weeks of culture on shoot induction medium. Bars denoted with different letters indicate significant differences among treatments according to Tukey's honestly significant difference test (Tukey's HSD) at a *p*-value of less than 0.05.

3.4. Rooting of Microshoots and Acclimation

For rooting, the individually harvested shootlets were transferred to MS basal medium containing an auxin such as IAA at concentration of 0, 0.25, 0.5, 1.0, and 2.0 µM (Table 5). Adding IBA to the basal medium was found to be effective for root induction from in vitro regenerated shootlets of *P. amboinicus* (Figure 1D). On medium containing 0.5 IBA, the shootlets responded best, yielding 5.3 roots/shootlets and with a root induction frequency of 93.7% in 4 weeks. For acclimatization, the individual plantlets with a complete root system were removed from semi-solid agar rooting media, shifted to pots containing sterile planting materials, and grown for weeks inside a growth chamber. After four weeks, the surviving pants were moved to field condition. Out of 50 in vitro-produced plantlets,

47 survived, resulting in a survival rate of 94 percent with no phenotypic changes between the plants.

Table 5. Effect of IBA on in vitro root induction from microshoots of *Plectranthus amboinicus*.

IBA (μM)	Regeneration %	Roots Number per Shootlet
0.0	36.67 ± 2.02 d	0.73 ± 2.02 c
0.25	93.76 ± 2.71 a	5.30 ± 0.44 a
0.5	87.10 ± 2.33 a	4.13 ± 0.45 ab
1.0	72.36 ± 2.30 b	3.87 ± 0.30 b
2.0	60.50 ± 1.88 c	1.06 ± 0.35 c

Data are expressed as mean ± SE recorded after 4 weeks of culture on shoot induction medium. Tukey's honestly significant difference test (Tukey's HSD) analysis was performed for comparisons of significant differences between treatments at a *p*-value of less than 0.05.

3.5. Flow Cytometric Profile of Plants

Flow cytometric examination of in vitro grown plants is a considerably faster and more reliable technique of ensuring ploidy genetic uniformity. Thus, comparative ploidy levels were determined in this study using a Muse Cell Analyzer in isolated nuclei from *P. amboinicus* micropropagated plants and a parent plant. The histograms produced from the nuclei of all plant sources, as previously stated, reveal a unimodal fluorescence peak of nuclear DNA. The cytometric profile indicated that all plant sources had nearly identical G0/G1 positions (Figure 4). As a result of the experimental findings, it was discovered that there were no differences between their usual fluorescence peak in the histogram of all plant sources, indicating that in vitro propagated *P. amboinicus* plants-maintained homogeneity of ploidy level, which ensures the successful clonal multiplication of this plant without compromising the genetic integrity.

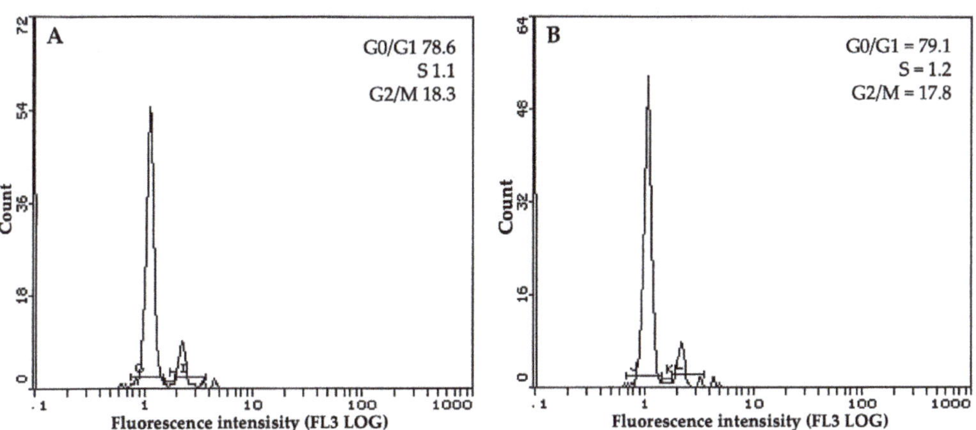

Figure 4. Flow cytometric profiles of tissue culture raised plants (**A**) and donor plants (**B**) of *Plectranthus amboinicus*.

3.6. Molecular Characterization

DAMD and ISSR primers were used for molecular characterization of in vitro propagated plants growing in the greenhouse and compared with the DNA profile of the donor plant. Five DAMD primers were selected and screened out, where all the primers generated a total of 48 bands after PCR amplification with an average of 9.6 bands per DAMD primer (Table 1). The bands were monomorphic across all in vitro propagated plants of *P. amboinicus* and 100% similar to that of donor plants. The DNA banding profiles of the *P. amboinicus* genotypes using DAMD primers HVR and 33.6 are presented in Figure 5.

Similarly, all nine ISSR primers tested also produced very clear, monomorphic bands within all in vitro propagated plants with a total of 150 scorable bands. The number of bands for each ISSR primer varied from 9 to 17 with an average of 11.4 bands per primer (Table 2, Figure 6).

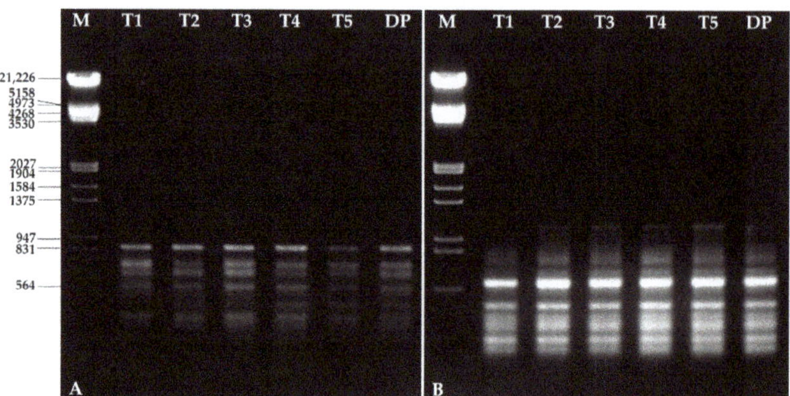

Figure 5. DAMD-PCR profiles of tissue culture and donor plants of *Plectranthus amboinicus*: (**A**) profile generated using primer HVB5 and (**B**) profile generated using primer 33.6. Lane M = Lambda DNA/EcoRI+HindIII marker; lanes T1–T5 = randomly selected regenerated plants; lane DP = donor plant.

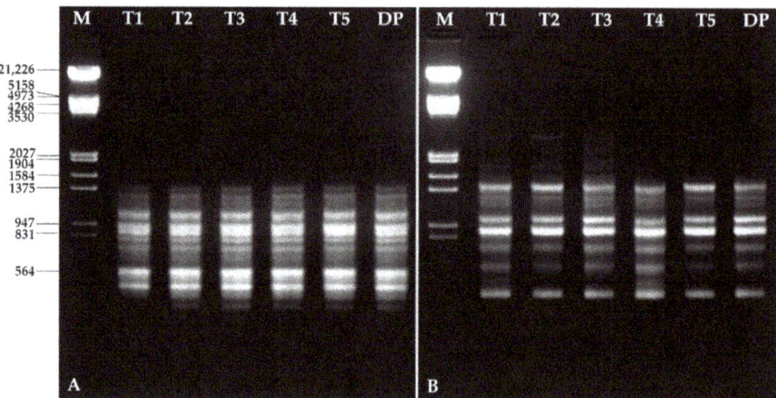

Figure 6. ISSR-PCR profiles of tissue culture and donor plants of *Plectranthus amboinicus*: (**A**) profile generated using primer UBC827 and (**B**) profile generated using primer UBC855. Lane M = Lambda DNA/EcoRI+HindIII marker; lanes T1–T5 = randomly selected regenerated plants; lane DP = donor plant.

4. Discussion

The most widely used approach for multiple shoot induction and plant regeneration is in vitro axillary bud proliferation, which is also thought to be the best way to ensure that the regenerated plants will have the same genetic makeup as the donor plants. Normally, dormant axillary buds are forced to develop into multiple shoots by proper application of growth regulators. Addition of cytokinins, viz., BA, Kin, TDZ, 2iP, and m-Topolin, in the medium led to the successful formation of multiple shoots in many plant species, including *P. edulis* [37,38], *P. amboinicus* [39,40], *Ruta graveolens* [33], *Pongamia pinnata* [59], *Rumex pictus* [60], *Mentha* × *piperita* [61], *Atropa acuminata* [62], *Cannabis sativa* [63], *Campomanesia xanthocarpa* [64], *Zingiber officinale* [65], *Cicer arietinum* [66,67], and *Humulus lupulus* [68].

The type of cytokinin used in the current research had a significant influence on the induction of multiple buds from nodal sections, with BA being shown to be more efficient than Kin. Apart from that, until BA concentration was optimized, both the regeneration frequency and the mean number of shoots generated per nodal sections continued to increase. Similar findings have been made for shoot organogenesis in a number of therapeutic plants in the past [69–73]. From the nodal explants of *P. amboinicus*, the concentration of BA at 5.0 μM produced the greatest number of shoots, compared to the other concentrations and PGR tested in this investigation. The current findings are comparable with those of Arumugam et al. [40] and Belete and Balcha [37], where 0.5 mg/L and 1.5 μM BA alone was proved to be more effective than kinetin for shoot induction in *P. amboinicus* and *P. edulis*, respectively. The presence of high BA in the culture medium, on the other hand, showed inhibitory effects on both shoot response and proliferation, which was consistent with the findings in *Bambusa ventricose* [74] and *Sapium sebiferum* [72].

The balance between auxin and cytokinin is critical during the whole micropropagation process, since the combination of the two regulates growth and development in a plant species. As a relatively high ratio of cytokinins is combined with a low ratio of auxins, they have a synergistic effect on cell division, resulting in a higher frequency of shoot bud induction and a greater number of shoots/explants when compared to cytokinin alone in the same experiment [75,76]. It is important to maintain a proper balance of plant growth regulators in the culture medium in combination with phytohormones produced by the plant when regenerating tissues in vitro, as this is one of the primary factors responsible for the induction, differentiation, and proliferation of shoots in the growth media. Similarly, in the present study, low auxin and high cytokinin concentrations also showed the greatest potential for inducing and multiplying shoots in *P. amboinicus*, and MS medium augmented with 5.0 μM BA in conjunction with 2.5 μM NAA was found to be the most effective treatments for direct shoot regeneration and multiplication in the microenvironment. In contrast, BA in combination with TDZ promoted a high number of shoots in *P. bourneae* [35]. The role of auxin and cytokinins in increasing axillary buds proliferation and ending apical dominance is thought to be the cause of the rise in the number of shoots observed in this investigation. The combination of auxin with cytokinin was also found to be effective for shoot multiplication and elongation in many plant species, including *P. amboinicus* [39,40], *Cassia alata* [77], *Artemisia abrotanum* [78], *Syzygium cumini* [79], *Manihot esculenta* [80], *Hildegardia populifolia* [81], *Asparagus cochinchinensis* [82], *Basella rubra* [83], and *Sapium sebiferum* [72].

The ability of microshoots to induce roots is critical because it has a direct impact on their survival in the greenhouse and their ability to adapt to their new environment. In the great majority of species, auxins are significant factors in root production because they stimulate the development of adventitious roots, which are essential for rooting [84]. The micropropagated shootlets of *P. amboinicus* were effectively rooted with IBA supplied in MS agar medium. The effects of IBA containing agar rooting media are in agreement with earlier approaches for inducing roots in *P. amboinicus* [39,40] and in several medicinal plants, including *Ruta graveolens* [33], *Hemidesmus indicus* [85], *Rauvolfia tetraphylla* [86], *Sapium sebiferum* [72], *Artemisia vulgaris* [87], and *Asystasia gangetica* [88].

Preserving the genetic consistency in tissue culture plants is critical for any micropropagation system prior to commercialization or in germplasm conservation. The PCR-based SPAR markers technique is one of the most significant approaches with increased use in the assessment of the genetic stability of tissue culture plants, since it uses only a small amount of genomic DNA, avoids the use of radioactivity and DNA blotting, and is susceptible to automation. In this study, DAMD and ISSR markers, as well flow cytometry, were used to confirm the genetic integrity of *P. amboinicus* micropropagated plants. Because micropropagation is known to induce somaclonal variation in micropropagated plants, the use of multiple markers has long been advocated for a better evaluation of genetic uniformity of in vitro plantlets [47]. Both DMAD and ISSR analyses showed a 100% monomorphic banding pattern, indicating the absence of variability among tissue culture plantlets of *P. amboinicus*. The appraisal of genetic constancy in a variety of micropropagated medicinal plants such as *Mentha arvensis* [46], *Pittosporum eriocarpum* [47], *Bacopa monnieri* [44], *Curcuma zedoaria* [48], *Ruta chalepensis* [50], *Rauvolfia serpentina* [89], *Nepenthes khasiana* [90] and *Ficus carica* [91] has been documented.

5. Conclusions

The present study describes a comprehensive and dependable approach for in vitro shoot regeneration, multiplication, and ex vitro establishment from axillary node explants of *P. amboinicus*, which will provide an alternative method for its mass multiplication and conservation as well as a source of material for commercial use and medicinal demands. For the first time, genetic homogeneity and true-to-type character of in vitro raised plants were confirmed by SPAR markers such as DAMD and ISSR, as well as flow cytometric analyses, ensuring the availability of legitimate raw materials for commercial production of the plant and its biologically active molecules.

Author Contributions: Conceptualization, M.F.; methodology, M.F. and A.A.A.; validation, M.F. and A.A.A.; formal analysis, M.F. and A.A.A.; investigation, M.F. and A.A.A.; resources, M.F. and A.A.A.; data curation, M.F.; writing—original draft preparation, M.F.; writing—review and editing, M.F. and A.A.A.; project administration, M.F. and A.A.A.; funding acquisition, M.F. and A.A.A. All authors have read and agreed to the published version of the manuscript.

Funding: This study was supported by the Researchers Supporting Project (RSP-2021/86), King Saud University, Riyadh, Saudi Arabia.

Institutional Review Board Statement: Not applicable.

Informed Consent Statement: Not applicable.

Data Availability Statement: Not applicable.

Acknowledgments: The authors are thankful to the Researchers Supporting Project (RSP-2021/86), King Saud University, Riyadh, Saudi Arabia for financial support.

Conflicts of Interest: The authors declare no conflict of interest.

References

1. Arumugam, G.; Swamy, M.K.; Sinniah, U.R. *Plectranthus amboinicus* (Lour.) Spreng: Botanical, Phytochemical, Pharmacological and Nutritional Significance. *Molecules* **2016**, *21*, 369. [CrossRef]
2. Lukhoba, C.W.; Simmonds, M.S.J.; Paton, A.J. Plectranthus: A review of ethnobotanical uses. *J. Ethnopharmacol.* **2006**, *103*, 1–24. [CrossRef]
3. Hsu, K.-P.; Ho, C.-L. Antimildew Effects of Plectranthus amboinicus Leaf Essential Oil on Paper. *Nat. Prod. Commun.* **2019**, *14*, 1–6. [CrossRef]
4. Senthilkumar, A.; Venkatesalu, V. Chemical composition and larvicidal activity of the essential oil of *Plectranthus amboinicus* (Lour.) Spreng against *Anopheles stephensi*: A malarial vector mosquito. *Parasitol. Res.* **2010**, *107*, 1275–1278. [CrossRef]
5. Khare, R.S.; Kundu, S. *Coleus aromaticus* Benth-A Nutritive Medicinal Plant Of Potential Therapeutic Value. *Int. J. Pharma Bio. Sci.* **2011**, *2*, 488–500.
6. Menéndez Castillo, R.A.; Pavón González, V. *Plecthranthus amboinicus* (Lour.) spreng. *Rev. Cuba. Plantas Med.* **1999**, *4*, 110–115.
7. Singh, G.; Singh, O.P.; Prasad, Y.; De Lampasona, M.; Catalan, C. Studies on essential oils, Part 33: Chemical and insecticidal investigations on leaf oil of *Coleus amboinicus* Lour. *Flavour Fragr. J.* **2002**, *17*, 440–442. [CrossRef]
8. Murthy, P.S.; Ramalakshmi, K.; Srinivas, P. Fungitoxic activity of Indian borage (*Plectranthus amboinicus*) volatiles. *Food Chem.* **2009**, *114*, 1014–1018. [CrossRef]
9. Negi, P.S. Plant extracts for the control of bacterial growth: Efficacy, stability and safety issues for food application. *Int. J. Food Microbiol.* **2012**, *156*, 7–17. [CrossRef]
10. Aguiar, J.J.S.; Sousa, C.P.B.; Araruna, M.K.A.; Silva, M.K.N.; Portelo, A.C.; Lopes, J.C.; Carvalho, V.R.A.; Figueredo, F.G.; Bitu, V.C.N.; Coutinho, H.D.M.; et al. Antibacterial and modifying-antibiotic activities of the essential oils of *Ocimum gratissimum* L. and *Plectranthus amboinicus* L. *Eur. J. Integr. Med.* **2015**, *7*, 151–156. [CrossRef]

11. Vijayakumar, S.; Vinoj, G.; Malaikozhundan, B.; Shanthi, S.; Vaseeharan, B. Plectranthus amboinicus leaf extract mediated synthesis of zinc oxide nanoparticles and its control of methicillin resistant *Staphylococcus aureus* biofilm and blood sucking mosquito larvae. *Spectrochim. Acta Part A Mol. Biomol. Spectrosc.* **2015**, *137*, 886–891. [CrossRef]

12. Oliveira, R.d.A.G.d.; Lima, E.d.O.; Souza, E.L.d.; Vieira, W.L.; Freire, K.R.; Trajano, V.N.; Lima, I.O.; Silva-Filho, R.N. Interference of *Plectranthus amboinicus* (Lour.) Spreng essential oil on the anti-*Candida* activity of some clinically used antifungals. *Rev. Bras. Farmacogn.* **2007**, *17*, 186–190. [CrossRef]

13. Asiimwe, S.; Borg-Karlsson, A.-K.; Azeem, M.; Mugisha, K.M.; Namutebi, A.; Gakunga, N.J. Chemical composition and toxicological evaluation of the aqueous leaf extracts of *Plectranthus amboinicus* Lour. Spreng. *Int. J. Pharm. Sci. Invent.* **2014**, *3*, 19–27.

14. Hattori, M.; Nakabayashi, T.; Lim, Y.A.; Miyashiro, H.; Kurokawa, M.; Shiraki, K.; Gupta, M.P.; Correa, M.; Pilapitiya, U. Inhibitory effects of various ayurvedic and Panamanian medicinal plants on the infection of herpes simplex virus-1 in vitro and in vivo. *Phytother. Res.* **1995**, *9*, 270–276. [CrossRef]

15. Kusumoto, I.T.; Nakabayashi, T.; Kida, H.; Miyashiro, H.; Hattori, M.; Namba, T.; Shimotohno, K. Screening of various plant extracts used in ayurvedic medicine for inhibitory effects on human immunodeficiency virus type 1 (HIV-1) protease. *Phytother. Res.* **1995**, *9*, 180–184. [CrossRef]

16. Hamidi, J.A.; Ismaili, N.; Ahmadi, F.; Lajisi, N.H. Antiviral and cytotoxic activities of some plants used in Malaysian indigenous medicine. *Pertanika. J. Trop. Agric. Sci.* **1996**, *19*, 129–136.

17. Bhattacharjee, P.; Majumder, P. Investigation of phytochemicals and anti-convulsant activity of the plant *Coleus amboinicus* (Lour.). *Int. J. Green Pharm.* **2013**, *7*, 211.

18. Gurgel, A.P.A.D.; da Silva, J.G.; Grangeiro, A.R.S.; Oliveira, D.C.; Lima, C.M.P.; da Silva, A.C.P.; Oliveira, R.A.G.; Souza, I.A. In vivo study of the anti-inflammatory and antitumor activities of leaves from *Plectranthus amboinicus* (Lour.) *Spreng* (Lamiaceae). *J. Ethnopharmacol.* **2009**, *125*, 361–363. [CrossRef]

19. Ramalakshmi, P.; Subramanian, N.; Saravanan, R.; Mohanakrishnan, H.; Muthu, M. Anticancer effect of *Coleus amboinicus* (Karpooravalli) on human lung cancer cell line (A549). *Int. J. Dev. Res.* **2014**, *4*, 2442–2449.

20. Silitonga, M.; Ilyas, S.; Hutahaean, S.; Sipahutar, H. Levels of apigenin and immunostimulatory activity of leaf extracts of bangunbangun (*Plectranthus amboinicus* Lour.). *Int. J. Biol.* **2015**, *7*, 46. [CrossRef]

21. Chiu, Y.-J.; Huang, T.-H.; Chiu, C.-S.; Lu, T.-C.; Chen, Y.-W.; Peng, W.-H.; Chen, C.-Y. Analgesic and antiinflammatory activities of the aqueous extract from *Plectranthus amboinicus* (Lour.) *Spreng.* Both in vitro and in vivo. *Evid. Based Complement. Alternat. Med.* **2012**, *2012*, 508137. [CrossRef] [PubMed]

22. Kumaran, A. Antioxidant and free radical scavenging activity of an aqueous extract of *Coleus aromaticus*. *Food Chem.* **2006**, *97*, 109–114. [CrossRef]

23. Praveena, B.; Pradeep, S.N. Antioxidant and antibacterial activities in the leaf extracts of Indian borage (*Plectranthus amboinicus*). *Food Sci. Nutr.* **2012**, *2012*, 146–152. [CrossRef]

24. Cartaxo, S.L.; Souza, M.M.d.A.; de Albuquerque, U.P. Medicinal plants with bioprospecting potential used in semi-arid northeastern Brazil. *J. Ethnopharmacol.* **2010**, *131*, 326–342. [CrossRef] [PubMed]

25. Hole, R.C.; Juvekar, A.R.; Roja, G.; Eapen, S.; D'Souza, S.F. Positive inotropic effect of the leaf extracts of parent and tissue culture plants of Coleus amboinicus on an isolated perfused frog heart preparation. *Food Chem.* **2009**, *114*, 139–141. [CrossRef]

26. Santos, F.A.V.; Serra, C.G.; Bezerra, R.J.A.C.; Figueredo, F.G.; Edinardo; Matias, F.F.; Menezes, I.R.A.; Costa, J.G.M.; Coutinho, H.D.M. Antibacterial activity of *Plectranthus amboinicus* Lour. (Lamiaceae) essential oil against *Streptococcus mutans*. *Eur. J. Integr. Med.* **2016**, *8*, 293–297. [CrossRef]

27. Shubha, J.R.; Bhatt, P. *Plectranthus amboinicus* leaves stimulate growth of probiotic *L. plantarum*: Evidence for ethnobotanical use in diarrhea. *J. Ethnopharmacol.* **2015**, *166*, 220–227. [CrossRef]

28. Patel, R.; Mahobia, N.K.; Gendle, R.; Kaushik, B.; Singh, S.K. Diuretic activity of leaves of *Plectranthus amboinicus* (Lour.) *Spreng* in male albino rats. *Pharmacognosy Res.* **2010**, *2*, 86–88. [CrossRef] [PubMed]

29. Jain, A.K.; Dixit, A.; Mehta, S.C. Wound healing activity of aqueous extracts of leaves and roots of *Coleus aromaticus* in rats. *Acta Pol. Pharm.* **2012**, *69*, 1119–1123.

30. Bhat, P.; Hegde, G.; Hegde, G.R. Ethnomedicinal practices in different communities of Uttara Kannada district of Karnataka for treatment of wounds. *J. Ethnopharmacol.* **2012**, *143*, 501–514. [CrossRef]

31. Selvakumar, P.; naveena, B.E.; prakash, S.D. Studies on the antidandruff activity of the essential oil of *Coleus amboinicus* and *Eucalyptus globulus*. *Asian Pac. J. Trop. Dis.* **2012**, *2*, S715–S719. [CrossRef]

32. Bhojwani, S.S.; Dantu, P.K. Micropropagation. In *Plant Tissue Culture: An Introductory Text*; Bhojwani, S.S., Dantu, P.K., Eds.; Springer: New Delhi, India, 2013. [CrossRef]

33. Faisal, M.; Ahmad, N.; Anis, M.; Alatar, A.A.; Qahtan, A.A. Auxin-cytokinin synergism in vitro for producing genetically stable plants of *Ruta graveolens* using shoot tip meristems. *Saudi J. Biol. Sci.* **2018**, *25*, 273–277. [CrossRef] [PubMed]

34. Sreedevi, E.; Anuradha, M.; Pullaiah, T. Plant regeneration from leaf-derived callus in *Plectranthus barbatus* Andr.[Syn.: *Coleus forskohlii* (Wild.) Briq.]. *Afr. J. Biotechnol.* **2013**, *12*, 2441–2448. [CrossRef]

35. Thaniarasu, R.; Kumar, T.S.; Rao, M. In vitro Propagation of *Plectranthus bourneae* Gamble? An Endemic Red Listed Plant. *Plant Tissue Cult. Biotechnol.* **2015**, *25*, 273–284. [CrossRef]

36. Thaniarasu, R.; Senthil Kumar, T.; Rao, M.V. Mass propagation of *Plectranthus bourneae* Gamble through indirect organogenesis from leaf and internode explants. *Physiol. Mol. Biol. Plants* **2016**, *22*, 143–151. [CrossRef]

37. Belete, K.; Balcha, A. Micropropagation of *Plectranthus edulis* (Vatke) Agnew from shoot tip and nodal explants. *Afr. J. Agric. Res.* **2015**, *10*, 6–13. [CrossRef]

38. Tsegaw, M.; Feyissa, T. Micropropagation of *Plectranthus edulis* (Vatke) Agnew from meristem cultur. *Afr. J. Biotechnol.* **2014**, *13*, 3682–3688. [CrossRef]

39. Rahman, Z.A.; Noor, E.S.M.; Ali, M.S.M.; Mirad, R.; Othman, A.N. In Vitro Micropropagation of a Valuable Medicinal Plant, *Plectranthus amboinicus*. *Am. J. Plant Sci.* **2015**, *6*, 7. [CrossRef]

40. Arumugam, G.; Sinniah, U.R.; Swamy, M.K.; Lynch, P.T. Micropropagation and essential oil characterization of *Plectranthus amboinicus* (Lour.) Sprengel, an aromatic medicinal plant. *In Vitro Cell. Dev. Biol. Plant* **2020**, *56*, 491–503. [CrossRef]

41. Mahmoud, D.S.; Sayed, L.M.; Diab, M.; Fahmy, E.M. In Vitro Propagation Of *Plectranthus barbatus* Andrews As Important Medicinal Plant. *Arab Univ. J. Agri. Sci.* **2019**, *27*, 511–517. [CrossRef]

42. Fonseka, D.; Wickramaarachchi, W.; Situge, C. Mass production of *Plectranthus zeylanicus*—A valuable medicinal and aromatic plant with a future value. *Int. J. Minor. Fruits Med. Aromat. Plants* **2019**, *5*, 15–20.

43. Kujeke, G.T.; Chitendera, T.C.; Masekesa, R.T.; Mazarura, U.; Ngadze, E.; Rugare, J.T.; Matikiti, A. Micropropagation of Livingstone Potato (*Plectranthus esculentus* N.E.Br). *Adv. Agric.* **2020**, *2020*, 8364153. [CrossRef]

44. Faisal, M.; Alatar, A.A.; El-Sheikh, M.A.; Abdel-Salam, E.M.; Qahtan, A.A. Thidiazuron induced in vitro morphogenesis for sustainable supply of genetically true quality plantlets of Brahmi. *Ind. Crops Prod.* **2018**, *118*, 173–179. [CrossRef]

45. Faisal, M.; Alatar, A.A.; Ahmad, N.; Anis, M.; Hegazy, A.K. An Efficient and Reproducible Method for in vitro Clonal Multiplication of *Rauvolfia tetraphylla* L. and Evaluation of Genetic Stability using DNA-Based Markers. *Appl. Biochem. Biotechnol.* **2012**, *168*, 1739–1752. [CrossRef]

46. Faisal, M.; Alatar, A.A.; Hegazy, A.K.; Alharbi, S.A.; El-Sheikh, M.; Okla, M.K. Thidiazuron induced in vitro multiplication of *Mentha arvensis* and evaluation of genetic stability by flow cytometry and molecular markers. *Ind. Crops Prod.* **2014**, *62*, 100–106. [CrossRef]

47. Thakur, J.; Dwivedi, M.D.; Sourabh, P.; Uniyal, P.L.; Pandey, A.K. Genetic Homogeneity Revealed Using SCoT, ISSR and RAPD Markers in Micropropagated *Pittosporum eriocarpum* Royle- An Endemic and Endangered Medicinal Plant. *PLoS ONE* **2016**, *11*, e0159050. [CrossRef]

48. Jena, S.; Ray, A.; Sahoo, A.; Sahoo, S.; Dash, B.; Kar, B.; Nayak, S. Rapid plant regeneration in industrially important *Curcuma zedoaria* revealing genetic and biochemical fidelity of the regenerants. *3 Biotech.* **2019**, *10*, 17. [CrossRef]

49. Rohela, G.K.; Jogam, P.; Bylla, P.; Reuben, C. Indirect Regeneration and Assessment of Genetic Fidelity of Acclimated Plantlets by SCoT, ISSR, and RAPD Markers in *Rauwolfia tetraphylla* L.: An Endangered Medicinal Plant. *BioMed Res. Int.* **2019**, *2019*, 3698742. [CrossRef]

50. Qahtan, A.A.; Faisal, M.; Alatar, A.A.; Abdel-Salam, E.M. High-Frequency Plant Regeneration, Genetic Uniformity, and Flow Cytometric Analysis of Regenerants in *Rutachalepensis*, L. *Plants* **2021**, *10*, 2820. [CrossRef]

51. Sirikonda, A.; Jogam, P.; Ellendula, R.; Kudikala, H.; Mood, K.; Allini, V.R. In vitro micropropagation and genetic fidelity assesment in *Flemingia macrophylla* (Willd.) Merr: An ethnomedicinal plant. *Vegetos* **2020**, *33*, 286–295. [CrossRef]

52. Thakur, M.; Rakshandha; Sharma, V.; Chauhan, A. Genetic fidelity assessment of long term in vitro shoot cultures and regenerated plants in Japanese plum cvs Santa Rosa and Frontier through RAPD, ISSR and SCoT markers. *S. Afr. J. Bot.* **2021**, *140*, 428–433. [CrossRef]

53. Faisal, M.; Abdel-Salam, E.M.; Alatar, A.A.; Qahtan, A.A. Induction of somatic embryogenesis in *Brassica juncea* L. and analysis of regenerants using ISSR-PCR and flow cytometer. *Saudi, J. Biol. Sci.* **2021**, *28*, 1147–1153. [CrossRef] [PubMed]

54. Abdolinejad, R.; Shekafandeh, A.; Jowkar, A.; Gharaghani, A.; Alemzadeh, A. Indirect regeneration of *Ficus carica* by the TCL technique and genetic fidelity evaluation of the regenerated plants using flow cytometry and ISSR. *Plant. Cell Tissue Organ. Cult.* **2020**, *143*, 131–144. [CrossRef]

55. Parzymies, M.; Pogorzelec, M.; Głębocka, K.; Śliwińska, E. Genetic Stability of the Endangered Species *Salix lapponum* L. Regenerated In Vitro during the Reintroduction Process. *Biology* **2020**, *9*, 378. [CrossRef] [PubMed]

56. Murashige, T.; Skoog, F. A revised medium for rapid growth and bio assays with tobacco tissue cultures. *Physiol. Plant.* **1962**, *15*, 473–497. [CrossRef]

57. Galbraith, D.W. Simultaneous flow cytometric quantification of plant nuclear DNA contents over the full range of described angiosperm 2C values. *Cytometry A* **2009**, *75*, 692–698. [CrossRef]

58. Doyle, J.J.; Doyle, J.L. A rapid DNA isolation procedure for small quantities of fresh leaf tissue. *Phytochem. Bull.* **1987**, *19*, 11–15.

59. Tan, S.N.; Tee, C.S.; Wong, H.L. Multiple shoot bud induction and plant regeneration studies of *Pongamia pinnata*. *Plant. Biotechnol.* **2018**, *35*, 325–334. [CrossRef]

60. El-Shafey, N.; Sayed, M.; Ahmed, E.; Hammouda, O.; Khodary, S.E. Effect of growth regulators on micropropagation, callus induction and callus flavonoid content of *Rumex pictus* Forssk. *Egypt, J. Bot.* **2019**, *59*, 293–302. [CrossRef]

61. Vaidya, B.N.; Asanakunov, B.; Shahin, L.; Jernigan, H.L.; Joshee, N.; Dhekney, S.A. Improving micropropagation of *Mentha × piperita* L. using a liquid culture system. *In Vitro Cell. Dev. Biol. Plant* **2019**, *55*, 71–80. [CrossRef]

62. Rajput, S.; Agrawal, V. Micropropagation of *Atropa acuminata* Royle ex Lindl. (a critically endangered medicinal herb) through root callus and evaluation of genetic fidelity, enzymatic and non-enzymatic antioxidant activity of regenerants. *Acta Physiol. Plant.* **2020**, *42*, 160. [CrossRef]
63. Wróbel, T.; Dreger, M.; Wielgus, K.; Słomski, R. Modified Nodal Cuttings and Shoot Tips Protocol for Rapid Regeneration of *Cannabis sativa* L. *J. Nat. Fibers* **2022**, *19*, 536–545. [CrossRef]
64. Machado, J.S.; Degenhardt, J.; Maia, F.R.; Quoirin, M. Micropropagation of *Campomanesia xanthocarpa* O. Berg (Myrtaceae), a medicinal tree from the Brazilian Atlantic Forest. *Trees* **2020**, *34*, 791–799. [CrossRef]
65. Zahid, N.A.; Jaafar, H.Z.E.; Hakiman, M. Micropropagation of Ginger (*Zingiber officinale* Roscoe) 'Bentong' and Evaluation of Its Secondary Metabolites and Antioxidant Activities Compared with the Conventionally Propagated Plant. *Plants* **2021**, *10*, 630. [CrossRef]
66. Kumari, P.; Singh, S.; Yadav, S.; Tran, L.S.P. Pretreatment of seeds with thidiazuron delimits its negative effects on explants and promotes regeneration in chickpea (*Cicer arietinum* L.). *Plant Cell Tissue Organ. Cult. (PCTOC)* **2018**, *133*, 103–114. [CrossRef]
67. Kumari, P.; Singh, S.; Yadav, S.; Tran, L.-S.P. Influence of different types of explants in chickpea regeneration using thidiazuron seed-priming. *J. Plant Res.* **2021**, *134*, 1149–1154. [CrossRef] [PubMed]
68. Clapa, D.; Hârța, M. Establishment of an Efficient Micropropagation System for Humulus lupulus L. cv. Cascade and Confirmation of Genetic Uniformity of the Regenerated Plants through DNA Markers. *Agronomy* **2021**, *11*, 2268. [CrossRef]
69. Faisal, M.; Ahmad, N.; Anis, M. An efficient micropropagation system for *Tylophora indica*: An endangered, medicinally important plant. *Plant. Biotechnol. Rep.* **2007**, *1*, 155–161. [CrossRef]
70. Sahoo, N.; Thirunavoukkarasu, M.; Behera, P.R.; Satpathy, G.B.; Panda, P.K. Direct shoot organogenesis from hypocotyl explants of *J. atropha curcas* L. an important bioenergy feedstock. *GCB Bioenergy* **2012**, *4*, 234–238. [CrossRef]
71. Panigrahi, J.; Gantait, S.; Patel, I.C. An Efficient In Vitro Approach for Direct Regeneration and Callogenesis of *Adhatoda vasica* Nees, a Potential Source of Quinazoline Alkaloids. *Natl. Acad. Sci. Lett.* **2017**, *40*, 319–324. [CrossRef]
72. Jinyan, H. A high throughput plant regeneration system from shoot stems of *Sapium sebiferum* Roxb., a potential multipurpose bioenergy tree. *Ind. Crops Prod.* **2020**, *154*, 112653. [CrossRef]
73. Hussain, S.A.; Anis, M.; Alatar, A.A. Efficient In Vitro Regeneration System for *Tecoma stans* L., Using Shoot Tip and Assessment of Genetic Fidelity Among Regenerants. *Proc. Indian Natl. Sci. Acad. (B Biol. Sci.)* **2020**, *90*, 171–178. [CrossRef]
74. Wei, Q.; Cao, J.; Qian, W.; Xu, M.; Li, Z.; Ding, Y. Establishment of an efficient micropropagation and callus regeneration system from the axillary buds of *Bambusa ventricosa*. *Plant Cell Tissue Organ. Cult.* **2015**, *122*, 1–8. [CrossRef]
75. Fatima, N.; Ahmad, N.; Anis, M. Enhanced in vitro regeneration and change in photosynthetic pigments, biomass and proline content in *Withania somnifera* L. (Dunal) induced by copper and zinc ions. *Plant. Physiol. Biochem.* **2011**, *49*, 1465–1471. [CrossRef] [PubMed]
76. Swain, D.; Lenka, S.; Hota, T.; Rout, G.R. Micro-propagation of *Hypericum gaitii* Haines, an endangered medicinal plants: Assessment of genetic fidelity. *Nucleus* **2016**, *59*, 7–13. [CrossRef]
77. Ahmed, M.R.; Anis, M.; Alatar, A.A.; Faisal, M. In vitro clonal propagation and evaluation of genetic fidelity using RAPD and ISSR marker in micropropagated plants of *Cassia alata* L.: A potential medicinal plant. *Agrofor. Syst.* **2017**, *91*, 637–647. [CrossRef]
78. Bolyard, M. In vitro regeneration of *Artemisia abrotanum* L. by means of somatic organogenesis. *In Vitro Cell. Dev. Biol. Plant* **2018**, *54*, 127–130. [CrossRef]
79. Naaz, A.; Hussain, S.A.; Naz, R.; Anis, M.; Alatar, A.A. Successful plant regeneration system via de novo organogenesis in *Syzygium cumini* (L.) Skeels: An important medicinal tree. *Agrofor. Syst.* **2019**, *93*, 1285–1295. [CrossRef]
80. Chinnadurai, V.; Viswanathan, P.; Kalimuthu, K.; Vanitha, A.; Ranjitha, V.; Pugazhendhi, A. Comparative studies of phytochemical analysis and pharmacological activities of wild and micropropagated plant ethanol extracts of *Manihot esculenta*. *Biocatal. Agric. Biotechnol.* **2019**, *19*, 101166. [CrossRef]
81. Upadhyay, A.; Shahzad, A.; Ahmad, Z. In vitro propagation and assessment of genetic uniformity along with chemical characterization in *Hildegardia populifolia* (Roxb.) Schott & Endl.: A critically endangered medicinal tree. *In Vitro Cell. Dev. Biol. Plant* **2020**, *56*, 803–816. [CrossRef]
82. Kim, Y.-G.; Okello, D.; Yang, S.; Komakech, R.; Rahmat, E.; Kang, Y. Histological assessment of regenerating plants at callus, shoot organogenesis and plantlet stages during the in vitro micropropagation of *Asparagus cochinchinensis*. *Plant. Cell Tissue Organ. Cult.* **2021**, *144*, 421–433. [CrossRef]
83. Kumar, S.S.; Giridhar, P. In vitro micropropagation of *Basella rubra* L. through proliferation of axillary shoots. *Plant Cell Tissue Organ Cult.* **2021**, *144*, 477–483. [CrossRef]
84. De Klerk, G.-J. Rooting of microcuttings: Theory and practice. *In Vitro Cell. Dev. Biol. Plant* **2002**, *38*, 415–422. [CrossRef]
85. Yadav, V.; Shahzad, A.; Ahmad, Z.; Sharma, S.; Parveen, S. Synthesis of nonembryonic synseeds in *Hemidesmus indicus* R. Br.: Short term conservation, evaluation of phytochemicals and genetic fidelity of the regenerants. *Plant. Cell Tissue Organ. Cult.* **2019**, *138*, 363–376. [CrossRef]
86. Hussain, S.A.; Ahmad, N.; Anis, M. Synergetic effect of TDZ and BA on minimizing the post-exposure effects on axillary shoot proliferation and assessment of genetic fidelity in *Rauvolfia tetraphylla* (L.). *Rend. Lincei Sci. Fis. Nat.* **2018**, *29*, 109–115. [CrossRef]
87. Jogam, P.; Sandhya, D.; Shekhawat, M.S.; Alok, A.; Manokari, M.; Abbagani, S.; Allini, V.R. Genetic stability analysis using DNA barcoding and molecular markers and foliar micro-morphological analysis of in vitro regenerated and in vivo grown plants of *Artemisia vulgaris* L. *Ind. Crops Prod.* **2020**, *151*, 112476. [CrossRef]

88. Dilkalal, A.; AS, A.; TG, U. In vitro regeneration, antioxidant potential, and genetic fidelity analysis of *Asystasia gangetica* (L.) T. Anderson. *In Vitro Cell. Dev. Biol. Plant* **2021**, *57*, 447–459. [CrossRef]

89. Faisal, M.; Alatar, A.A.; Ahmad, N.; Anis, M.; Hegazy, A.K. Assessment of Genetic Fidelity in *Rauvolfia serpentina* Plantlets Grown from Synthetic (Encapsulated) Seeds Following In Vitro Storage at 4 °C. *Molecules* **2012**, *17*, 5050–5061. [CrossRef]

90. Devi, S.P.; Kumaria, S.; Rao, S.R.; Tandon, P. Single primer amplification reaction (SPAR) methods reveal subsequent increase in genetic variations in micropropagated plants of *Nepenthes khasiana* Hook. f. maintained for three consecutive regenerations. *Gene* **2014**, *538*, 23–29. [CrossRef]

91. Parab, A.R.; Lynn, C.B.; Subramaniam, S. Assessment of genetic stability on in vitro and ex vitro plants of *Ficus carica* var. black jack using ISSR and DAMD markers. *Mol. Biol. Rep.* **2021**, *48*, 7223–7231. [CrossRef]

horticulturae

Article

Improvement of In Vitro Seed Germination and Micropropagation of *Amomum tsao-ko* (Zingiberaceae Lindl.)

Quyet V. Khuat [1,2,*], Elena A. Kalashnikova [1], Rima N. Kirakosyan [1], Hai T. Nguyen [3], Ekaterina N. Baranova [4,5] and Marat R. Khaliluev [1,4,*]

[1] Agronomy and Biotechnology Faculty, Russian State Agrarian University—Moscow Timiryazev Agricultural Academy, Timiryazevskaya 49, 127550 Moscow, Russia; kalash0407@mail.ru (E.A.K.); mia41291@mail.ru (R.N.K.)

[2] Biology and Agricultural Engineering Faculty, Hanoi Pedagogical University 2, Nguyen Van Linh, Phuc Yen 15000, Vietnam

[3] Biotechnology Faculty, Vietnam National University of Agriculture, Gia Lam, Hanoi 12406, Vietnam; nthaicnsh@vnua.edu.vn

[4] All-Russia Research Institute of Agricultural Biotechnology, Timiryazevskaya 42, 127550 Moscow, Russia; greenpro2007@rambler.ru

[5] N.V. Tsitsin Main Botanical Garden of Russian Academy of Sciences, Botanicheskaya 4, 127276 Moscow, Russia

* Correspondence: khuatvanquyet@hpu2.edu.vn or khuatquyetst@gmail.com (Q.V.K.); marat131084@rambler.ru (M.R.K.); Tel.: +7(965)173-96-65 (Q.V.K.); +7(499)-977-31-41 (M.R.K.)

Citation: Khuat, Q.V.; Kalashnikova, E.A.; Kirakosyan, R.N.; Nguyen, H.T.; Baranova, E.N.; Khaliluev, M.R. Improvement of In Vitro Seed Germination and Micropropagation of *Amomum tsao-ko* (Zingiberaceae Lindl.). *Horticulturae* **2022**, *8*, 640. https://doi.org/10.3390/horticulturae8070640

Academic Editor: Sergio Ruffo Roberto

Received: 15 June 2022
Accepted: 11 July 2022
Published: 15 July 2022

Publisher's Note: MDPI stays neutral with regard to jurisdictional claims in published maps and institutional affiliations.

Abstract: Black cardamom (*Amomum tsao-ko* Crevost & Lemarié) is a spice plant of great commercial value in Vietnam, but with limited propagation ability. Its seeds are characterized by a thick and hard seed coat, a small endosperm, and a small embryo, which are the causes of the physical dormancy of the seeds and low germination. Attempts in this study to improve the germination rate and achieve uniform germination included mechanical scarification, immersion in hot or cold water, acid scarification, and the application of plant growth regulators. Although immersion of seeds in cold water and application of plant growth regulators (PGRs) (gibberellic acid (GA$_3$) and 1-naphtaleneacetic acid (NAA)) showed positive effects on seed germination and subsequent seedling growth, mechanical scarification provided the highest germination rate of black cardamom seeds (68.0%) and significantly shortened germination time (53.7 days) compared to control (16.0% and 74.7 days). On the other hand, an efficient micropropagation protocol has been established using shoot tip explants derived from in-vitro-grown seedlings. Murashige and Skoog (MS) medium supplemented with 4.0 mg/L 6-benzylaminopurine (BAP) + 0.5 mg/L NAA proved to be most suitable for rapid multiplication and rooting, providing a mean of 5.4 shoots per explant, 6.8 cm shoot length, and 16.2 roots per explant after 7 weeks of culture. Well-rooted black cardamom plantlets have been successfully adapted to ex vitro conditions. "Fasco" bio-soil was more suitable for acclimatization, with a 48.9% survival rate, 23.3 cm plant length, and 5.7 leaves per plant after 3 months of planting. Improved germination and multiplication protocols can be used to improve propagation performances and to develop elite of black cardamom planting material.

Keywords: black cardamom; seed dormancy; mechanical scarification; shoot micropropagation

1. Introduction

Amomum tsao-ko Crevost & Lemarié, belonging to the ginger family (Zingiberaceae), is a valuable medicinal plant in Vietnam and is commonly known as "black cardamom", "do ho", or "sa nhan coc" [1,2]. It is a particularly shade-loving and moisture-loving species, so it can only be grown under the forest canopy, at an altitude of 1300–2200 m, with an average annual temperature of 13–15.3 °C, usually frequent fog, annual rainfall of 3500–3800 mm, and more than 90% humidity [1,3–5]. Therefore, it is distributed and cultivated in high ranges with damp soil rich in humus and under the partial shade of evergreen forests belonging to the Northern Midland and Mountainous region of Vietnam (Lai Chau, Lao Cai, Yen Bai, Ha Giang, Tuyen Quang, and Cao Bang provinces). Additionally, black cardamom is also grown in China (Yunnan province) and Laos (Phongsaly province) [1,2,6]. In Vietnamese traditional medicine, black cardamom fruit is utilized to treat dyspepsia, nausea, malaria, bad breath, infections, etc. [1,7,8]. Black cardamom is also a spice used in many traditional Vietnamese dishes, e.g., Vietnamese beef noodle, chicken herbal dish, hotpot, etc.

For the above-mentioned reasons, black cardamom has become one of the main medicine crops for local people in the Northern Midland and Mountainous region of Vietnam. The area of black cardamom cultivation is constantly expanding. In Vietnam, local people use mainly seeds and rhizome segments to cultivate black cardamom. Because plants grown from rhizome segments are susceptible to diseases caused by viruses, fungi, or bacteria, they often provide a lower yield and fruit quality than plants grown from seeds. Additionally, the destructive harvesting of the rhizomes for vegetative propagation appears not to be workable because there is always the possibility of losing the mother plant during this process. However, the practice of seed propagation of local people illustrates that cardamom seeds germinate slowly and unevenly, and seedlings are slow growth. Studies on other species from the ginger family such as *Alpinia malaccensis* Roscoe [9], *Alpinia galanga* Willd. [10], korarima (*Aframomum corrorima* P.C.M. Jansen) [11,12], large cardamom (*Amomum subulatum* Roxb.) [13], and green cardamom (*Elettaria cardamomum* Maton) [14] have all shown that seed germination of these species was not fast and/or that many seeds do not germinate due to the presence of some kind of dormancy, possibly associated with the hard and impermeable nature of the seed coat. Additionally, in the above-mentioned articles, various seed treatments (including mechanical scarification, soaking in hot or cold water, acid scarification, and soaking in plant growth regulators) were applied to break seed dormancy. Presently, similar studies in black cardamom are lacking. Quyet et al. (2021) [15] found that surface disinfection with 0.1% $HgCl_2$ for 10 min showed the best seed disinfection effect, and MS medium diluted to 1/16 concentration was the best for in vitro germination of black cardamom seeds. Additionally, the authors applied a mechanical scarification treatment to improve the germination rate of dried seeds. However, the germination rate was reported to be quite low (33.3%). Determining the cause and applying more seed treatments to establish the best seed treatments in order to further improve the germination rate of these seeds is necessary.

Currently, the technique involving propagation using plant tissue promises to produce many disease-free and uniform-quality crops in a short time compared with traditional methods. In vitro propagation has been reported in various species of the genus *Amomum*, e.g., green cardamom [16–18], large cardamom [19–21], siam cardamom (*Amomum krevanh* Pierre) [22], white amomum (*Amomum villosum* Lour.) [23,24], purple amomum (*Amomum longiligulare* T.L. Wu) [25], and *Amomum* sp. [26]. Rhizome buds are used by most researchers as a source of explants when carrying out in vitro propagation of these species. Regarding black cardamom, Quyet et al. (2021) [27] carried out clonal micropropagation and also used rhizome buds as an explant source. In this study, the authors evaluated the separate effects of two cytokinins, 6-benzylaminopurine (BAP) and kinetin, in the fast multiplication phase and two auxins, indole-3-butyric acid (IBA) and 1-naphtaleneacetic acid (NAA), for root induction. Eyob S. (2009) [11], when carrying out the micropropagation of korarima, a species of the ginger family, used two different explant sources including shoot tip derived from in vitro grown seedlings and rhizome buds. The author found that the highest survival and number of shoots were obtained from the shoot tips of in vitro seedlings compared with rhizome buds. Reghunath (1989) [28] applied the shoot tips of green cardamom to perform micropropagation experiments through callus culture. The results of this study showed that Murashige and Skoog (MS) medium supplemented with the combination of 4.0 mg/L NAA or 1.0 mg/L 2,4-dichlorophenoxyacetic acid (2,4-D) and 1 mg/L BAP provided the highest callus induction rate after 28 days of culture. On the other hand, the combined effect of cytokines and auxins in micropropagation has been reported to be effective in many species of the genus *Amomum*, including green cardamom, large cardamom, and purple amomum. These were not reported in the previous study [27]. Therefore, similar studies on black cardamom are needed to determine the most effective micropropagation protocol for black cardamom.

Therefore, the overall goal of the present study was to investigate the effects of different seed treatment methods on the seed germination and seedling growth of black cardamom under in vitro conditions and determine the effect of different PGRs on in vitro shoot multiplication and root formation of seedling shoot tips in order to develop a protocol that produces a constant supply of viable and clean clonally propagated plant materials for crop production.

2. Materials and Methods

2.1. Seed Material

Matured capsules of black cardamom were collected at the end of December 2020 from the cardamom forest in Tam Duong district, Lai Chau province, Vietnam, around 22°23′04.5″ N latitude and 103°32′44.0″ E longitude. After harvesting, they were rinsed with water, preserved by vacuum packing, and transferred to the biotechnology laboratory of the Russian State Agrarian University—Moscow Timiryazev Agricultural Academy. At the laboratory, seeds were extracted from the capsules, rinsed thoroughly with tap water to remove the aril, and used immediately for experiments (Figure 1).

Figure 1. Black cardamom plants in Lai Chau province, Vietnam. (**a**) Black cardamom plants and their growing environment; (**b**) inflorescence shoots; (**c**) bloomed inflorescences; (**d**) fresh capsules; (**e**) fresh seeds; (**f**) dried ripe capsules (Photo by Ma A Chang, 2020).

A voucher specimen (No. QF 001) was deposited at the herbarium of Hanoi Pedagogical University N°2, Vietnam. Botanical identification was achieved by Dr. Tam H. M.

2.2. Morphological and Anatomical Characteristics of Black Cardamom Seeds

The characteristics of the shape, color, and size of seeds were examined via stereomicroscope "Stemi" DV4 (Carl Zeiss, Oberkochen, Germany). Detailed morphology and anatomy of seeds were evaluated using a scanning electron microscope (SEM). Black cardamom seeds were fixed in 2.5% glutaraldehyde in a 0.1 M Sorenson buffer, pH 7.2. After that, they were washed with buffer and dehydrated through ethanol series (30%, 50%, 70%, 96%, and 2 × 100%). In the next step, carbon dioxide (CO_2) for critical-point drying (Hitachi (Tokyo, Japan) HCP-2 critical point dryer) was applied. Dried seeds were mounted on an SEM stub with carbon-conductive tabs and coated with gold and palladium using an Eiko IB-3 ion coater (Eiko, Tokyo, Japan). Samples were observed and photographed under a JSM-6380LA SEM at 20 kV (JEOL, Tokyo, Japan).

2.3. Experimental Design and Treatments

Germination of black cardamom seeds is erratic, delayed, and poor under the traditional type of sowing. Attempts made to improve germination percentage and also to achieve uniform germination include mechanical scarification, immersion in hot or cold water, acid scarification, and soaking in PGRs (Table 1). Seeds without damage to the uniform size and color were selected randomly to conduct the experiments.

Table 1. Summary of black cardamom seed treatments.

Treatment	Abbreviation	Description
Mechanical scarification	ME	Soak in tap water for 24 h and scarify them manually by cutting 1–1.5 mm of the seed coat at the opposite site of hilum by a sterile scalpel.
Hot water	HW2m	Soak in hot water at 100 °C for 2 min.
	HW4m	Soak in hot water at 100 °C for 4 min.
Cold water	CW	Soak in cold water at 4 ± 1 °C for 24 h.
Acid scarification	NAS10m	Soak in 50% nitric acid (HNO_3) for 10 min.
	NAS15m	Soak in 50% nitric acid (HNO_3) for 15 min.
	HAS10m	Soak in 25% hydrochloric acid (HCl) for 10 min
	HAS15m	Soak in 25% hydrochloric acid (HCl) for and 15 min.
PGRs	GA$_3$24h	Soak in gibberellic acid (GA$_3$), 200 ppm, for 24 h.
	NAA24h	Soak in 1-naphthylacetic acid (NAA) 200 ppm for 24 h.

Control (without treatment). ME treatment are performed after disinfection seed step. Other treatments are performed before disinfection seed step. Each treatment was replicated three times with 25 seeds for each. To avoid contamination, only one seed was cultured per glass vessel.

2.4. Obtaining of Aseptic Donor Seeds and Culture Conditions

Above-mentioned treated seeds (Table 1) were rinsed with tap water. After that, seeds were washed in liquid soap for 10 min and then rinsed directly under running tap water. In the next step, seeds were disinfected in 70% ethanol for 30 s, followed by immersion in 0.1% (*w/v*) aqueous mercuric chloride for 10 min [15]. After surface disinfection, seeds were rinsed 4–5 times with sterile distilled water and transferred to the culture vessels containing basal MS medium [29] with macronutrients diluted to 1/16 strength for germination and growth [15]. The pH of the medium was adjusted to 5.6–5.8 by 1N NaOH before being autoclaved at 121 °C and 1.1 atm for 20 min. The cultures were maintained in a culture room at 25 ± 2 °C during a long-day photoperiod (16 h of light: 8 h of dark) with cool white fluorescent light (2000–2500 lux).

2.5. Data Recording and Germination Assessment

Seeds were considered germinated when the healthy, white radical had emerged through the integument. Data were scored from 30 to 90 days after culture. The contamination and germination seeds were counted every 10 days until the 90th day.

The following germination parameters were determined:

(1) Germination percentage (GP), the number of germinated seeds as a percentage of the total number of tested seeds, is given as:

$$GP = (\text{germinated seeds/total tested seeds}) \times 100\%;$$

(2) Mean germination time (MGT, days) is given according to Scott et al. [30] as:

$$MGT = \sum Tk\, Nk/S,$$

where T_k is the number of days since the beginning of the experiment, N_k the number of seeds germinated per day, and S is the total number of seeds germinated.

(3) Germination rate index (GRI) was calculated for each treatment using the following equation:

$$GRI = (G_1/1) + (G_2/2) + \dots + (G_i/i),$$

where G is the germination day 1, 2, …, and i represents the corresponding day of germination [31].

Seedling length was measured using a ruler scale on the 110th day of culture.

2.6. Culture Media for Clonal Propagation

Seedlings (2–3 cm in length) were used as an explant source (shoot tips) for clonal propagation. They then were placed on agar-solidified MS medium supplemented with various concentrations of PGRs (auxin: 0.5 and 1.0 mg/L NAA, 1.0 mg/L and 2.0 mg/L 2,4-D); cytokinin: 1.0–4.0 mg/L BAP; see Table 3 for details). The experiment was arranged completely randomly and repeated three times, with 12 explants per treatment. Data were scored after 7 weeks. The length of shoots was measured using a ruler scale.

2.7. Plantlets Adaptation to Ex Vitro Conditions

After twelve weeks, in-vitro-propagated plantlets (4.0–5.0 cm in length, with 3–4 leaves) were acclimatized by transferring the culture vessels to the greenhouse for 5 days and then opening the cap of the culture vessels for 2 days. In the next step, they were taken out of the culture medium, and their basal portions were rinsed thoroughly in running tap water to remove agar and traces of medium. After that, these plantlets (n = 30 for each variant) were treated with 0.5% (*w/v*) Bavistin solution (10 min) to prevent fungal contamination and planted directly on 2 different types of soil including "Garden Star" soil universal (consisting of peat and mineral fertilizers) and "Fasco" bio-soil (consisting of high- and lowland peat, sand, bio-humus, dolomite flour, and complete mineral fertilizer). Survival rate of the plantlets and their morphometric characteristics were recorded after 1 and 3 months of soil adaptation.

2.8. Statistical Analysis

Mean values of all data were calculated using Microsoft Office Excel 2013 packages. Analysis of variance (ANOVA) was performed using Statistica, version 10.0, and means were compared using Duncan's multiple range test at a significance level of $\alpha = 0.05$.

3. Results

3.1. Morphological and Anatomical Characteristics of Black Cardamom Seeds

Black cardamom seeds are conical, polyhedral, 0.35–0.7 cm in diameter, brown or black, and covered with greyish-white membranous aril (Figure 2a–c). They have a pungent, slightly spicy, and aromatic taste. The weights of 100 fresh and dried seeds were found as 9.5 and 6.1 g, respectively.

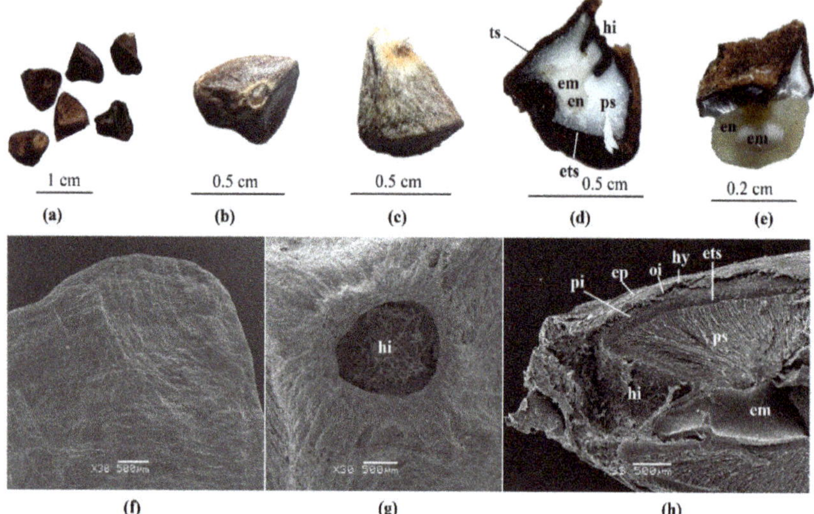

Figure 2. Black cardamom seeds: (**a,b**) morphological characteristics of seeds; (**c**) seed covered by arillus; (**d,e**) anatomical characteristics of seeds under the stereomicroscope: ts—testa, ets—endotesta, hi—hilum, ps—perisperm, en—endosperm, em—embryo; (**f–h**) anatomical characteristics of seeds under the SEM: ep—epidermal cells of testa, hy—hypodermis, oi—oil cell layer, pi—pigment layer. Scale bars (**f–h**) = 500 μm.

The longitudinal section of the black cardamom seed was observed by stereo microscope and SEM (Figure 2d–h). From the outside to the inside of the seed, the observed structures include:

(1) Testa (or seed coat): very hard, representing an effective barrier against water and air, and consisting of: epidermis, comprising one layer of cells, longitudinally elongated; hypodermis, comprising one layer of cells, tangentially elongated; below the hypodermis, a layer of large parenchymatous cells containing volatile oil; below this layer, the pigment layer comprising several layers of brown cells; endotesta, comprising one layer of palisade sclerenchymatous cells, brown.
(2) Perisperm: well-developed, composed of parenchymatous cells, white, and encircling the endosperm and embryo.
(3) Endosperm: small, greyish-white, and partially enclosing the embryo.
(4) Embryo: linear and fully developed.

3.2. In Vitro Germination Steps of Untreated Black Cardamom Seeds

In the in vitro germination of untreated black cardamom seed (control), the radicle emerges from the hilum of the seed and forms the primary root at around the 75th day of culture (Figure 3a). After that, the coleoptile emerges, grows to approximately 0.2–0.3 cm, then stops growing and will be pierced by the first leaf of the plumule (Figure 3c,d). The first leaf will grow, expand, and be supplemented by additional leaves (Figure 3e). At the other end of the embryonic axis, the primary root soon dies while adventitious roots (roots that arise directly from the shoot system) emerge, develop, and will ultimately produce a fibrous root system (Figure 3b).

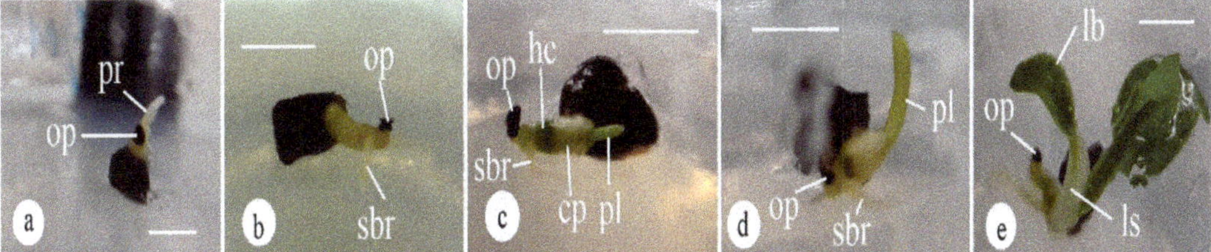

Figure 3. In vitro germination steps of untreated black cardamom seed: (**a**,**b**) radicle emerges (at 75th day of culture): op—operculum, pr—primary root, sbr—shoot-borne root; (**c**) coleoptile emerges and is pierced by the primary leaf (at 85th day of culture): cp—coleoptile, hc—hypocotyl, pl—primary leaf; (**d**,**e**) the first leaf grows, expands, and is supplemented by additional leaves (at 85th and 110th day of culture, respectively): lb—leaf blade, ls—leaf sheath. Scale bars = 0.5 cm.

3.3. Effect of Seed Treatments on the Germination and Seedling Growth of Black Cardamom under In Vitro Conditions

Efficiency of seed surface disinfection by 0.1% $HgCl_2$ for 10 min was significantly different between seed treatments after 10 days of culture (Table 2).

Table 2. Different seed treatment effects on seed germination and seedling growth of black cardamom cultured in vitro.

Treatment	Contamination Free (%)	GP [2] (%)	MGT [3] (days)	GRI [4]	Seedling Length (cm)	No. of Leaves
Control	68.0 ± 2.3 [1] c	16.0 ± 2.1 c	74.7 ± 2.9 b	0.05 ± 0.01 e	2.09 ± 0.11 bc	1.83 ± 0.15 b
ME	69.3 ± 2.7 c	68.0 ± 4.0 a	53.7 ± 0.7 d	0.34 ± 0.02 a	3.05 ± 0.07 a	3.07 ± 0.02 a
HW2m	66.7 ± 2.7 c	8.0 ± 2.3 d	85.0 ± 2.9 a	0.02 ± 0.01 f	1.87 ± 0.08 cd	1.62 ± 0.25 b
HW4m	68.0 ± 4.6 c	5.3 ± 1.0 d	88.3 ± 1.7 a	0.02 ± 0.00 f	1.66 ± 0.08 d	1.58 ± 0.28 b
CW	68.0 ± 2.3 c	28.0 ± 2.2 b	58.9 ± 1.4 cd	0.13 ± 0.01 b	2.25 ± 0.05 b	2.70 ± 0.1 a
NAS10m	78.7 ± 1.3 b	16.0 ± 2.0 c	61.4 ± 3.3 c	0.07 ± 0.01 de	2.17 ± 0.07 bc	1.75 ± 0.14 b
NAS15m	84.0 ± 2.3 ab	22.7 ± 1.3 bc	61.1 ± 2.0 c	0.10 ± 0.01 cd	2.23 ± 0.1 b	1.80 ± 0.17 b
HAS10m	82.7 ± 1.3 ab	22.7 ± 1.3 bc	62.3 ± 1.5 c	0.10 ± 0.01 cd	2.10 ± 0.17 bc	2.62 ± 0.09 a
HAS15m	86.7 ± 1.3 a	17.3 ± 1.0 c	62.5 ± 3.8 c	0.08 ± 0.01 de	2.07 ± 0.11 bc	2.58 ± 0.14 a
GA$_3$24h	66.7 ± 4.8 c	29.3 ± 1.2 b	61.4 ± 0.8 c	0.13 ± 0.01 b	2.93 ± 0.1 a	2.79 ± 0.16 a
NAA24h	62.7 ± 2.7 c	25.3 ± 1.1 b	60.6 ± 2.2 cd	0.11 ± 0.01 bc	2.81 ± 0.16 a	2.95 ± 0.07 a
LSD$_{0.05}$	5.3	4.7	6.8	0.02	0.31	0.47

Treatments: control; ME—soak in water for 24 h + scarify by scalpel; HW2m and HW4m—soak in hot water at 100 °C for 2 and 4 min, respectively; CW—soak in cold water for 24 h; NAS10m and NAS15m—soak in 50% HNO_3 for 10 and 15 min, respectively; HAS10m and HAS15m—soak in 25% HCl for 10 and 15 min, respectively; GA$_3$24h—soak in 200 ppm GA$_3$ for 24 h; NAA24h—soak in 200 ppm NAA for 24 h. Means followed by a different letter are significantly different at an alpha level of 0.05 according to the Duncan's multiple range test. Percentage values were arcsin $\sqrt{X}$ transformed prior to statistical analysis. [1] Mean ± standard error. [2] GP (%) = germination percentage. [3] MGT (days) = mean germination time. [4] GRI = germination rate index.

Acid seed scarification had a significantly higher percentage of contamination-free seed compared with other treatments. Among these, seeds were dipped into 25% HCl for 15 min and disinfected with 0.1% $HgCl_2$ for 10 min in the next step and gave the highest percentage of contamination-free seeds, reaching 86.7%. In the other seed treatments, there was not a statistically significant difference in the percentage of contamination free seed compared to the control.

Results also showed that there was a significant difference in germination of black cardamom seeds under different seed treatments at $\alpha = 0.05$ (Table 2). Applying mechanical scarification (ME) provided the highest mean germination percentage (reaching 68.0%) and the lowest mean germination time (reaching 53.7 days) after 90 days of culture. In this treatment, it took only approximately 50 days for 50% seed germination, while all other treatments had a lower germination percentage of 50% (Figure 4). Treatment of seed immersion in cold water (CW) or PGRs (GA$_3$24h and NAA24h) also showed significantly higher mean germination percentages than the control. Acid seed scarification methods (NAS10m, NAS15m, HAS10m, and HAS15m) did not show a significant effect on the germination percentage of black cardamom seeds, but they significantly reduced the seed germination time compared to the control. Hot-water-soaking treatments (HW2m and HW4m) appeared to have a negative effect on seed germination (Table 2). In addition, after the 90th day of culture, seed germination was not observed in any of the treatments.

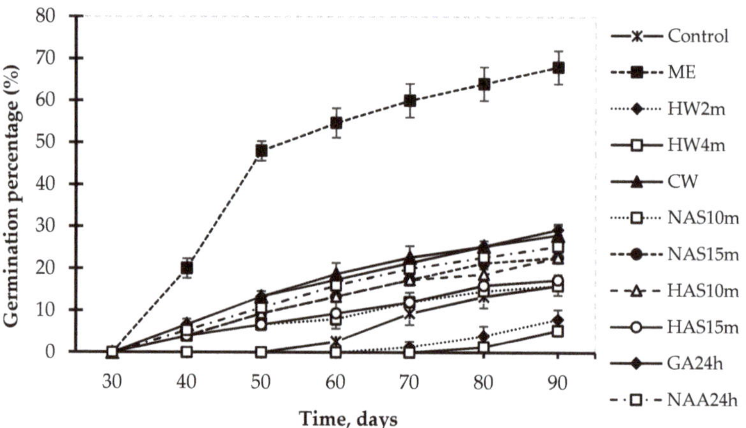

Figure 4. In vitro black cardamom seed germination patterns under different treatments.

Generally, the ME, GA$_3$24h, and NAA24h treatments significantly improved in vitro seedling length and number of leaves after 20 days of subsequent culture from the 90th day. In the other seed treatments, there were not statistically significant differences in mean seedling length and mean number of leaves compared to the control (Table 2, Figure 5).

Figure 5. Seedling growth of black cardamom at the 110th day under different seed treatments: T0—Control; T1—ME (soak in water for 24 h + scarify by scalpel); T2—HW2m (soak in hot water at 100 °C for 2 min); T3—HW4m (soak in hot water at 100 °C for 4 min); T4—CW (soak in cold water for 24 h); T5—NAS10m (soak in 50% HNO$_3$ for 10 min); T6—NAS15m (soak in 50% HNO$_3$ for 15 min); T7—HAS10m (soak in 25% HCl for 10 min); T8—HAS15m (soak in 25% HCl for 15 min); T9—GA$_3$24h (soak in 200 ppm GA$_3$ for 24 h); T10—NAA24h (soak in 200 ppm NAA for 24 h). Scale bars = 1 cm.

3.4. Effect of Different PGRs on Shoot Multiplication, Elongation, and Rooting in Black Cardamom

After 7 weeks of culture on the MS medium supplemented with various concentrations of PGRs (auxin and cytokinin), the growth responses of the explants (shoot tip explants derived from in vitro grown seedlings) were different (Table 3).

Table 3. Effect of different PGRs in MS culture medium on shoot multiplication, elongation, and rooting in black cardamom.

Type and Concentration of PGRs (mg/L)			Shoots Number (units)	Shoot Length (cm)	Root Number (units)	Callus Induction (%)	Shoot and Root Quality
BAP	NAA	2,4-D					
0.0	0.0	0.0	0.6 ± 0.1 d	3.0 ± 0.1 d	2.7 ± 0.2 d	0.0 e	+
1.0	0.0	0.0	3.4 ± 0.3 c	5.1 ± 0.1 c	5.6 ± 0.2 c	0.0 e	++
1.0	4.0	0.0	0.0 e	0.0 e	0.0 e	0.0 e	
1.0	0.0	1.0	0.0 e	0.0 e	0.0 e	83.3 ± 5.3 a	
1.0	0.0	2.0	0.0 e	0.0 e	0.0 e	58.3 ± 8.3 b	
2.0	0.5	0.0	3.7 ± 0.5 c	5.1 ± 0.1 c	5.4 ± 0.3 c	0.0 e	++
2.0	1.0	0.0	3.8 ± 0.4 c	5.5 ± 0.2 b	5.8 ± 0.2 c	0.0 e	++
3.0	0.5	0.0	4.0 ± 0.4 bc	5.8 ± 0.2 b	5.9 ± 0.5 c	0.0 e	++
3.0	1.0	0.0	4.1 ± 0.3 bc	5.8 ± 0.1 b	6.3 ± 0.3 c	0.0 e	+++
4.0	0.5	0.0	5.4 ± 0.3 a	6.8 ± 0.3 a	16.2 ± 0.8 a	0.0 e	+++
4.0	1.0	0.0	4.9 ± 0.2 ab	5.8 ± 0.2 b	8.9 ± 0.7 b	0.0 e	++
$LSD_{0.05}$			0.9	0.5	1.2	9.5	

Means followed by the same letter are not significantly different at $\alpha = 0.05$ according to the Duncan's multiple range test. Values of callus induction were arcsin $\sqrt{X}$ transformed prior to statistical analysis. The data have been recorded on a per-explant basis and recorded after 7 weeks of culture. +: small shoots, thin and light green leaves, and thin and few root-hairs roots; ++: medium shoots, thin and light green leaves, and medium and a few root-hairs roots; +++: fat shoots, thick and dark green leaves, and fat and many root-hairs roots.

Results indicated that the combination of BAP and NAA showed a high response for both shoot, leaf, and root induction after seven weeks of culture. MS medium supplemented with 4.0 mg/L BAP and 0.5 mg/L NAA exhibited the overall best response (mean shoot number: 5.4 ± 0.3, mean shoot length: 6.8 ± 0.3 cm, and mean root number: 16.2 ± 0.8) (Table 3). Shoots developed in clusters, stout pseudo-stems, thick and dark green leaves, and fat and many root-hairs roots (Figure 6c,e–g). The second-highest response was obtained using MS culture medium supplemented with 4.0 mg/L BAP and 1.0 mg/L NAA (mean shoot number: 4.92 ± 0.22, mean shoot length: 5.8 ± 0.2 cm, and mean root number: 8.9 ± 0.7) (Table 3). The in vitro response of seedling shoot tips on medium without PGRs (control) was low (mean shoot number: 0.6 ± 0.1, mean shoot length: 3.0 ± 0.1 cm, and mean root number: 2.7 ± 0.2) (Table 3). It can be seen that MS medium supplemented with a combination of BAP at high concentration and NAA is very suitable not only for shoot multiplication but also for root formation, and, hence, a separate constitution for root initiation is not required, as is often the case.

On the other hand, the combination of high-concentration auxin (including NAA and 2,4-D) with lower-concentration cytokinin (BAP) also showed different effects on callus induction of explants. The treatments related to 2,4-D produced a high callus induction rate (Table 3). However, produced callus were white and did not initiate shoot organogenesis in any of the explants when they were subjected to callogenesis induction treatments (Figure 6d). At the combination of 4.0 mg/L NAA with 1.0 mg/L BAP, callus induction was not observed; the explants were increasingly dark brown to blackish and, finally, rotten.

Figure 6. Different stages of in vitro micropropagation of black cardamom through seedling shoot tip culture on MS medium supplemented with BAP (4 mg/L) and NAA (0.5 mg/L) (except Figure (**d**)): (**a**) in vitro seedling, the source of explant; (**b**) explant (seedling shoot tip) after 1 week; (**c,e,g**) cluster of multiple shoots after 3, 5, and 7 weeks; (**f**) rooting after 7 weeks; (**d**) callus induction on MS medium supplemented with BAP (1 mg/L) and 2,4-D (1 mg/L) after 7 weeks. Scale bars = 1 cm.

3.5. Plantlets Adaptation to Ex Vitro Conditions

After being acclimatized to changes in temperature, humidity, and water loss, the in-vitro-propagated plantlets (4.0–5.0 cm in length, with 3–4 leaves (Figure 7a)) were directly planted on two soil types. The results presented in Table 4 show that the mean survival rate of in vitro plantlets on the two soil types reached 75.6–83.3% after a month of planting and 41.1–48.9% after 3 months of planting (Figure 7b,d). Among the two soil types studied, "Fasco" bio-soil produced a higher survival rate of 83.3% after a month of planting and 48.9% after 3 months of planting. The plantlets grew on this soil type better (mean length of 23.3 cm and a mean number of leaves of 5.7). The corresponding results when planting on "Garden Star" soil were all lower (Figure 7c,e).

Figure 7. Effect of acclimatization of in-vitro-propagated black cardamom plantlets after transfer in the greenhouse conditions: (**a**) in-vitro-propagated plantlets; plantlets after 1 (**b**) and 3 (**d**) months of planting on "Fasco" bio-soil; plantlets after 1 (**c**) and 3 (**e**) months of planting on "Garden Star" soil. Scale bar = 2 cm.

Table 4. Effect of different soil types on efficiency of black cardamom plantlet adaptation and their morphometric characteristics.

Soil Type	Survival Rate (%)		Plantlet Length (cm) after 3 Months	No. of Leaves after 3 Months
	After a Month	**After 3 Months**		
Garden Star	75.6 ± 4.8 [1]	41.1 ± 4.2	13.2 ± 0.6	4.3 ± 0.2
Fasco	83.3 ± 1.9	48.9 ± 4.8	23.3 ± 0.5	5.7 ± 0.2

[1] Mean ± standard error.

4. Discussion

The objectives of the study reported in this paper were to find solutions to improve the germination rate of black cardamom seeds, thereby creating an explant source for micropropagation to create many disease-free crops with outstanding characteristics of the selected mother plant to provide for mass production.

Seed coat hardness is an important factor that affects germination in seeds [32]. Seed dormancy has been reported in several species of the Zingiberaceae, including *Alpinia malaccensis* Roscoe [9], *Alpinia galanga* Willd. [10], korarima [11,12], large cardamom [13], and green cardamom [14]. In black cardamom, the morphological and anatomical characteristics of seeds established in the study, including the very hard testa, small endosperm (i.e., possessing low food storage), and small embryo, can be responsible for the difficulty of seed germination and the slow growth of seedlings. The strong inhibitory effect of the seed coat on seed germination may be caused by several possible mechanisms, including mechanical constraint, prevention of water and oxygen uptake, and retention or production of chemical inhibitors [33–35]. The integument breaking or softening, for instance, is needed to remove dormancy imposed by seed coat hardness or impermeability. Several authors (Alamgir and Hossain 2005 [36], 2005 [37]; Azad et al., 2006 [38], 2006 [39], 2010 [40], 2010 [41], 2011 [42]) have discussed different methods of pre-sowing treatments for seed germination in order to break dormancy, enhance the rate of germination, and speed up the germination process. Our studies showed that the seed treatments had a significant effect on the germination and growth parameters of black cardamom seeds. Among the applied seed treatments, mechanical scarification produced the best effect on seed germination parameters and subsequent seedling development. The findings of our previous study [15] also displayed similar results, but the germination rate of seeds was significantly improved in this experiment. The difference in the type of seeds used and the time of sowing in our two experiments is probably the cause for that difference. While in the first report we used dried seeds that were collected in October and dried in the open sun for 3 days, in this experiment we used fresh seeds that were collected in December. According to local people's experience, the best time to sow black cardamom seeds is at the end of December, and they must be sown immediately after harvest because they will lose their ability to germinate when stored for a long time. In a study on the effect of moisture in seeds on the germination of green cardamom seeds, Sangakkara U. R. (1990) [43] found that reducing the moisture content of seeds significantly reduced and even destroyed the viability of them. Robert (1973) [44] termed them recalcitrant seeds. Bearing similar physiological characteristics to green cardamom seeds may be responsible for the difference in results in our two experiments. Further studies are needed to confirm this issue. Copeland (1995) [45], Hartmann (1997) [46], Missanjo E. (2014) [47], and Botseleng B. (2014) [48] also confirmed that mechanical scarification produced the best effect in breaking the dormancy of seeds. However, recent studies on breaking dormancy of some Zingiberaceous species have shown that the effect of chemical scarification treatments is better than mechanical scarification treatments. According to Radhamani et al. [49], the treatments of green cardamom seeds in India with 25% sulfuric acid (H_2SO_4) for ten minutes and 80% absolute alcohol for 30 min were the most effective treatments in breaking seed dormancy. Dahanayake [14] reported that soaking in 50%nitric acid (HNO_3) for 15 min was the most effective for green cardamom seeds' dormancy in Sri Lanka. In our study, acid scarification treatments were not effective in increasing the germination rate of black cardamom seeds. Similarly, the report of Seid et al. [50] also showed that acid scarification treatments were not effective in breaking dormancy of green cardamom seeds. According to this report, soaking in 80% alcohol for 30 min was the most effective method for breaking the dormancy of green cardamom seed. Endogenous gibberellins have been widely studied in relation to the breaking of seed dormancy in various species. GA_3 has been exogenously applied as a substitute for stratification and has increased germination in many plant species, e.g., *Fagus sylvatica* L. [51], *Crataegus pseudoheterophylla* Pojark. [52], and *Juniperus polycarpos* K. Koch [53]. In a previous study on korarima, combined treatment of 50% H_2SO_4 for 60 min and 250 mg/L GA_3 for 24 h yielded the highest dormancy-breaking effect [11]. In this study, the positive effects of PGRs on the seed germination efficiency and growth of black cardamom seedlings were also noted. Hot water treatments have been reported to enhance germination of hard-coated seeds by elevating water and O_2 permeability of the testa [54]. Rivai et al. [9] showed that treatment with hot water at 75 °C for 5 min gave the highest germination rate of *Alpinia malaccensis* Roscoe seeds. However, in our study, the hot water treatment seemed to have a negative effect on the seed germination and growth of black cardamom seedlings. This result is similar to that of Okunlola et al. (2010) [55] regarding African locust bean (*Parkia biglobosa* Benth.). Cold water treatment of black cardamom gave a fair germination percentage and a reduced mean germination time when compared to hot water and control treatments. This also implies that soaking in cold water could also reduce the dormancy period in seeds of black cardamom when compared to control. This result concurs with earlier reports of Emerhi and Nwiisuator (2010) [56] and Falemara et al. (2014) [57] that show that soaking in cold water is a feature that enhances germination in seeds of tropical trees. The subsequent increase in the germination percentage, decrease in mean germination time, and increase in germination index when subjected to different treatment methods are indications that the hard seed coat is responsible for the dormancy in black cardamom.

The next step in our study was to apply shoot tip explants derived from in-vitro-grown seedlings to establish an efficient in vitro propagation procedure that could be used for commercial purposes. In a previous study on large cardamom [19], when rhizome segments were cultured on an MS medium supplemented with BAP or NAA alone or in a combination of BAP + IBA, BAP + NAA, or IBA + NAA, they produced shoots and roots simultaneously. These researchers reported 90% survival of plants after transfer to soil [19]. In the present investigation, the MS medium supplemented with 4.0 mg/L BAP and 0.5 mg/L NAA was found to be suitable for large-scale multiplication of black cardamom. The use of MS medium supplemented with BAP at high concentration and NAA significantly shortened the time of in vitro plantlet generation (7 weeks) compared with our previous study (14 weeks). Earlier reports on the clonal micropropagation of other Zingiberaceous species such as green cardamom, ginger, and turmeric, indicated similar results [16,58–62]. According our previous study [27], MS medium supplemented with 1 mg/L BAP without auxin was best for shoot multiplication of rhizome buds of black cardamom after six weeks of culture (mean shoot number and mean shoot length were 4.5 and 5.5 cm, respectively). However, in this study, the growth response of shoot tip explants derived from in vitro grown seedlings to this medium was moderate (mean shoot number: 3.4 ± 0.3, mean shoot length: 5.1 ± 0.1 cm, and mean root number: 5.6 ± 0.2) (Table 3).

"Fasco" and "Garden Star" are two commonly used soils in the Russian Federation because they have a composition suitable for most crops grown here. The main components of these two soils are peat and mineral fertilizers. However, in the composition of "Fasco" bio-soil, there is also bio-humus—an organic fertilizer containing many beneficial microorganisms for the soil, which helps to increase the fertility and aeration of the soil. This difference may explain the more positive effect of "Fasco" bio-soil on survival and subsequent plantlets growth compared with "Garden Star" soil.

5. Conclusions

This study is one of the first reports of the morphological, anatomical, and germinating characteristics of black cardamom seeds collected in Vietnam. Seed coat ultrastructural features of the black cardamom using SEM were first reported. Because of the consistency in seed morphological features, SEM is considered useful when it comes to phylogenetic information and a great potential source of taxonomy. Additionally, we identified black cardamom seeds as physical dormancy seeds and the hard seed coat as the major cause. Applying mechanical scarification treatment before sowing has been shown to be the most effective for improving seed germination rates. Methods of immersion in cold water or PGRs before sowing are also recommended, although the effect of dormancy breaking is not as high as that of the mechanical scarification method. Traditional methods of propagation of black cardamom (by seeds and rhizomes) do not currently meet the increasing demand for large-scale production of crops. Therefore, the developed in vitro propagation protocol can be an effective solution for rapid multiplication of high-yielding elite plants to meet the needs of expanding cultivation of this important crop, but it needs to be tested for large-scale multiplication and field planting. It can be said that this investigation carries great commercial significance, as black cardamom is used as a spice in many Asian countries, as a flavoring agent, and for pharmaceutical purposes by other industries.

Author Contributions: Q.V.K. and R.N.K.—experiments on black cardamom culture in vitro. E.N.B.—preparation of plant material for SEM; Q.V.K. and H.T.N.—statistical analysis of experimental data; E.A.K.—manuscript conceptualization. Q.V.K., E.A.K. and M.R.K.—writing of the manuscript; Q.V.K. and M.R.K.—reviewing and editing of the manuscript; all authors approved the final manuscript for publication and agreed to be accountable for all aspects of the manuscript. All authors have read and agreed to the published version of the manuscript.

Funding: The results of Sections 3.1–3.3 were supported by assignments 122020300187-2 (Tsitsin Main Botanical Garden of Russian Academy of Sciences) and 0431-2022-0003 (All-Russia Research Institute of Agricultural Biotechnology) of the Ministry of Science and Higher Education of the Russian Federation, respectively. Additionally, Sections 3.4 and 3.5 were supported by the Ministry of Science and Higher Education of the Russian Federation in accordance with agreement № 075-15-2022-746, dated 13 May 2022 (internal number MK-3084.2022.1.4) on providing a grant in the form of subsidies from the federal budget of the Russian Federation under grants from the President of the Russian Federation for state support of young Russian scientists—PhDs— and leading scientific schools of the Russian Federation.

Institutional Review Board Statement: Not applicable.

Informed Consent Statement: Not applicable.

Data Availability Statement: Data sharing is not applicable to this article.

Acknowledgments: The authors are grateful to Ma A Chang for provided photos of black cardamom and Tam H. M. for botanical identification.

Conflicts of Interest: The authors declare no conflict of interest.

References

1. Binh, N.Q. Zingiberaceae. In *Flora of Vietnam*; Publishing House for Science and Technology: Hanoi, Vietnam, 2017; pp. 304–318.
2. Ho, P.H. *An Illustrated Flora of Vietnam*; Young Publishing House: Hanoi, Vietnam, 2000; Volume 3.
3. Hoat, N.B.; Thuan, N.D. *Techniques for Growing, Using, and Processing Medicinal Plants*; Agricultural Publisher: Hanoi, Vietnam, 2005.
4. Thom, C.T.; Lai, P.T.; To, N.T. *Techniques for Growing some Medicinal Plants*; Labor Publishing House: Hanoi, Vietnam, 2006.
5. Hai, P.T. *Module Curriculum for Production of Non-Timber Forest Products*; Information and Media Publisher: Hanoi, Vietnam, 2013.
6. Lamxay, V.; Newman, M. A revision of *Amomum* (Zingiberaceae) in Cambodia, Laos and Vietnam. *Edinb. J. Bot.* **2012**, *69*, 99–206. [CrossRef]
7. Loi, D.T. *Vietnamese Medicinal Plants and Herbs*; Medical Publishing House: Hanoi, Vietnam, 2001; pp. 423–424.
8. Moi, L.D. *Essential–Oil Plant Resources in Vietnam*; Agricultural Publisher: Hanoi, Vietnam, 2002; Volume 2.
9. Rivai, R.R.; Wardani, F.F.; Devi, M.G. Germination and breaking seed dormancy of *Alpinia malaccensis*. *Nusant. Biosci.* **2015**, *7*, 67–72. [CrossRef]
10. Baradwaj, R.G.; Rao, M.V.; Baskin, C.C.; Kumar, T.S. Non-deep simple morphophysiological dormancy in seeds of the rare *Alpinia galanga*: A first report for Zingiberaceae. *Seed Sci. Res.* **2016**, *26*, 165–170. [CrossRef]
11. Eyob, S. Promotion of seed germination, subsequent seedling growth and in vitro propagation of korarima (*Aframomum corrorima* (Braun) PCM Jansen). *J. Med. Plants Res.* **2009**, *3*, 652–659.
12. Dawid, J. Effect of seed treatment and nursery potting media on emergence and seedling growth of korarima (*Aframomum cororima* (Braun) PCM Jansen). *Int. J. Res. Stud. Sci. Eng. Technol.* **2019**, *6*, 20–30.
13. Bhowmick, T.P.; Chattopadhyay, S.B. Germination of seeds of larger cardamom. *Sci. Cult.* **1960**, *26*, 185–186.
14. Dahanayake, N. Application of seed treatments to increase germinability of cardamom (*Elettaria cardamomum*) seeds under in vitro conditions. *Sabaragamuwa Univ. J.* **2015**, *14*, 101–107. [CrossRef]
15. Quyet, V.K.; Kalashnikova, E.A.; Kirakosyan, R.N.; Hai, N.T. In vitro germination of *Amomum tsao-ko* Crevost & Lemarié seeds. *IOP Conf. Ser. Earth Environ. Sci.* **2021**, *848*, 012207.
16. Nadgauda, R.S.; Mascarenhas, A.F.; Madhusoodanan, K.J. Clonal multiplication of *Elettaria cardamomum* Maton. by tissue culture. *J. Plant. Crops* **1983**, *11*, 60–64.
17. Malhotra, E.V.; Kamalapriya, M.; Bansal, S.; Meena, D.P.S.; Agrawal, A. Improved protocol for micro-propagation of genetically uniform plants of commercially important cardamom (*Elettaria cardamomum* Maton). *Vitr. Cell. Dev. Biol.-Plant.* **2021**, *57*, 409–417. [CrossRef]

18. Tyagi, R.K.; Goswami, R.; Sanayaima, R.; Singh, R.; Tandon, R.; Agrawal, A. Micropropagation and slow growth conservation of cardamom (*Elettaria cardamomum* Maton). *Vitr. Cell. Dev. Biol.-Plant.* **2009**, *45*, 721–729. [CrossRef]

19. Sajina, A.; Mini, P.M.; John, C.Z.; Babu, K.N.; Ravindran, P.N.; Peter, K.V. Micro-propagation of large cardamom *Amomum subulatum* Roxb. *J. Spices Arom. Crops.* **1997**, *6*, 145–148.

20. Pradhan, S.; Pradhan, S.; Basistha, B.C.; Subba, K.B. In vitro micropropagation of *Amomum subulatum* (Zingiberaceae), a major traditional cash crop of Sikkim Himalaya. *Int. J. Life Sci. Biotech. Pharma Res.* **2014**, *3*, 169–180.

21. Purohit, S.; Nandi, S.K.; Paul, S.; Tariq, M.; Palni, L.M.S. Micropropagation and genetic fidelity analysis in *Amomum subulatum* Roxb: A commercially important Himalayan plant. *J. Appl. Res. Med. Arom. Plants* **2017**, *4*, 21–26. [CrossRef]

22. Wannakrairoj, S.; Tefera, W. Thidiazuron and other plant bioregulators for axenic culture of Siam cardamom (*Amomum krervanh* Pierre ex Gagnep.). *Agric. Nat. Resour.* **2012**, *46*, 335–345.

23. Ping, J.L. Micropropagation of *Amomum villosum* Lour. *Subtrop. Plant Sci.* **2004**, *33*, 37–38.

24. Hong, H.; Na, T.L. Tissue culture and plantlet regeneration of *Amomum villosum*. *Plant Physiol. Commun.* **2005**, *1*, 57–61.

25. Phuc, N.D.; Thanh, T.N.; Chau, D.T.T.; Phuong, T.T.B. In vitro propagation of *Amomum longiligulare* TL Wu. *J. Biotech.* **2011**, *9*, 689–698.

26. Truong, T.B.P.; Than, T.B.K.; Nguyen, D.T.; Nguyen, T.T.L.; Nguyen, T.T. In vitro propagation of *Amomum* sp. in A Luoi district, Thua Thien Hue province. *Hue Univ. J. Sci. Nat. Sci.* **2017**, *126*, 37–52.

27. Quyet, V.K.; Hai, N.T.; Hai, N.T.L.; Hang, P.T.T.; Kirakosya, R.; Kalashnikova, E. Plant regeneration of *Amomum tsaoko* Crevost & Lemarié in vitro. *IOP Conf. Ser. Earth Environ. Sci.* **2021**, *677*, 042065.

28. Reghunath, B.R. In Vitro Studies on the Propagation of Cardamom (*Elettaria cardamomum* Maton). Ph.D. Thesis, Kerala Agricultural University, Trichur, India, 1989.

29. Murashige, T.; Skoog, F. A revised medium for rapid growth and bioassays with tobacco tissue culture. *Physiol. Plant.* **1962**, *15*, 473–497. [CrossRef]

30. Scott, S.J.; Jones, R.A.; Williams, W. Review of data analysis methods for seed germination 1. *Crop Sci.* **1984**, *24*, 1192–1199. [CrossRef]

31. Esechie, H.A. Interaction of salinity and temperature on the germination of sorghum. *J. Agron. Crop Sci.* **1994**, *172*, 194–199. [CrossRef]

32. Aref, I.M.; Ali, H.; Atta, E.; Shahrani, T.; Ismail, A. Effects of seed pretreatment and seed source on germination of five Acacia spp. *Afr. J. Biotech.* **2011**, *10*, 15901–15910. [CrossRef]

33. Bewley, J.D.; Black, M. *Seeds, Physiology of Development and Germination*; Plenum Press: New York, NY, USA, 1994.

34. Schmidt, L.H. *Guide to Handling of Tropical and Subtropical Forest Seed*; DANIDA Forest Seed Centre: Humleback, Denmark, 2000; pp. 1–45.

35. Taiz, L.; Zeiger, E. Abscisic Acid: A Seed Maturation and Antistress Signal. In *Plant Physiology*, 3rd ed.; Sinauer Associates, Inc.: Sunderland, MA, USA, 2002; Volume 23, pp. 538–558.

36. Alamgir, M.; Hossain, M.K. Effect of pre-sowing treatments on *Albizia procera* (Roxb.) Benth seeds and initials development of seedlings in the nursery. *J. For. Environ.* **2005**, *3*, 53–60.

37. Alamgir, M.; Hossain, M.K. Effect of pre-sowing treatments on germination and initials seedling development of *Albizia saman* in the nursery. *J. For. Res.* **2005**, *16*, 200–204. [CrossRef]

38. Azad, M.S.; Islam, M.W.; Matin, M.A.; Bari, M.A. Effect of pre-sowing treatment on seed germination of *Albizia lebbeck* (L.) Benth. *South Asian J. Agricult.* **2006**, *1*, 32–34.

39. Azad, M.S.; Matin, M.A.; Islam, M.W.; Musa, Z.A. Effect of pre-sowing treatment on seed germination of Lohakath (*Xylia kerrii* Craib & Hutch.). *Khulna Univ. Stud.* **2006**, *7*, 33–36.

40. Azad, M.S.; Musa, Z.A.; Matin, A. Effect of pre-sowing treatments on seed germination of *Melia Azedarach*. *J. For. Res.* **2010**, *21*, 193–196. [CrossRef]

41. Azad, M.S.; Paul, N.K.; Matin, M.A. Do pre-sowing treatments affect seed germination in *Albizia richardiana* and *Lagerstroemia speciosa*. *Front. Agricult. China* **2010**, *4*, 181–184. [CrossRef]

42. Azad, M.S.; Manik, M.R.; Hasan, M.S.; Matin, M.A. Effect of different pre-sowing treatments on seed germination percentage and growth performance of *Acacia auriculiformis*. *J. For. Res.* **2011**, *22*, 183–188. [CrossRef]

43. Sangakkara, U.R. Relationship between storage and germinability of cardamom (*Elatteria cardamomum* Maton). *J. Agron. Crop Sci.* **1990**, *164*, 16–19. [CrossRef]

44. Roberts, E.H. Predicting the storage life of seeds. *Seed Sci. Technol.* **1973**, *1*, 499–514.

45. Copeland, L.O.; McDonald, M.F. *Principles of Seed Science and Technology*; Kluwer Academic Publishers: Norwell, MA, USA, 2012; p. 488.

46. Hartmann, H.T.; Kester, D.E.; Davies, F.T.; Geneve, R.L. The Biology of Propagation by Cuttings. *Plant Propag. Princ. Pract.* **1997**, *6*, 276–328.

47. Missanjo, E.; Chioza, A.; Kulapani, C. Effects of different pretreatments to the seed on seedling emergence and growth of Acacia polyacantha. *Int. J. For. Res.* **2014**, *2014*, 583069.

48. Botsheleng, B.; Mathowa, T.; Mojeremane, W. Effects of pre-treatments methods on the germination of Pod mahogany (*Afzelia quanzensis*) and mukusi (*Baikiaea plurijuga*) seeds. *Int. J. Innov. Res. Sci. Eng. Technol.* **2014**, *3*, 8108–8113.

49. Radhamani, J.; Malik, S.K.; Chandel, K.P.S. Structural changes associated with dormancy breaking treatments in cardamom. *Ind. J. Plant Genet. Res.* **1991**, *4*, 27–33.
50. Seid, H.; Mulualem, T.; Shiferaw, G.; Atilaw, A.; Herko, B. Enhancement of seed germination and seedling growth of cardamom (*Elleteria cardamomum*) at Tepi south-western part of Ethiopia. *Acad. Res. J. Agric. Sci. Res.* **2019**, *7*, 303–306.
51. Nicolas, C.; Nicolas, G.; Rodriguez, D. Antagonistic effects on abscisic acid and gibberellic acid on the breaking of dormancy of *Fagus sylvatica* seeds. *Physiol. Plant.* **1996**, *96*, 244–250. [CrossRef]
52. Ahmadloo, F.; Kochaksaraei, M.T.; Azadi, P.; Hamidi, A.; Beiramizadeh, E. Effects of pectinase, BAP and dry storage on dormancy breaking and emergence rate of *Crataegus pseudoheterophylla* Pojark. *New For.* **2015**, *46*, 373–386. [CrossRef]
53. Daneshvar, A.; Tigabu, M.; Karimidoost, A.; Oden, P.C. Stimulation of germination in dormant seeds of *Juniperus polycarpos* by stratification and hormone treatments. *New For.* **2016**, *47*, 751–761. [CrossRef]
54. Airi, S.; Bhatt, I.D.; Bhatt, A.; Rawal, R.S.; Dhar, U. Variations in seed germination of *Hippophae salicifolia* with different presoaking treatments. *J. For. Res.* **2009**, *20*, 27–30. [CrossRef]
55. Okunlola, A.I.; Adebayo, R.A.; Orimogunje, A.D. Methods of braking seed dormancy on germination and early seedling growth of African locust bean (*Parkia biglobosa* Benth). *J. Hort. For.* **2011**, *3*, 1–6.
56. Emerhi, E.A.; Nwiisuator, D. Germination of *Baillonella toxisperma* (Pierrre) seeds as influenced by seed pretreatment. In Proceedings of the 2nd Biennial National Conference of the Forests and Forest Products Society, Akure, Nigeria, 26–29 April 2010; pp. 323–328.
57. Falemara, B.C.; Chomini, M.S.; Thlama, D.M.; Udenkwere, M. Pre-germination and dormancy response of *Adansonia digitata* L seeds to pre-treatment techniques and growth media. *Eur. J. Bot. Plant Sci. Pathol.* **2014**, *2*, 13–23.
58. Nadgauda, R.S.; Mascrenhas, A.F.; Hendre, R.R.; Jagannathan, V. Rapid clonal multiplication of turmeric *Curuma longa* L. plants by tissue culture. *Ind. J. Exp. Biol.* **1978**, *16*, 120–122.
59. Kumar, K.B.; Kumar, P.P.; Singh, P.; Balachandran, S.M.; Iyer, R.D. Development of clonal plantlets from immature panicles of cardamom. *J. Plant. Crops* **1985**, *13*, 31–34.
60. Bhagyalakshmi; Singh, N.S. Meristem culture and micropropagation of a variety of ginger (*Zingiber officinale* Rosc.) with a high yield of oleoresin. *J. Hortic. Sci. Biotechnol.* **1988**, *63*, 321–327. [CrossRef]
61. Nirmal Babu, K.; Samsudeen, K.; Ratnambal, M.J. In vitro plant regeneration from leaf-derived callus in ginger (*Zingiber officinale* Rosc.). *Plant Cell Tiss. Organ Cult.* **1992**, *29*, 71–74. [CrossRef]
62. Salvi, N.D.; George, L.; Eapen, S. Micropropagation and field evaluation of micropropagated plants of turmeric. *Plant Cell Tiss. Organ Cult.* **2002**, *68*, 143–151. [CrossRef]

Article

In Vitro Propagation of *Aconitum violaceum* Jacq. ex Stapf through Seed Culture and Somatic Embryogenesis

Abdul Hadi [1,2], Seema Singh [1], Shah Rafiq [1,*], Irshad A. Nawchoo [2], Nasir Aziz Wagay [3], Eman A. Mahmoud [4], Diaa O. El-Ansary [5], Hanoor Sharma [6], Ryan Casini [7], Kowiyou Yessoufou [8] and Hosam O. Elansary [9,*]

1   Plant Tissue Culture and Research Laboratory, Department of Botany, University of Kashmir, Srinagar 190006, India; abhadi.scholar@kashmiruniversity.net (A.H.); seemakuniv@gmail.com (S.S.)
2   Plant Reproductive Biology, Genetic Diversity and Phytochemistry Research Laboratory, Department of Botany, University of Kashmir, Srinagar 190006, India; irshadnawchoo@uok.edu.in
3   Department of Botany, Government Degree College, Baramulla 193101, India; nasir.wagay1989@rediffmail.com
4   Department of Food Industries, Faculty of Agriculture, Damietta University, Damietta 34511, Egypt; emanmail2005@yahoo.com
5   Precision Agriculture Laboratory, Department of Pomology, Faculty of Agriculture (El-Shatby), Alexandria University, Alexandria 21545, Egypt; diaaagri@hotmail.com
6   Microbiology and Immunology Department, Wright State University, Dayton, OH 45435, USA; hanoor.sharma@clearlabs.com
7   College of Public Health, University of California, 2121 Berkeley Way, Berkeley, CA 94704, USA; ryan.casini@berkeley.edu
8   Department of Geography, Environmental Management, and Energy Studies, University of Johannesburg, APK Campus, Johannesburg 2006, South Africa; kowiyouy@uj.ac.za
9   Plant Production Department, College of Food & Agriculture Sciences, King Saud University, Riyadh 11451, Saudi Arabia
*   Correspondence: shahrafiq.scholar@kashmiruniversity.net (S.R.); helansary@ksu.edu.sa (H.O.E.)

Citation: Hadi, A.; Singh, S.; Rafiq, S.; Nawchoo, I.A.; Wagay, N.A.; Mahmoud, E.A.; El-Ansary, D.O.; Sharma, H.; Casini, R.; Yessoufou, K.; et al. In Vitro Propagation of *Aconitum violaceum* Jacq. ex Stapf through Seed Culture and Somatic Embryogenesis. *Horticulturae* 2022, 8, 599. https://doi.org/10.3390/horticulturae8070599

Academic Editor: Sergio Ruffo Roberto

Received: 9 June 2022
Accepted: 28 June 2022
Published: 4 July 2022

Publisher's Note: MDPI stays neutral with regard to jurisdictional claims in published maps and institutional affiliations.

Abstract: *Aconitum violaceum* Jacq. ex Stapf is a threatened medicinal plant with restricted global distribution. The highest frequency of seed germination was recorded on Murashige and Skoog's (MS) basal medium, supplemented with 0.5 mg L^{-1} kinetin with a germination rate of 77.32% and mean germination time of 27 days. Among the various plant growth regulators examined, 0.1 mg L^{-1} kinetin (Kn) + 0.5 mg L^{-1} indole-3-acetic acid (IAA) proved to be effective for maximum embryogenic callus production (51.0%) within 31 days of inoculation. The conversion rate of somatic embryos into complete plantlets was highest in the MS medium augmented with 0.1 mg L^{-1} Kn + 0.5 mg L^{-1} IAA (68.00%), with an average root initiation time of 25 days. The rooted plantlets were subsequently hardened into jiffy pots with a combination of loamy soil, coco-peat, and vermicompost (1:1:1 *v/v*), and then transplanted into a greenhouse with a 60% survival rate. To our knowledge, this is the first study on direct in vitro propagation and embryogenic callus induction from seeds. The established regeneration protocol could be employed to propagate *A. violaceum* on a large scale in a short time. This would contribute significantly to its rapid propagation and germplasm conservation, and establish a framework for the domestication of this highly valued threatened medicinal plant.

Keywords: acclimatization; callusing; multiple shooting; seed germination; micropropagation

1. Introduction

Aconitum violaceum Jacq. ex Stapf is a biennial herbaceous medicinal plant of the Ranunculaceae family and is endemic to the north-western Himalayan region of India, Pakistan, and Nepal [1–3]. In India, it is primarily found in the alpine and subalpine regions of the north-western Himalayas, at an elevation range of 3000–4000 m asl and shares its position with vulnerable plant species [4,5]. The toxic alkaloids of *A. violaceum* can be easily converted into less harmful alkaloids by heating or by using an alkaline treatment. They can then be employed in Ayurvedic and Unani medicines after they have been detoxified [6]. Traditionally, it is used to treat boils [7–9], asthma, high fever [7–12], gastric troubles [8–12], sciatic pains [8,11,13], intestinal worms [7,12,13], renal pain [7,11,12], and snake and scorpion bites [6]. This is due to its wide spectrum of biological activities, such as antioxidant [14], antimicrobial [14], anti-inflammatory [15–17], anti-malarial [15–18], anti-proliferative [16–20], analgesic, and antipyretic properties [14,15,17].

A. violaceum faces a great threat of extinction in its natural habitat due to various factors, such as physiological seed dormancy of >5 months [4,21,22], and long-term burying of seeds beneath snow (more than five months). Due to its specific ecological requirement, seed germination and seedling establishment in *A. violaceum* are quite challenging. *A. violaceum* grows along the margins of irrigation canals and at the edges of alpine streams; consequently, the majority of the seeds are susceptible to being carried away from their natural environment by running water, heavy rainfall, and floods, exposing them to adverse environmental conditions [4,21]. The plant maintains its spatial continuity through its rhizome. Therefore, uprooting whole plants due to collection, overgrazing, and premature harvesting, alongside construction of high-altitude roads, dams, cemented water channels, and human settlements, also contribute to the decline of this species from its natural habitat. Beetles and aphids are another potential hazard to this species as they consume the flowers and other reproductive portions of the plant, thereby reducing the species' sexual potential [22]. In addition, the species has a unique niche, which may limit its area of occupancy and dispersion [4].

When properly applied, mass propagation methods may help to reduce the extinction risk of vulnerable species by using procedures such as in vitro propagation [22]. In vitro regeneration techniques have been established for some *Aconitum* species such as *A. chasmanthum* Stapf ex Holmes [22], *A. ferox* [23], *A. heterophyllum* [24], *A. nagarum* [25], and *A. vilmorinianum* [26], but no such attempts have been made for *A. violaceum*. Giri et al. [24] regenerated complete plantlets from the somatic embryos of leaf and petiole explants in *Aconitum heterophyllum*. Hatano et al. [27] regenerated complete plantlets from the somatic embryo of anther explant in *Aconitum carmichaeli*. High concentrations of NAA (10 mg L^{-1}) and darkness promote somatic embryo development in *Ranunculus sceleratus* [28]. After surveying extensive literature from various databases (DOAJ, Google Scholar, PubMed, Scihub, ScienceDirect, etc.) only a few studies have reported on the in vitro propagation of *A. violaceum* to date. Rawat et al. [29] attempted in vitro propagation from the nodal explant. Given the aforementioned issues, the current research on *A. violaceum* was carried out to develop: (1) direct in vitro seed germination protocols; (2) embryogenic callus production; and (3) complete plant regenerations from somatic embryos.

Due to its immense medicinal and economic value, establishing in vitro propagation protocols for *A. violaceum* would not only provide elite clones for pharmaceutical uses and facilitate rapid propagation and germplasm conservation, but it would also help to domesticate the plant and preserve wild populations.

2. Materials and Methods

2.1. Collection of Plant Material

The seeds of *A. violaceum* Jacq. ex Stapf were collected in the month of August (10–28) as soon as the follicle burst for seed dispersal, from the villages of Khawous and Numsuru (3220 m asl to 3340 m asl; 34°15.905 N to 34°12.705 N; 75°96.040 E to 75°96.720 E) in Ladakh, India. The plant specimen was identified and authenticated by the Center for Biodiversity and Taxonomy (CBT), Department of Botany, University of Kashmir, Hazratbal-Srinagar. A voucher specimen (accession number 3738-KASH) was deposited at the Kashmir University Herbarium (KASH). Seeds were collected in dry plastic and glass bottles and brought to the plant tissue culture laboratory, Department of Botany, University of Kashmir, where they were stored between 4 °C and 6 °C with 55 to 65% relative humidity for experimental purposes.

2.2. Chemicals

MS (Murashige and Skoog 1962) medium and plant growth regulators (PGRs), such as 2,4-dichlorophenoxyacetic acid (2,4-D), indole-3-acetic acid (IAA), kinetin (Kn), 6-benzylaminopurine (BAP), and 1-naphthaleneacetic acid (NAA), were purchased from Hi-Media, India. Mercuric chloride ($HgCl_2$), sodium hypochlorite (NaClO), and sucrose were also purchased from Hi-Media, India, while agar (plant agar) was purchased from Sigma Aldrich, India.

2.3. Explant Selection, Culture Conditions, and Establishment of In Vitro Cultures

A. violaceum seeds were used as explants to start in vitro cultures. The seeds were immersed in tap water for 96 h (h) before inoculation on the MS medium. Prior to culture, seeds were rinsed for about 2 h under running tap water, surface sterilized with 1% NaClO (v/v) for 7 min with occasional agitation, washed 4–5 times with double-distilled autoclaved water, and then dipped in 70% ethanol for about 30–45 s. They were then washed with aseptic (autoclaved, double-distilled) water 3–4 times. The seeds were placed in the folds of sterile filter paper to absorb the leftover moisture. Seeds were then cultured in 30 mL borosilicate culture vials on the MS basal medium containing 0.8% (w/v) agar and sucrose (3% w/v), fortified with different PGRs at different concentrations (0.05–2.0 mg L^{-1}), either individually or in combination. They were then incubated at $10 \pm 2\,^{\circ}C$ with 50–55% humidity for a 12–12 h photoperiod (42–60 $\mu mol\, m^{-2}\, s^{-1}$) in a plant growth chamber. Under the laminar air flow hood, all studies, from surface sterilization to inoculation, were carried out successfully. Data were represented as a mean of 50 replicates per repetition. Observations were recorded from each non-contaminated vial (experimental unit). Mean germination time (MGT) in days was calculated by following the methods of Darrudi et al. [30] using the following equation:

$$\text{MGT} = \frac{\Sigma(n \times D)}{N}$$

where
'n' = number of newly germinated seeds after each incubation period
'D' = number of days since the experiment began, and
'N' = total number of seeds germinated at the end of the experiment.

Seed germination rate was calculated using the following equation:

$$\% \text{ seed germination} = (nx/Na) \times 100$$

where
'nx' = total germinated seeds
'Na' = number of seeds used at the beginning of the experiment.

2.4. Embryogenic and Non-Embryogenic Callus Production

Mature and immature seeds of *A. violaceum*, taken from the wild plants, were cultured on the MS basal medium augmented with various concentrations of PGRs, such as Kn (0.05 mg L^{-1} to 3.0 mg L^{-1}), IAA (0.1 mg L^{-1} to 1.5 mg L^{-1}), and 2,4-D (0.05–2.5 mg L^{-1}) with sucrose 3% (w/v) and 0.8 % agar for production of embryogenic and non-embryogenic calluses. The pH of the medium was adjusted to 5.8 ± 0.02 before autoclaving at $121\,^{\circ}C$. Immature and mature seeds were surface sterilized with 1% sodium hypochlorite (NaClO) (v/v) for 7 min with occasional stirring, then rinsed 3–5 times with sterile water before being immersed in 70% ethanol for about 30 s and washed 3–4 times with sterile (autoclaved, double-distilled) water. The remaining moisture was removed by putting the seeds in the folds of sterile filter paper. Cultures were incubated at $10 \pm 2\,^{\circ}C$ in a 16/8-h light/dark cycle or in completely dark conditions on racks fitted with cool fluorescent tube lights of 60.0 $\mu mol\, m^{-2}\, s^{-1}$ illuminance and RH 55%. The frequency of embryogenic callus induction was studied after 6 to 8 weeks (wk) of incubation. Consequently, embryogenic and non-embryogenic calluses sub-cultured within 1 culture were used for whole plant regeneration. The experiment was conducted at least 3 times with a total of 10 replicates for each treatment (1 vial was considered as 1 replicate).

2.5. Effect of Light Requirement on Percentage of Somatic Embryo Formation

The effect of light conditions on the production of somatic embryos was investigated by keeping 50% of embryogenic calluses in the dark (the culture vials were wrapped in aluminum foil) and the other 50% were kept in a 16/8 h photoperiod for 10–30 days.

2.6. Whole Plant Regeneration from Somatic Embryos

To accelerate complete plant development from somatic embryos, the effects of Kn and IAA on further conversion of somatic embryos were studied. The somatic embryos were sub-cultured on the MS medium augmented with different concentrations of Kn (0.05–1.0 mg L^{-1}) and IAA (0.1–1.0 mg L^{-1}), either individually or in combination. After 8 weeks of culture, the average number of shoots and roots developed by somatic embryos were evaluated. All cultures were incubated at 10 ± 2 °C in a 16/8-h light/dark cycle with 60% RH. Each treatment consisted of 10 somatic embryos in 30 mL borosilicate culture vials and 150 mL conical flasks, and the experiment was repeated 3 times.

2.7. In Vitro Rooting

Multiple shoots developed directly from non-embryogenic and embryogenic calluses were sub-cultured on the MS basal medium and enriched with different combinations and concentrations of auxins and cytokinins. IAA concentrations were kept slightly higher compared to Kn. The effect of light and various concentrations of PGRs on the development of roots were also examined.

2.8. Hardening and Acclimatization of In Vitro Plantlets

Media traces from the roots of *A. violaceum* regenerated by different methods from seed explants were removed by thoroughly washing under tap water. The plantlets were then transplanted into jiffy pots composing a mixture of sterilized loamy soil, coco-peat, and vermicompost (1:1:1). The plantlets in the jiffy pots were enclosed in transparent polybags for 2 wks to ensure adequate humidity and were maintained in a growth chamber. After 3 wks, the hardened plantlets were transferred into a net-shade greenhouse for further acclimatization. Well-established plantlets were then transferred into the Kashmir University Botanical Garden (KUBG) and were kept in the shade with occasional watering. The phenotypic data, including the plant height, leaf number, floral bud numbers, and survival rate, was determined for up to 8 wks.

2.9. Experimental Design and Statistical Analysis

The data were evaluated using a factorial design that was completely randomized (CRD). Substantial variations between each treatment were assessed using Duncan's multiple range test (DMRT) of one-way ANOVA using SPSS statistical software version 23 (IBM Corp. Released 2015. IBM SPSS Statistics for Windows, version 23.0. Armonk, NY, USA: IBM Corp.). Results were shown as mean value $\pm$ SEM (standard error mean) for each experiment. Graphs were prepared in origin pro (version 9, 2021; developed by originLab corporation, Northampton, MA, USA).

3. Results

3.1. Seed Germination

Seed germination rate was enhanced when the seeds were immersed in tap water for 96 h in winter and early spring, prior to culture. Among the various PGRs used, the highest rate of in vitro seed germination was recorded in winter and spring on the MS basal medium enriched with 0.5 mg L^{-1} Kn and 1.0 mg L^{-1} Kn with a MGT of 27.22 ± 0.70 and 26.88 ± 0.16 days, and percentage germination of 77.32 ± 0.38 and 75.33 ± 0.14. The number of shoot buds was also recorded as maximum (2.0 ± 0) in this treatment. The shoot length of each seed was recorded highest in the MS medium augmented with 1.0 mg L^{-1} Kn with an average length of 4.7 ± 0.11 cm (Table 1, Figure 1). Kn at low concentrations (<0.01 mg L^{-1}) and high concentrations (>1.5 mg L^{-1}) reduced seed germination percentage. Likewise, high concentrations of IAA (>1.5 mg L^{-1}) also reduced the rate of seed germination. Thus, concentrations of Kn and IAA ranging between 0.1 and 1.5 mg L^{-1} were proven to be ideal for seed germination in *A. violaceum*. None of the seeds germinated in the control condition (MS basal) without PGRs (plant growth regulators). The data were recorded for up to eight weeks, which was represented by mean value ± SEM (standard error mean).

Figure 1. Direct in vitro seed germination of *Aconitum violaceum* Jacq. ex Stapf on the MS basal medium enriched with different concentrations of PGRs, either individually or in combination: (**A**) 0.1 mg L^{-1} Kn; (**B**) 0.35 mg L^{-1} Kn; (**C**) 0.5 mg L^{-1} Kn; (**D**) 1.0 mg L^{-1} Kn; (**E**) 1.5 mg L^{-1} Kn; (**F**) 0.1 mg L^{-1} IAA; (**G**) 0.5 mg L^{-1} IAA; (**H**) 1.0 mg L^{-1} IAA; (**I**) 0.1 mg L^{-1} BAP; (**J,K**) germinated seeds with well-formed roots before hardening; and (**L**) hardened plantlets in the jiffy pots composing a mixture of loamy soil, coco-peat, and vermicompost (1:1:1 *v/v*). Scale bar (**A–I**) represents 5 mm; (**J–L**) represents 1 cm.

Table 1. Effects of plant growth regulators (IAA, Kn, BAP, and NAA) on direct in vitro seed germination of wild-growing *Aconitum violaceum* Jacq. ex Stapf. in Ladakh, India.

MS Medium + PGRs (mg L^{-1})				MGT (days, Mean ± SEM)	Percentage Germination (Mean ± SEM)	Number of Shoot Buds Formed Per Seed (Mean ± SEM)	Shoot Length (cm, Mean ± SEM)	Remark
IAA	Kn	BAP	NAA					
-	-	-	-	0.0 ± 0 a	0.0 ± 0 a	0.0 ± 0 a	0.0 ± 0 a	No seed germination
0.1	-	-	-	35.44 ± 0.18 de	27.54 ± 0.18 d	1.0 ± 0 b	2.83 ± 0.03 cd	Healthy growth of seedlings
0.5	-	-	-	34.11 ± 0.09 cd	25.99 ± 0.25 d	1.33 ± 0.33 b	2.9 ± 0.05 c,d	Poor seed germination
1.0	-	-	-	37.88 ± 0.12 ef	24.44 ± 0.10 d	1.0 ± 0 b	2.56 ± 0.03 bc	Poor seed germination
1.5	-	-	-	0.0 ± 0 a	0.0 ± 0 a	0.0 ± 0 a	0.0 ± 0 a	No response
-	0.1	-	-	31.33 ± 0.30 c	36.22 ± 0.28 e	1.33 ± 0.33 b	3.8 ± 0.2 e	Healthy seed germination
-	0.35	-	-	28.00 ± 0.38 b	45.99 ± 0.33 f	1.33 ± 0.33 b	3.96 ± 0.20 ef	Healthy growth of seedlings
-	0.5	-	-	27.22 ± 0.70 b	77.32 ± 0.38 g	2.0 ± 0 c	4.3 ± 0.3 f	Healthy growth of seedlings
-	1.0	-	-	26.88 ± 0.16 b	75.33 ± 0.14 g	2.0 ± 0 c	4.7 ± 0.11 g	Healthy growth of seedlings
-	1.5	-	-	31.88 ± 0.10 c	6.64 ± 0.15 b	1.0 ± 0 b	2.26 ± 0.13 b	Poor seed germination
-	2.0	-	-	0.0 ± 0 a	0.0 ± 0 a	0.0 ± 0 a	0.0 ± 0 a	No response
-	-	0.05	-	39.22 ± 0.45 f	12.6 ± 0.13 c	1.33 ± 033 b	3.0 ± 0.05 d	Poor seed germination
-	-	0.1	-	44.77 ± 0.15 g	16.13 ± 0.35 c	1.0 ± 0 b	2.5 ± 0.11 b,c	Poor seed germination
-	-	-	0.05	40.11 ± 0.44 f	12.82 ± 0.04 c	1.0 ± 0 b	2.6 ± 0.05 bcd	Poor seed germination

Data were collected for up to eight weeks. The same letters in each column indicate that the data were not substantially different according to Duncan's multiple range test (DMRT) at $p \leq 0.05$. Data were evaluated from 50 replicates (per repetition) and repeated three times. Abbreviations: MS: Murashige and Skoog. PGRs: plant growth regulators. IAA: indole-3-acetic acid. Kn: kinetin. BAP: 6-benzylaminopurine. NAA: 1-napthaleneacetic acid. MGT: mean germination time. SEM: standard error mean.

3.2. Effects of Kn, IAA, and 2, 4-D on Embryogenic and Non-Embryogenic Callus Formation

The impacts of Kn and IAA on embryogenic callus production from immature and mature seed cultures were explored. Various concentrations of Kn, either individually or in combination with IAA, were found to be efficient in inducing a nodular mass of embryogenic callus. However, the highest proportion of embryogenic callus production was achieved on the MS basal medium supplemented with 0.1 mg L^{-1} Kn + 0.5 mg L^{-1} IAA, with a percent culture response of 51.0 ± 1.0% and a mean initiation time of 31.0 ± 0.57 days, followed by 0.1 mg L^{-1} Kn alone (35.53 ± 2.23). At concentrations ranging between 0.1 and 1.5 mg L^{-1} Kn and IAA, the proportion of somatic embryo formation is high. Raising the concentrations further dramatically reduces somatic embryo development and promotes callus induction. The MS media enriched with 2,4-D at concentrations 0.1 to 0.5 mg L^{-1} produced only non-embryogenic calluses (Figure 4A). Establishment of embryogenic callus induction was examined in this study (Table 2, Figure 2A–L). When the immature seeds were incubated on the MS medium augmented with various doses of Kn and IAA (0.1 mg L^{-1} to 0.5 mg L^{-1}), either individually or in combination, two different types of calluses were generated simultaneously. Light green transparent calluses developed from immature and mature seeds after four to six weeks of culture, whereas creamy nodular calluses were mainly produced after six to eight weeks of incubation (Figure 2A,B). During callus proliferation, nodular calluses developed into many globular staged embryos (Figure 2C,D) after eight weeks of culture. Globular stage somatic embryos further transformed into subsequently staged embryos (Figure 2E,H).

Figure 2. Direct somatic embryogenesis and complete plant regeneration from immature and mature seed cultures of *Aconitum violaceum* in the MS medium fortified with various concentrations of PGRs, either separately or in combination: (**A,B**) embryogenic and non-embryogenic callus formation from the immature and mature seed cultures, enriched with 0.1 mg L^{-1} Kn + 0.5 mg L^{-1} IAA and 0.1 mg L^{-1} Kn; (**C,D**) globular-shaped somatic embryo development; (**E–H**) torpedo and subsequent stages of somatic embryo formation; (**I**) direct germination of somatic embryos on the MS medium, augmented with Kn 0.1 + IAA 0.5 mg L^{-1}; (**J**) multiple shoot and root development from the cultured somatic embryos; (**K**) well-developed rooted plantlets before hardening; (**L**) hardening of plantlet in jiffy pots. Scale bar (**A–I**) represents 5 mm; (**J–L**) represents 1 cm.

Table 2. Effects of Kn, IAA, and 2, 4-D on embryogenic and non-embryogenic callus formation from the immature and mature seeds of *Aconitum violaceum*.

MS Medium + PGRs (mg L^{-1})			Callus Initiation Time (Days, Mean ± SEM)	Callus Proliferation Rate	% Culture Response (Mean ± SEM)	Callus Attributes to Various Treatments
Kn	IAA	2,4-D				
-	-	-	0.0 ± 0 a	-	0.0 ± 0 a	No callus formation
0.1	-	-	46.66 ± 0.88 d	High	35.53 ± 2.23 d	Embryogenic; friable, transparent or light greenish, proliferative
0.1	0.3	-	51.66 ± 1.20 e	Moderate	24.0 ± 2.0 b	Embryogenic; friable, creamy or light green, proliferative
0.1	0.5	-	31.0 ± 0.57 b	High	51.0 ± 1.0 e	Embryogenic; friable, healthy creamy or greenish, proliferative
0.5	0.1	-	42.33 ± 1.20 cd	Low	28.88 ± 2.93 bc	Compact, light brown
-	-	0.1	0.0 ± 0 a	-	0.0 ± 0.0 a	No callus formation
-	-	0.5	41.66 ± 3.52 c	Moderate	30.88 ± 2.44 cd	Compact, light brown

Data are expressed as mean value ± SEM. The same letters within each column represent that the data were not considerably different at $p \leq 0.05$ according to DMRT (one-way ANOVA). Each treatment is represented by 10 replicates (one vial is considered one replicate) in three repetitions. Abbreviations: MS: Murashige and Skoog. PGRs: plant growth regulators. Kn: kinetin. IAA: indole-3-acetic acid. 2,4-D: 2,4-dichlorophenoxyacetic acid. SEM: standard error mean.

3.3. Effect of Light on Percentage of Somatic Embryo Formation

The MS medium cultured vials augmented with 0.1 mg L^{-1} Kn and 0.5 mg L^{-1} IAA, incubated in the dark, produced more somatic embryos (77.34 ± 0.05%) than the cultured vials fortified with the same PGRs (0.1 mg L^{-1} Kn and 0.5 mg L^{-1} IAA) incubated in the light (56.21 ± 0.07). Likewise, other concentrations showed similar patterns of somatic embryo formation. Thus, it was demonstrated that darkness had a significant impact on somatic embryo formation in *Aconitum violaceum* (Figure 3). In all trials, cultures that were incubated in the dark for a minimum of 10–25 days may facilitate the induction of somatic embryos.

3.4. Complete Plant Development from Somatic Embryos

To regenerate complete plantlets from somatic embryos, the effects of Kn and IAA on further differentiation of somatic embryos were studied. Somatic embryos of different stages were sub-cultured on the MS basal medium augmented with various concentrations of Kn (0.05–1.0 mg L^{-1}) and IAA (0.1–1.0 mg L^{-1}), either individually or in combination (Figure 2I). However, somatic embryo conversion into plantlets was recorded only in 0.1 mg L^{-1} Kn, 0.1 mg L^{-1} Kn + 0.3 mg L^{-1} IAA, and 0.1 mg L^{-1} Kn + 0.5 mg L^{-1} IAA. The highest percentage of whole plant regeneration from somatic embryos was achieved on Kn 0.1 + IAA 0.5 mg L^{-1} with a percentage culture response of 68.00 ± 1.52%. These observations clearly illustrate that the 0.1 mg L^{-1} Kn + 0.5 mg L^{-1} IAA treatment had a more stimulatory effect than other treatments employed to convert somatic embryos into normal plantlets (Table 3, Figures 2J and 4F,G). The healthy plantlets were transferred into earthen/plastic pots consisting of a mixture of garden soil, peat moss, and vermicompost in a ratio of (1:1:1) and kept in a plant growth chamber (Figures 2L and 4H–J). These findings clearly demonstrate that complete plant regeneration via somatic embryogenesis could be possible.

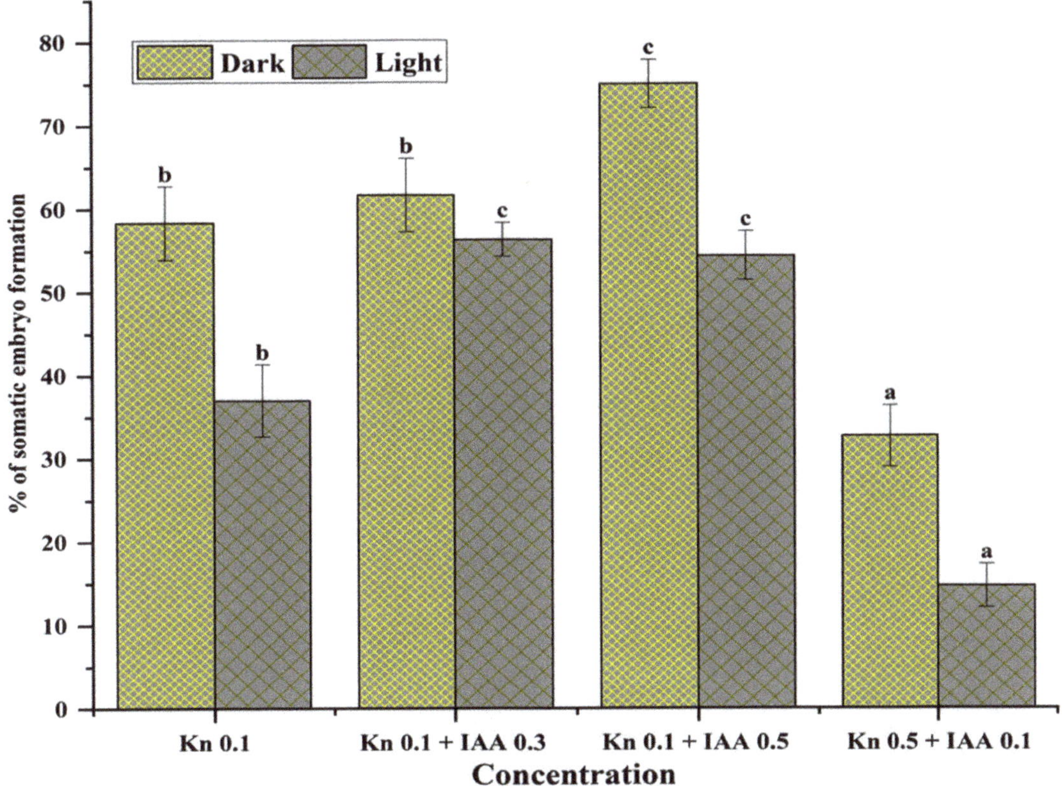

Figure 3. Influence of light conditions and concentrations of Kn and IAA (mg L^{-1}) on somatic embryo formation from the embryogenic callus of *Aconitum violaceum*. Mean value ± standard error mean followed by different letters in each column represents that the observations were substantially different according to DMRT (one-way ANOVA) at $p \leq 0.05$. Abbreviations: Kn: kinetin. IAA: indole-3-acetic acid.

Table 3. Effect of Kn and IAA on whole plant regeneration from somatic embryos and in vitro rooting in *Aconitum violaceum*.

MS Medium + PGRs (mg L^{-1})		Initiation of Shoot (Days, Mean ± SEM)	Initiation of Root Days (Mean ± SEM)	% Culture Response (Mean ±SEM)	Root Length (cm) (Mean ± SEM)	Shoot Length (cm) (Mean ± SEM)	Number of Shoots (Mean ± SEM)
Kn	IAA						
-	-	0.0 ± 0 a	0.0 ± 0 a	0.0 ± 0 a	0.0 ± 0 a	0.0 ± 0 a	0.0 ± 0 a
0.1	-	26 ± 1.0 d	34.66 ± 0.66 d	56.00 ± 2.08 b	4.33 ± 0.35 b	8.16 ± 0.44 b	5.33 ± 0.66 b
0.1	0.3	21.33 ± 0.88 c	28.33 ± 0.88 c	56.33 ± 4.91 b	4.16 ± 0.23 b	9.0 ± 1.0 b	5.66 ± 0.88 b
0.1	0.5	16.66 ± 0.66 b	25. 00 ± 1.15 b	68.00 ± 1.52 c	4.66 ± 0.20 b	10.33 ± 0.6 c	9.33 ± 1.76 c

The data were evaluated for up to 8 weeks. The results were displayed as mean value ± SEM. The same letters within each column represent that data are statistically different at $p \leq 0.05$ according to DMRT (one-way ANOVA). Abbreviations: MS: Murashige and Skoog. PGRs: plant growth regulators. Kn: kinetin. IAA: indole-3-acetic acid.

Figure 4. Callus development and plant regeneration from the seeds of *Aconitum violaceum* in MS basal medium enriched with various concentrations of PGRs, either alone or in combination: (**A**) callus induction in MS medium + 0.5 mg L^{-1} 2,4-D; (**B**) callus induction in MS medium + 0.1 mg L^{-1} Kn; (**C**) multi-shoot formation in MS medium + 0.1 mg L^{-1} Kn; (**D**) root induction when sub-culture in MS medium is augmented with 0.1 mg L^{-1} Kn + 0.5 mg L^{-1} IAA; (**E**) embryogenic callus production and root induction in MS medium fortified with 0.1 mg L^{-1} Kn; (**F**,**G**) multiple shoot and root development from embryogenic callus when sub-cultured in MS medium enriched with 0.1 mg L^{-1} Kn + 0.5 mg L^{-1} IAA; (**H**,**I**) hardened plants in the pots containing loamy soil, coco-peat, and vermicompost (1:1:1 *v/v*); (**J**) plant acclimatized in green house. Scale bar (**A–G**) represents 5 mm; (**H–J**) represents 1 cm.

3.5. In Vitro Rooting

The in vitro developed plantlets, through various methods (i.e., somatic embryogenesis and callusing), developed roots through sub-culturing in the MS basal medium enriched with the same concentrations (concentrations at which somatic embryos were developed) or different concentrations of Kn and IAA (Table 3, Figure 4D–G). The multi-shoot plantlets from sub-culturing in the MS medium, enriched with rooting hormone, developed roots within four weeks of inoculation. In most cases, more than three roots were formed per shoot in each treatment. The highest proportion of root formation occurred on the MS basal medium fortified with 0.1 mg L^{-1} Kn + 0.5 mg L^{-1} IAA and 30 g L^{-1} sucrose, with a percentage response of 68.00 ± 1.52% and average root length 4.66 ± 0.23, followed by 0.1 mg L^{-1} Kn + 0.3 mg L^{-1} IAA with a percentage response of 56.33 ± 4.91% and average root length of 5.66 ± 0.88. Among all the PGRs tested, the Kn and IAA combination was proven to be most effective for healthy root development and formation of maximum root–shoot ratio. The root induction ratio (number of roots formed per shoot) is further enhanced by dim light or complete darkness. Several cultured vials were wrapped in aluminum foil to reduce the intensity of light; these vials developed roots rapidly compared to vials that were kept in light (16/8-h photoperiod). Further root induction in *A. violaceum* was facilitated by a low concentration of PGRs. Increasing the concentration of PGRs promotes further callusing on the developed roots.

3.6. Hardening and Acclimatization of In Vitro Plantlets

The plantlets were then transferred into jiffy pots, filled with a mixture of sterilized garden soil/loamy soil, coco-peat/peat moss, and vermicompost (1:1:1). The plantlets in the jiffy pots were enclosed in transparent polybags for two weeks to ensure adequate humidity and were kept in the growth chamber. After the third week, the hardened plantlets were transferred into the greenhouse for further acclimatization. Healthy and well-established plantlets were then transplanted into the Kashmir University Botanical Garden (KUBG) and kept in the shade with occasional watering. After three to five weeks, well-established plantlets produced three to five leaves (Figure 5B). Transplanted plantlets attained a maximum height of 25.33 ± 1.76 after eight weeks (Table 4; Figure 5D). Overall, 55% of the transplanted plantlets survived and reached the budding stage (Figure 5D).

Figure 5. Hardening and acclimatization of in vitro raised plantlets of *Aconitum violaceum* Jacq. ex Stapf: (**A,B**) three to four week old hardened plantlets; (**C**) five to six week old acclimatized plants; (**D**) eight week old acclimatized plants. Scale bar represents 1 cm.

Table 4. Morphological data of tissue culture-raised plantlets of *Aconitum violaceum* after acclimatization in the field.

Week	Plant Height (Mean ± SEM)	Number of Leaves (Mean ± SEM)	Number of Floral Buds (Mean ± SEM)
1st	2.33 ± 0.33 a	2.66 ± 0.33 a	0.0 ± 0 a
2nd	3.33 ± 0.66 a	3.0 ± 0.57 a	0.0 ± 0 a
3rd	6.0 ± 1.52 ab	4.33 ± 0.33 a	0.0 ± 0 a
4th	9.0 ± 1.15 b	6.0 ± 0.57 b	0.0 ± 0 a
5th	12.66 ± 0.88 c	8.66 ± 0.33 c	0.0 ± 0 a
6th	15.66 ± 0.88 c	9.66 ± 0.66 c	0.0 ± 0 a
7th	20.66 ± 1.45 d	9.66 ± 0.66 c	6.33 ± 0.66 b
8th	25.33 ± 1.76 e	9.66 ± 0.66 c	8.0 ± 0.57 c

Data represented as mean value ± SEM (standard error mean). The same letters in each column represent that the data were not considerably different according to DMRT (one-way ANOVA) at $p \leq 0.05$.

4. Discussion

Aconitum species are primarily harvested on a large scale for their rhizome, which has excellent pharmacological properties and is widely used in indigenous medicines, indirectly resulting in the decline of several *Aconitum* species from their native habitat [31]. Seed-based propagation is the most efficient, affordable, and practicable technique for most species to be cultivated on a large scale for commercial purposes [31]. *Aconitum* species are mostly wild-growing at high elevations, so their propagation through seeds at low altitudes is constrained by ecological factors such as soil fertility, soil textures, pH, humidity, temperature, phytosociology, nature of vegetation, and dormancy of seeds [4,32,33]. Thus, providing pre-sowing treatments (chilling, water soaking, hot water treatment, or with different chemicals etc.) can enhance its cultivation at lower elevations [32]. Soaking seeds in cold water or hot water is often used in seed germination of *Aconitum* spp [33] and *Iris* spp [34]. Cold scarification is the most efficient strategy for breaking dormancy in Ranunculaceae [35–37]. The seeds stratified at 4 °C for 4–10 days enhanced the germination rate in *A. chasmanthum* and *A. nagarum* when cultured on the MS basal medium [22,25]. Similarly, *A. violaceum* seeds also require chilling treatment for successful germination. *A. violaceum* seeds were cultured in different seasons to determine the organogenic response. However, the seeds sown in winter and early spring showed a better germination rate of >77% and gave peculiar organogenic responses (i.e., non-embryogenic and embryogenic callus formation, somatic embryos development, multiple shooting and rooting, and direct seed germination) on the MS basal medium supplemented by different PGRS (auxins and cytokinins), either alone or in combination. Similar results were achieved by Deb and Lunghu with *A. nagarum* [25]. Among the different PGRs used, Kn and IAA were found to be best for direct seed germination, embryogenic callus induction, and multiple shoot regeneration in *A. violaceum*. Seeds stored at lower temperatures (1 °C to 6 °C) showed a better rate of germination during an in vitro culture. Similarly, seeds from *A. heterophyllum* that have chilled for 25–40 days showed a germination rate of 60% [38]. These studies suggest that cold storage and cold-frame sowing or culturing of *A. violaceum* seeds showed a high frequency of germination. Other studies have also suggested that a low temperature is the most significant factor in the germination and in vitro propagation of *Aconitum* spp, which was also verified in the current study. The proposed in vitro regeneration will be helpful in the domestication of understudied plant species at lower altitudes. In its natural habitat, *A. violaceum* regenerates with rhizomes and the new stocks grow healthy and vigorous. The seed also has the potential to germinate into new seedlings; however, the plantlets of germinated seeds are delicate and susceptible to environmental fluctuation and rarely reach maturity stage. Most of the seeds face juvenile mortality at early stages due to extreme diurnal climatic fluctuation in their habitat.

The key rationale is that plants reproduced from direct somatic embryogenesis are commonly more uniform than plants regenerated indirectly by callus tissues [39]. Secondary embryogenesis approaches allow for the rapid production of enormous populations in a short period of time [40]. Secondary somatic embryos could also be developed from the surfaces of somatic embryos [41]. In the current study, somatic embryos were generated from immature and mature seed cultures of *A. violaceum*. Prior to culture on the growth medium, all seeds were stored at a low temperature (1 °C to 6 °C). The seeds cultured in winter and early spring produced a nodular mass of embryogenic potential callus on the MS basal medium enriched with various concentrations of auxins and cytokinins. Similar findings were observed by Vandelook et al. [42] in *Aconitum lycoctonum*, where a low temperature (below 10 °C) was suitable for the growth and germination of embryos. Giri et al. [24] developed complete plantlets from somatic embryos of leaf and petiole explants of *Aconitum heterophyllum* on the MS basal media augmented with 1 mg L^{-1} 2,4-D and 0.5 mg L^{-1} Kn, or 5 mg L^{-1} NAA and 1 mg L^{-1} BAP. Among the various growth regulators studied, Kn and IAA were found to be the most effective for the direct production of an embryogenic callus from immature and mature seed cultures. At lower concentrations, Kn alone was able to induce a nodular mass of embryogenic callus, but the same was insufficient for the conversion of somatic embryos into complete plantlets. However, Kn in combination with IAA converts embryogenic calluses into somatic embryos and facilitates the regeneration of complete plantlets. Similarly, Kn also enhanced embryogenic callus induction in *Drimiopsis kirkii* [43], and in *Iris* species such as *Iris sanguinea* [44,45]. The addition of IAA in the range of 0.1–1.0 mg L^{-1} accelerated the rate of somatic embryo germination, which eventually reached to 68.00 ± 1.52%. The dark conditions promote somatic embryo formation and development of roots in in *A. violaceum*. Similar results were reported by Xu Kd [28] in *Ranunculus sceleratus*, where darkness enhanced the frequency of somatic embryo formation. Cold storage can enhance somatic embryo conversion frequency, potentially caused by epigenetic changes triggered by temperature stress [46]. Moreover, in vitro culturing of seeds on the MS basal medium fortified with various concentrations of 2,4-D induced callusing. Increasing the concentration of 2,4-D to >1.0 mg L^{-1} resulted in a decrease in the percentage of callus induction, which is consistent with the findings of Li et al. [47,48], who found that increasing the concentration of 2,4-D from 11.3 to 18 μM decreased callus induction in *Rosa hybrida*. Lower concentrations of PGRs and a low temperature promotes optimal growth and development of in vitro culture of *A. violaceum*. From the current investigation, we observed that the optimal ambient temperature in the growth chamber should be maintained at 8 °C–10 °C, and relative humidity of 50–55% should be maintained for complete regeneration of plantlets through somatic embryogenesis. Moreover, pre-soaking in cold water for 96 h enhances the rate of seed germination, embryogenic callus production, and other organogenic responses. Optimal application of Kn along with IAA is sufficient for direct seed germination, multiple shoot induction, somatic embryo induction, and complete plant regeneration from somatic embryos in a short time. In the tissue culture-raised plantlets, the mortality rate was highest in the juvenile stage. The acclimatized plantlets required slightly acidic, porous, and loamy soil for successful establishment. The plantlets grew well in semi-shaded places with temperature fluctuations ranging from 10 °C to 20 °C. A further rise in temperature would not be feasible for the survival of *A. violaceum*.

5. Conclusions

The present study is the first to report on the development of an in vitro propagation protocol for seed germination and somatic embryo formation from seeds of the threatened endemic plant species, *A. violaceum*. Furthermore, the study revealed that seeds are suitable explants for efficient multiplication and restoration of *A. violaceum* within a short period of time (approximately three to five months), starting from the initiation of seed germination or somatic embryo development to final tissue culture-raised plantlets. The regeneration protocols established here could be useful for mass multiplication and conservation of this important economic plant species. In addition, this work may be useful in the discovery of physiologically active secondary metabolites from in vitro-derived plantlets under controlled circumstances and their commercial utilization.

Author Contributions: Conceptualization, A.H. and S.S.; methodology, A.H., S.S., S.R. and I.A.N.; software, A.H., S.S., I.A.N., S.R. and N.A.W.; validation, A.H. and S.S.; formal analysis, A.H., S.S. and S.R; investigation, A.H., S.S. and S.R.; resources, E.A.M., D.O.E.-A., H.S., R.C., K.Y. and H.O.E.; data curation, writing—original draft preparation, A.H., S.S., I.A.N. and N.A.W.; writing—review and editing, A.H., S.S., I.A.N. and S.R.; funding acquisition, A.H., S.S., I.A.N., N.A.W., E.A.M., D.O.E.-A., H.S., R.C., K.Y. and H.O.E. All authors have read and agreed to the published version of the manuscript.

Funding: King Saud University (RSP-2021/118).

Institutional Review Board Statement: Not applicable.

Informed Consent Statement: Not applicable.

Data Availability Statement: Not applicable.

Acknowledgments: The first author acknowledges the Council of Scientific and Industrial Research (CSIR), Pusa, New Delhi, for providing financial assistance as SRF during this study. The authors extend their deep appreciation to the Researchers Supporting Project (RSP-2021/118), King Saud University, Riyadh, Saudi Arabia.

Conflicts of Interest: The authors declare no conflict of interest.

References

1. Ved, D.; Saha, D.; Ravikumar, K.; Haridasan, K. *Aconitum violaceum* . In *The IUCN Red List of Threatened Species*; International Union for Conservation of Nature and Natural Resources: Gland, Switzerland, 2015; p. e.T50126562A79581679. [CrossRef]
2. Kala, C.P. Status and conservation of rare and endangered medicinal plants in the Indian Trans-Himalaya. *Biol. Conserv.* **2000**, *93*, 371–379. [CrossRef]
3. Chaudhary, L.B.; Rao, R.R. Notes on the genus *Aconitum* L. (Ranunculaceae) in North West Himalaya (India). *Feddes Repert.* **1998**, *109*, 527–537. [CrossRef]
4. Hadi, A.; Singh, S.; Nawchoo, I.A.; Rafiq, S.; Ali, S. Impacts of Habitat Variability on the Phenotypic Traits of *Aconitum Violaceum* Jacq. Ex Stapf. At different altitudes and environmental conditions in the Ladakh Himalaya, India. *Plant Sci. Today* **2022**, *9*. [CrossRef]
5. CAMP. Threat assessment and management priorities of selected medicinal plants of Western Himalayan states, India. In Proceedings of the Conservation Assessment of Medicinal Plants Workshop, Shimla, Banglore, India, 22–26 May 2003.
6. Anonymous. The wealth of India. Dictionary of Indian raw material and industrial products. In *Raw Materials*; National Institute of Science Communication and Information resources (NISCAIR): New Delhi, India, 1998; Volume I CSIR, p. 253.
7. Sabir, S.; Arshad, M.; Hussain, M.; Sadaf, H.M.; Sohail; Imran, M.; Yasmeen, F.; Saboon; Chaudhari, S.K. A probe into biochemical potential of *Aconitum violaceum*: A medicinal plant from Himalaya. *Asian Pac. J. Trop. Dis.* **2016**, *6*, 502–504. [CrossRef]
8. Hadi, A.; Singh, S.; Ali, S.; Mehdi, M. Traditional uses of medicinal plants by indigenous tribes of Ladakh union territory. *Res. J. Agril. Sci.* **2022**, *13*, 68–77.
9. Pandey, M.R. Use of medicinal plants in traditional Tibetan therapy system in upper Mustang, Nepal. *Our Nat.* **2006**, *4*, 69–82. [CrossRef]
10. Ameri, A. The effects of Aconitum alkaloids on the central nervous system. *Prog. Neurobiol.* **1998**, *56*, 211–235. [CrossRef]
11. Bhattarai, S.; Chaudhary, R.P.; Quave, C.L.; Taylor, R.S.L. The use of medicinal plants in the trans-Himalayan arid zone of Mustang district, Nepal. *J. Ethnobiol. Ethnomed.* **2010**, *6*, 14. [CrossRef]
12. Lone, P.A.; Bhardwaj, A.K.; Shah, K.W.; Tabasum, S. Ethnobotanical survey of some threatened medicinal plants of Kashmir Himalaya, India. *J. Med. Plant Res.* **2014**, *8*, 1362–1373.
13. Gairola, S.; Sharma, J.; Bedi, Y.S. A cross-cultural analysis of Jammu, Kashmir and Ladakh (India) medicinal plant use. *J. Ethnopharmacol.* **2014**, *155*, 925–986. [CrossRef]
14. Khan, F.A.; Khan, S.; Khan, N.M.; Khan, H.; Khan, S. Antimicrobial and antioxidant role of the aerial parts of *Aconitum violaceum*. *J. Mex. Chem. Soc.* **2021**, *65*, 84–93. [CrossRef]
15. Yadav, S.; Verma, D.L. Acylated flavonol glycosides from the flowers of *Aconitum violaceum* Staph. *Nat. Sci.* **2010**, *8*, 239–243.
16. Miana, G.A.; Ikram, M.; Khan, M.I.; Sultana, F. Alkaloids of *Aconitum violaceum*. *Phytochemistry* **1971**, *10*, 3320–3322. [CrossRef]
17. Kunwar, R.M.; Shrestha, K.P.; Bussmann, R.W. Traditional herbal medicine in Far-west Nepal: A pharmacological appraisal. *J. Ethnobiol. Ethnomed.* **2010**, *6*, 35. [CrossRef] [PubMed]
18. Uprety, Y.; Asselin, H.; Boon, E.K.; Yadav, U.; Shrestha, K.K. Indigenous use and bio-efficacy of medicinal plants in the Rasuwa District, Central Nepal. *J. Ethnobiol. Ethnomed.* **2010**, *6*, 3. [CrossRef]
19. Dall'Acqua, S.; Shrestha, B.B.; Gewali, M.B.; Jha, P.K.; Carrara, M.; Innocenti, G. Diterpenoid alkaloids and phenol glycosides from *Aconitum naviculare* (Bruhl) Stapf. *Nat. Prod. Commun.* **2008**, *3*, 1985–1989. [CrossRef]
20. Gao, L.M.; Yan, L.Y.; He, H.Y.; Wei, X.M. Norditerpenoid alkaloids from *Aconitum spicatum* Stapf. *J. Integr. Plant Biol.* **2006**, *48*, 364–369. [CrossRef]
21. Mir, A.H.; Tyub, S.; Kamili, A.N. Ecology, distribution mapping and conservation implications of four critically endangered endemic plants of Kashmir Himalaya. *Saudi J. Biol. Sci.* **2020**, *27*, 2380–2389. [CrossRef]
22. Rafiq, S.; Wagay, N.A.; Bhat, I.A.; Kaloo, Z.A.; Rashid, S.; Lin, F.; El-Abedin, T.K.Z.; Wani, S.H.; Mahmoud, E.A.; Almutairi, K.F.; et al. In vitro Propagation of *Aconitum chasmanthum* Stapf Ex Holmes: An Endemic and Critically Endangered Plant Species of the Western Himalaya. *Horticulturae* **2021**, *7*, 586. [CrossRef]
23. Singh, M.; Chettri, A.; Pandey, A.; Sinha, S. In vitro propagation and phytochemical assessment of *Aconitum ferox* Wall: A threatened medicinal plant of Sikkim Himalaya. *Proc. Natl. Acad. Sci. India Sect. B Biol. Sci.* **2019**, *90*, 313–321. [CrossRef]
24. Giri, A.; Ahuja, P.S.; Ajay Kumar, P.V. Somatic embryogenesis and plant regeneration from callus cultures of *Aconitum heterophyllum* Wall. *Plant Cell Tissue Organ Cult.* **1993**, *32*, 213–218. [CrossRef]
25. Deb, C.R.; Langhu, T. Development of in vitro propagation protocol of *Aconitum nagarum* Stapf. *Plant Cell Biotechnol. Mol. Biol.* **2017**, *18*, 324–332.

26. Mou, Z.; Ye, F.; Shen, F.; Zhao, D. In vitro Germination and Micropropagation of *Aconitum vilmorinianum*: An Important Medicinal Plant in China. *Python-Int. J. Exp. Bot.* **2022**, 1–18. [CrossRef]

27. Hatano, K.; Shoyama, Y.; Nishioka, I. Somatic embryogenesis and plant regeneration from the anther of *Aconitum carmichaeli* Debx. *Plant Cell Rep.* **1987**, *6*, 446–448. [CrossRef]

28. Xu, K.; Wang, W.; Yu, D.; Li, X.-L.; Chetn, J.M.; Feng, B.J.; Zhao, Y.W.; Cheng, M.J.; Liu, X.X.; Li, C.W. NAA at a high concentration promotes efficient plant regeneration via direct somatic embryogenesis and SE-mediated transformation system in *Ranunculus sceleratus*. *Sci. Rep.* **2020**, *9*, 18321. [CrossRef]

29. Rawat, J.M.; Agnihotri, R.K.; Nautiyal, S.; Rawat, B.; Chandra, A. In vitro propagation, genetic and secondary metabolite analysis of *Aconitum violaceum* Jacq.: A threatened medicinal herb. *Acta Physiol. Plant.* **2013**, *35*, 2589–2599. [CrossRef]

30. Darrudi, R.; Hassandokht, M.R.; Nazeri, V. Effects of KNO3 and CaCl2 on seed germination of *Rheum khorasanicum* B. Baradaran & A. Jafari. *J. Appl. Sci. Res.* **2014**, *10*, 171–175.

31. Nidhi, S.; Vikas, S.; Barkha, K.; Dobriyal, A.K.; Jadon, V.S. Advancement in research on *Aconitum* sp. (Ranunculaceae) under different area: A review. *Biotechnology* **2010**, *9*, 411–427. [CrossRef]

32. Sharma, R.K.; Sharma, S.; Sharma, S.S. Seed germination behaviour of some medicinal plants of Lahaul and Spiti cold desert (Himachal Pradesh): Implications for conservation and cultivation. *Cur. Sci.* **2006**, *90*, 1113–1118.

33. Srivastava, N.; Sharma, V.; Kamal, B.; Jadon, V.S. *Aconitum*: Need for sustainable exploitation (with special reference to Uttarakhand). *Int. J. Green Pharm.* **2010**, *4*, 220–228. [CrossRef]

34. Li, C.; Zhao, X.; Dong, K.H.; Yang, J.F. Effects of diferent treatment methods on seed germination of *Iris lactea* Pall. *Anim. Husb. Feed. Sci.* **2013**, *34*, 23. [CrossRef]

35. Baskin, C.C.; Baskin, J.M. Deep complex morphophysiological dormancy in seeds of the mesic woodland herb Delphinium tricorne (Ranunculaceae). *Int. J. Plant Sci.* **1994**, *155*, 738–743. [CrossRef]

36. Walck, J.L.; Baskin, C.C.; Baskin, J.M. Seeds of *Thalictrum mirabile* (Ranunculaceae) require cold stratification for loss of non- deep simple morphophysiological dormancy. *Can. J. Bot.* **1999**, *77*, 1769–1776. [CrossRef]

37. Forbis, T.A.; Floyd, S.K.; DeQueiroz, A. The evolution of embryo size in angiosperms and other seed plants: Implications for the evolution of seed dormancy. *Evolution* **2002**, *56*, 2112–2125. [CrossRef]

38. Paramanick, D.; Panday, R.; Shukla, S.S.; Sharma, V. Primary pharmacological and other important findings on the medicinal plant "*Aconitum heterophyllum*" (aruna). *J. Pharmacopunct.* **2017**, *20*, 89. [CrossRef]

39. Maheswaran, G.; Williams, E.G. Direct somatic embryoid formation in immature embryos of *Trifolium repens, T. pretense* and *Medicago sativa*, and rapid clonal propagation of *T. repens*. *Ann. Bot.* **1984**, *54*, 201–211. [CrossRef]

40. Joseph, R.; Yeoh, H.H.; Loh, C.S. Induced mutations in cassava using somatic embryos and the identification of mutant plants with altered starch yield and composition. *Plant Cell Rep.* **2004**, *23*, 91–98. [CrossRef]

41. Chen, J.L.; Beversdorf, W.D.A. Combined use of microprojectile bombardment and DNA imbibition enhances transformation frequency of canola (*Brassica napus* L.). *Theoret. Appl. Genet.* **1994**, *88*, 187–192. [CrossRef]

42. Vandelook, F.; Lenaerts, J.; Jozef, A.V.A. The role of temperature in post dispersal embryo growth and dormancy break in seed of *Aconitum lycoctomum* L. *Flora-Morphol. Distrib. Funct. Ecol. Plants* **2009**, *204*, 536–542. [CrossRef]

43. Lan, T.H.; Hong, P.I.; Huang, C.C.; Chang, W.C.; Lin, C.S. High-frequency direct somatic embryogenesis from leaf tissues of *Drimiopsis kirkii* Baker (giant squill). *Vitr. Cell. Dev. Biol. Plant* **2009**, *45*, 44–47. [CrossRef]

44. Jevremovic, S.; Jeknic, Z.; Subotic, A. Micropropagation of *Iris* sp. *Methods Mol. Biol.* **2013**, *11013*, 291–303. [CrossRef] [PubMed]

45. Boltenkov, E.V.; Zarembo, E.V. In vitro regeneration and callogenesis in tissue culture of foral organs of the genus *Iris* (Iridaceae). *Biol. Bull.* **2005**, *32*, 138. [CrossRef]

46. Rhie, Y.H.; Lee, S.Y. Seed dormancy and germination of *Epimedium koreanum* Nakai. *Sci. Hortic.* **2020**, *272*, 109600. [CrossRef]

47. Montalban, I.; Garcia-Mendiguren, O.; Goicoa, T.; Ugarte, M.; Moncalean, P. Cold storage of initial plant material afects positively somatic embryogenesis in *Pinus radiata*. *New For.* **2015**, *46*, 309–317. [CrossRef]

48. Li, X.; Krasnyanski, S.; Korban, S.S. Somatic embryogenesis, secondary somatic embryogenesis, and shoot organogenesis in Rosa. *J. Plant Physiol.* **2002**, *159*, 313–319. [CrossRef]

 *horticulturae*

Article

Responses of Potato (*Solanum tuberosum* L.) Breeding Lines to Osmotic Stress Induced in In Vitro Shoot Culture

Alexandra Hanász [1,*], Judit Dobránszki [2], Nóra Mendler-Drienyovszki [3], László Zsombik [3] and Katalin Magyar-Tábori [3]

1   Kerpely Kálmán Doctoral School of Crop Production and Horticultural Sciences, University of Debrecen, Böszörményi St. 138, 4032 Debrecen, Hungary
2   Centre for Agricultural Genomics and Biotechnology, Faculty of the Agricultural and Food Science and Environmental Management, University of Debrecen, P.O. Box 12, 4400 Nyiregyhaza, Hungary; j.dobranszki@gmail.com
3   Research Institute of Nyíregyháza, Institutes for Agricultural Research and Educational Farm (IAREF), University of Debrecen, P.O. Box 12, 4400 Nyiregyhaza, Hungary; mendlernedn@gmail.com (N.M.-D.); zsombik@agr.unideb.hu (L.Z.); mtaborik@gmail.com (K.M.-T.)
*   Correspondence: hanasz.alexandra@agr.unideb.hu

Abstract: In vitro experiments were conducted to study the responses of potato (*Solanum tuberosum* L.) genotypes to osmotic stress. In vitro shoot cultures of 27 breeding lines and their drought-tolerant parents (referent lines: C103 and C107) were tested under osmotic stress induced by addition of PEG 6000 (Mw = 6000; 5.0, 7.5, 10.0%, w/v), D-mannitol (0.1, 0.2, 0.3 M) and PEG 600 (Mw = 600; 2.5, 5.0, 7.5%, w/v) to the Murashige-Skoog medium. Stress index (SI) was calculated from shoot length (SL) and root length (RL), root numbers (RN) and the rate of surviving shoots (SR) ($SI_{SL;RL;RN;SR}$ = Parameter$_{SL;RL;RN;SR}$ of treated shoots/Parameter$_{SL;RL;RN;SR}$ of control shoots × 100) to compare genotypes. In the average of each breeding line and concentration, the osmotic agents resulted in SI values of 40.1, 60.8, 82.6 and 76.0 for SI_{SL}, SI_{RL}, SI_{RN} and SI_{SR}, respectively. In general, all SI values of C103 and $SI_{RL,RN}$ of C107 were significantly higher than those of the breeding lines. Nine breeding lines were found to be promising based on their final ranking. According to the results, 7.5% and 10% PEG 6000 or 0.2 M and 0.3 M D-mannitol treatments proved to be suitable for the selection of osmotic stress-tolerant genotypes.

Keywords: polyethylene glycol; D-mannitol; tissue culture; stress index; morphological traits

Citation: Hanász, A.; Dobránszki, J.; Mendler-Drienyovszki, N.; Zsombik, L.; Magyar-Tábori, K. Responses of Potato (*Solanum tuberosum* L.) Breeding Lines to Osmotic Stress Induced in In Vitro Shoot Culture. *Horticulturae* 2022, 8, 591. https://doi.org/10.3390/horticulturae8070591

Academic Editor: Sergio Ruffo Roberto

Received: 26 May 2022
Accepted: 29 June 2022
Published: 1 July 2022

Publisher's Note: MDPI stays neutral with regard to jurisdictional claims in published maps and institutional affiliations.

1. Introduction

One of the most important abiotic stress factors is drought. Drought periods during the growing season are becoming more common due to climate change and global warming [1]. Stress caused by water deficit is even greater in areas free of permanent precipitation or where an adequate irrigation system is not available [2].

Water deficit affects almost all plant growth and developmental processes [3], can induce several morphological, physiological and biochemical changes [4], and can result in large yield losses of up to 50–70% in crop production [5].

Improving the drought tolerance and water use efficiency of varieties by using breeding methods can play an important role in reducing drought-related crop losses and thus contribute to a secure food supply for the growing population. This means that demand for drought-tolerant plant species and especially cultivars that are able to adapt to drought conditions is constantly increasing [6,7].

Potato (*Solanum tuberosum* L.) is one of the most important crops worldwide, because of its high productivity and nutritional values [8]. However, the potato crop often suffers from stress because it is sensitive to both drought [9,10] and high temperature [11]. Even though potato has difficulty tolerating water shortages due to its shallow roots [12,13], it basically manages water well.

By changes in physiology and morphology, plants are able to respond adaptively to altered environmental conditions, preceded and accompanied by cellular and molecular changes [14], so many processes are involved in the development of drought tolerance. This fact suggests that there may be large differences between species and cultivars in drought stress responses, and even variability has been found in the susceptibility of potato cultivars [15,16]. Several experiments were performed to identify traits that could be related to drought stress and to select genotypes with the desired characters in several crops [4,17] including potatoes [18–20]. Some of these traits can also be detected at the cellular and/or tissue level, such as the adjustment of osmotic pressure, which can be tested/modelled under laboratory conditions [21] or even in vitro [22,23]. Micropropagation is a commonly used process to produce virus- and disease-free plant propagating material [24], and in vitro conditions provide good opportunities to study different physiological processes and interactions and to select appropriate genotypes [25].

Osmotically active compounds such as the inert and non-penetrating polyethylene glycol (PEG) with various molecular sizes or the sugar alcohols such as D-mannitol and sorbitol are widely used to induce osmotic stress in plants [25]. Although sugars and their derivatives can significantly increase the osmotic pressure of the medium used, it must also be taken into account that they can be taken up and metabolized by plants. In addition, a high sugar content (8%) can induce tuber development in potatoes [26].

PEG is a neutral polymer, and its high molecular weight makes it suitable to imitate water deficit when added to the medium, because it cannot pass through the cell wall of the plant [27–30]. D-mannitol can be produced and metabolized by certain plant species [31]; however, it has been used efficiently in many osmotic stress tolerance experiments [25].

Numerous studies on osmotic stress tolerance of potato have been conducted in recent years. Gopal and Iwama [6] studied the effect of PEG and sorbitol on the morphological development of potato in vitro shoot cultures, and they found that in vitro tests gave results similar to those obtained under field conditions.

In addition to shoot culture, callus culture was also suitable for osmotic stress tolerance testing of genotypes, and results were well related to their field performance during drought [32]. The in vitro model provided a good basis for assessing the drought tolerance of genotypes also in other experiments [33,34]. Changes in morphological and physiological traits affected by osmotic stress are similar under in vivo conditions to those observed in in vitro tissue cultures. Survival rate, shoot length, fresh and dry weight, and number and length of roots are the most commonly affected characteristics [35–38]. Despite the fact that the growth parameters of the plantlets were strongly influenced by osmotic stress during the stress resistance experiments—in fact, they were most often inhibited—it was possible to differentiate the genotypes according to their stress tolerance only if a stress index was formed based on the measured growth parameters [32,39,40].

Although a stress index based on morpho-physiological characters alone could be used to distinguish potato genotypes [32,35,41–43], identification of quantitative trait loci (QTLs) for morphological characters [1,42] and using molecular and biochemical markers as a tool for drought tolerance study are becoming more and more widespread [1]. Our experiments were focused on the investigation of the osmotic stress tolerance of 27 potato breeding lines under in vitro conditions to reveal whether there were significant differences in SI counted from simple morphological parameters compared to the drought-tolerant referent lines. In addition, different types of osmotic agents were applied at three levels to find the effective method(s) to distinguish our breeding lines according to their osmotic stress tolerance and to select varieties that should be included in further studies. Targeted trait-specific selection started under in vitro conditions can significantly reduce the time required for development of a new variety, so the results of laboratory tests prior to the field experiment may provide a good basis for further stages of plant breeding.

2. Materials and Methods

2.1. Location of the Experiments and Plant Material

The study was conducted at the Centre for Agricultural Genomics and Biotechnology, Faculty of the Agricultural and Food Science and Environmental Management, University of Debrecen, Hungary. This study was carried out on 29 potato (*Solanum tuberosum* L.) genotypes including 27 breeding lines (C2, C3, C4, C5, C6, C8, C9, C10, C11, C12, C14, C17, C19, C20, C21, C22, C26, C28, C30, C32, C35, C37, C41, C42, C57, C58 and C63) and two drought-tolerant referent lines (C103 and C107). Selection of referent genotypes was based on their response to drought conditions in previous in vivo (field and greenhouse) experiments, in which they were proven to be drought-tolerant. C103 parental line belongs to the mid-early maturity group; its tuber is round oval, its skin is red, and the tuber flesh is yellow. C107 parental line is a late genotype with oblong, purple pink skinned tubers and white flesh. They are tolerant to late blight (*Phytophthora infestans*). All of the breeding lines tested originated from crossings of referent lines. In vitro culture establishment of breeding lines was initiated from seeds; they were surface sterilized by 3% NaOCl for 4 min, then they were washed with sterile distilled water three times. The seeds were placed onto medium containing Murashige and Skoog [44] (MS) salts and vitamins, 3% sucrose (VWR Chemicals, Radnor, PA, USA) and 0.7% agar-agar (Sigma-Aldrich, A1296, St. Louis, MO, USA) and incubated in dark at 24 °C for 5 days. Then, they were transferred to a growing room (at 22/15 ± 2 °C day/night temperature and 16 h daily illumination by 65 μmol m^{-2} s^{-1} PPF). In vitro shoot cultures of both referent lines and breeding lines were subcultured every 4 weeks on the same MS medium mentioned above.

2.2. Explant, Treatments and Culture Conditions

Single nodal cuttings, each containing an axillary bud from 4-week-old in vitro shoot cultures, were used as explant, and the shoot apex and basal part of the shoots were discarded. Explants were placed onto the basal MS medium supplemented with 3% sucrose and 0.7% agar and with polyethylene glycol (Alfa Aesar, Haverhill, MA, USA) with different molecular weights either in concentrations of 2.5, 5.0 and 7.5% or 5.0, 7.5 and 10% for PEG 600 and PEG 6000, respectively. Moreover, 0.1, 0.2 and 0.3 M D-mannitol (Merck, Burlington, VT, USA) were also applied in order to induce the osmotic stress, while the control medium was free of osmotic agent. Twenty explants per jar (450 mL, cylindrical shaped) were placed onto 50 mL of medium. Treatments consisted of five repetitions (five jars); thus, a total of 100 plantlets were observed. Experiments were repeated twice. Shoot cultures were grown under controlled conditions, in a culture room at 22/15 ± 2 °C day/night temperature and 16 h daily illumination by 65 μmol m^{-2} s^{-1} PPF for 4 weeks. At the end of experiments, the rate of survival (SR) was observed and the shoot length (SL) and the number and length of roots (RN and RL, respectively) on surviving explants were measured. Before analysis, results were expressed as percentages of the results obtained on the medium without stress agents (SI, stress index; [45]), to compare the responses to the different levels of osmotic stress in the breeding lines and referent lines.

$$SI_{SL} = \text{Shoot length of treated shoots (mm)/shoot length of control shoots (mm)} \times 100$$

$$SI_{RL} = \text{The longest root length of treated shoots (mm)/the longest root length of control shoots (mm)} \times 100$$

$$SI_{RN} = \text{Root number of treated shoots/root number of control shoots} \times 100$$

$$SI_{SR} = \text{Survival rate of treated shoots (per jar)/survival rate of control shoots (per jar)} \times 100$$

2.3. Statistical Analysis

SI calculated from morphological data (shoot and root parameters) were analyzed statistically by ANOVA followed by LSD and Tukey-B tests, using SPSS Statistics 27.0 (IBM, New York, NY, USA).

3. Results

Osmotic stress tolerance of 27 breeding lines and 2 drought-tolerant referent genotypes and their responses were compared to each other after SI values were calculated. We found that all morpho-physiological parameters were affected by treatments (Table 1), and differences were found between genotypes. Significant interactions between genotypes and type and concentration of osmotic agent were also detected.

Table 1. The main effect of osmotic agents calculated from SI values for morpho-physiological traits on the average of all potato breeding lines at different osmotic stress levels.

Traits	Level	D-Mannitol	PEG 600	PEG 6000	Mean
	1	68.9	55.7	45.1	
SL	2	39.1	32.3	39.3	
	3	23.7	18.3	38.2	
Mean		**43.9**	**35.4**	**40.9**	**40.1**
	1	90.5	76.4	61.5	
RL	2	70.0	47.5	64.9	
	3	48.3	30.1	58.0	
Mean		**69.6**	**51.3**	**61.5**	**60.8**
	1	93.9	80.8	96.6	
RN	2	88.3	62.2	99.4	
	3	75.8	41.5	105.2	
Mean		**86.0**	**61.5**	**100.4**	**82.6**
	1	98.5	85.9	87.8	
SR	2	92.5	56.1	84.8	
	3	78.5	21.7	78.4	
Mean		**89.8**	**54.6**	**83.7**	**76.0**
Total mean		<u>72.3</u>	<u>50.7</u>	<u>71.6</u>	<u>64.9</u>

Levels for D-mannitol: 1: 0.1 M, 2: 0.2 M, 3: 0.3 M; for PEG 600: 1: 2.5%, 2: 5%, 3: 7.5%; for PEG 6000: 1: 5%, 2: 7.5%, 3: 10%.

3.1. Effect of Osmotic Stress Induced by PEG 6000 on In Vitro Shoot Cultures

3.1.1. Changes in the SI of Survival Rate (SI_{SR})

SI values of survival rates (SI_{SR}) were significantly decreased or most frequently not affected by 5% PEG 6000 (Table 2, Figure S1a). Some breeding lines showed significantly lower SI_{SR} values than the referent lines. The raised level (7.5%) of PEG 6000 decreased significantly the SI_{SR} values very rarely, while other responses were similar to those observed at the level of 5% PEG (Table 2, Figure S1b). Increasing the PEG 6000 level to 10% resulted in significantly decreased SI_{SR} values for seven lines (C2, C17, C20, C28, C41, C42 and C57) compared to those observed at 7.5% (Table 2, Figure S1c). Although the C103 referent line showed the best SI_{SR} results along with two breeding lines (C8, C30), very similar results were found in eight breeding lines. However, the SI_{SR} value of the C107 referent line was significantly reduced, resulting in a number of breeding lines ahead of it.

Table 2. Survival rate SI (SI_{SR}) values of potato genotypes under osmotic stress induced by PEG 6000 added to the medium at levels of 5%, 7.5% and 10%.

Breeding Line	SI_{SR} (%)								
	5% PEG 6000			7.5% PEG 6000			10% PEG 6000		
C2	100	a A		99	a A		92	a–c B	▲
C3	87	a–d A		75	a–c A	▲□	75	a–e A	□
C4	97	ab A		75	a–c AB	▲□	60	d–f B	□
C5	93	a–c A		92	a A		83	a–e A	□
C6	63	e A	▲□	60	b–d A	▲□	74	a–e A	□
C8	100	a A		100	a A		100	a A	▲
C9	41	f A	▲□	48	d A	▲□	28	gh A	▲□
C10	95	a–c A		97	a A		92	a–c A	▲
C11	100	a A		99	a A		96	a A	▲
C12	99	a A		100	a A		96	a A	▲
C14	87	a–d A		100	a A		96	a A	▲
C17	64	e A	▲□	59	b–d A	▲□	17	h B	▲□
C19	72	de A	▲□	80	ab A	▲□	89	a–d A	
C20	100	a A		100	a A		93	ab B	▲
C21	84	a–d B	□	87	a B		98	a A	▲
C22	76	c–e B	▲□	99	a A		93	ab A	▲
C26	92	a–c A		92	a A		88	a–d A	
C28	90	a–d A		84	ab A	□	64	c–f B	□
C30	100	a A		100	a A		100	a A	▲
C32	98	ab A		96	a A		89	a–d A	
C35	89	a–d A		77	a–c A	▲□	86	a–e A	
C37	73	de A	▲□	50	d B	▲□	64	b–f B	□
C41	94	a–c A		79	ab A	▲□	44	fg B	▲□
C42	94	a–c A		96	a A		57	ef B	▲□
C57	98	ab A		84	ab A	□	62	c–f B	□
C58	79	b–e A	▲□	83	ab A	□	89	a–d A	
C63	98	ab A		54	cd B	▲□	80	a–e A	□
C103	95	a–c B		100	a A		100	a A	
C107	93	a–c A		96	a A		74	a–e A	

Symbols mark significant differences from the C107 referent line (▲), or from the C103 referent line (□) according to LSD test; the small letters indicate significantly different means between the breeding lines within a treatment, and the capital letters indicate significantly different means between treatment levels within a breeding line according to Tukey-B test.

3.1.2. Changes in the SI of Shoot Length (SI_{SL})

Each breeding line and both referent lines responded with decreased shoot length to the presence of PEG 6000 at all concentrations (Table 3, Figure S2a–c). In general, inhibition of shoot growth increased with increasing concentration of PEG 6000 to 7.5%, but higher concentrations did not result in further significant decrease. The best result (SI_{SL} 76.8) was obtained in C9 breeding line, although SI_{SL} values of some breeding lines were very similar (C2, C22, C30 and C41) (Table 3, Figure S2a). At raised PEG 6000 concentration (7.5%), five breeding lines (C2, C9, C12, C30 and C32) showed significantly higher SI_{SL} values compared to both referent lines (Table 3, Figure S2b). At the highest level of PEG 6000 (10%), only one breeding line (C22) reached significantly higher SI_{SL} values compared to both referent lines, but most of the breeding lines performed better than the C107 referent line (Table 3, Figure S2c).

Table 3. Shoot length SI values (SI_{SL}) of potato genotypes under osmotic stress induced by PEG 6000 added to the medium at levels of 5.0%, 7.5% and 10.0%.

Breeding Line	SI_{SL} (%)								
	5% PEG 6000			7.5% PEG 6000			10% PEG 6000		
C2	65.9	a–c A	▲□	62.4	a A	▲□	56.7	ab A	▲
C3	35.2	h–l B	▲	32.3	e–g B	□	43.5	c–f A	▲□
C4	34.5	h–l A	▲	33.9	e–g A	□	32.7	g–m A	▲□
C5	24.9	lm A	□	26.9	fg A	□	30.3	j–o A	▲□
C6	32.0	i–l B		41.5	c–e A	▲	42.0	d–g A	▲□
C8	62.2	b–d A	▲□	22.8	g B	▲□	35.4	e–l C	▲□
C9	76.8	a A	▲□	64.1	a A	▲□	65.1	a A	▲□
C10	40.9	f–j A	▲	31.6	e–g B	□	28.3	k–o B	□
C11	41.4	f–i A	▲	36.5	d–f B	□	41.8	d–h A	▲□
C12	52.3	d–f A	▲□	49.2	bc A	▲□	57.8	ab A	▲
C14	34.7	h–l B	▲	40.3	c–e A	▲	34.2	f–m B	▲□
C17	28.6	j–m A	□	25.6	fg AB	□	21.9	no B	□
C19	46.1	f–h A	▲□	46.9	b–d A	▲	30.5	j–o B	▲□
C20	19.6	m C	□	31.4	e–g B	□	37.2	d–j A	▲□
C21	42.1	f–i A	▲	31.0	e–g B	□	33.0	f–m B	▲□
C22	73.7	ab A	▲□	46.3	b–d B	▲	44.8	c–e B	▲□
C26	49.5	e–g A	▲□	47.9	bc A	▲	31.1	i–n B	▲□
C28	40.5	f–j A	▲	27.3	fg B	□	25.1	l–o B	□
C30	67.8	a–c A	▲□	62.5	a A	▲□	41.5	d–i B	▲□
C32	45.0	f–h A	▲□	54.2	ab A	▲□	44.9	c–e A	▲□
C35	60.2	c–e A	▲□	32.8	e–g C	□	47.5	cd B	▲
C37	38.9	g–j A	▲	34.0	e–g B	□	31.3	h–n B	▲□
C41	68.0	a–c A	▲□	48.1	bc B	▲	40.4	d–j B	▲□
C42	46.9	f–h A	▲□	40.1	c–e B	▲	34.5	e–m B	▲□
C57	46.9	f–h A	▲□	40.9	c–e A	▲	20.7	o B	□
C58	38.1	g–k A	▲	34.8	ef A	□	37.0	d–k A	▲□
C63	52.3	d–f A	▲□	26.0	fg C	□	37.0	d–k B	▲□
C103	37.5	g–k B		42.6	c–e B		52.3	ab A	
C107	26.1	lm B		31.5	e–g A		24.4	l–o B	

Symbols mark significant differences from the C107 referent line (▲), or from the C103 referent line (□) according to LSD test; the small letters indicate significantly different means between the breeding lines within a treatment, and the capital letters indicate significantly different means between treatment levels within a breeding line according to Tukey-B test.

3.1.3. Changes in the SI of Root Length (SI_{RL})

Application of 5% PEG 6000 to medium resulted in decreased root length in each breeding line and in referent lines (Table 4, Figure S3a). SI_{RL} values of 10 breeding lines were significantly higher than those of the referent lines. At the level of 7.5% PEG 6000, increased SI_{RL} values were found in some breeding lines, and the SI_{RL} values of four breeding lines (C10, C14, C22 and C30) were significantly higher than those of both referent lines (Table 4, Figure S3b). When 10% PEG 6000 was added to medium, the SI_{RL} value only of C22 was higher than that of referent line C103, but the SI_{RL} value of C107 was lower than that of the majority of breeding lines (Table 4, Figure S3c).

Table 4. Root length SI (SI_{RL}) values of potato genotypes under osmotic stress induced by PEG 6000 added to the medium at levels of 5.0%, 7.5% and 10%.

Breeding Line	SI_{RL} (%)								
	5% PEG 6000			7.5% PEG 6000			10% PEG 6000		
C2	50.7	f–j B		69.5	c–h A		46.8	e–i B	□
C3	56.6	d–j B		52.4	h–n B	▲□	63.6	b–e A	▲□
C4	51.1	f–j A		53.7	g–m A	▲□	51.3	d–h A	□
C5	69.4	c–f B	▲□	84.2	cd A	▲	85.1	ab A	▲
C6	48.1	h–j B		71.4	c–h A		51.8	d–h B	□
C8	83.8	a–c A	▲□	58.4	f–l B	□	72.3	b–d C	▲□
C9	87.7	ab A	▲□	62.8	e–j B		63.2	b–e B	□
C10	88.2	ab A	▲□	87.9	bc A	▲□	57.7	c–h B	□
C11	47.4	ij A	□	33.7	no B	▲□	48.0	e–i A	□
C12	47.2	ij A	□	76.7	c–f B		58.8	c–f C	▲□
C14	88.1	ab A	▲□	111.8	a B	▲□	57.0	c–g C	□
C17	44.0	f A	□	25.2	o B	▲□	17.3	k B	▲□
C19	67.5	c–h A	▲□	54.8	g–l B	▲□	39.2	f–k C	□
C20	66.6	c–h A	▲	45.0	j–o B	▲□	44.8	e–j B	□
C21	57.3	d–j A		65.4	d–i A		51.4	d–h B	□
C22	74.7	a–d B	▲□	90.0	bc A	▲□	95.9	a A	▲□
C26	49.8	g–j A	□	51.1	i–n A	▲□	24.1	jk B	▲□
C28	54.2	e–j A		41.0 B	l–o	▲□	34.5	g–k B	□
C30	91.0	a B	▲□	106.3	ab A	▲□	81.5	ab B	▲
C32	45.9	ij A	□	53.8	g–m A	▲□	36.9	f–k B	□
C35	89.0	ab A	▲□	61.5	e–k C	□	75.6	a–c B	▲
C37	55.6	e–j A		39.8	m–o B	▲□	50.3	d–h A	□
C41	52.1	f–j A		37.5	m–o B	▲□	25.7	i–k B	▲□
C42	63.6	d–i A		72.2	c–h A		32.6	h–k B	▲□
C57	45.8	d–f A		55.8	f–l A	□	20.9	k B	▲□
C58	63.2	d–j B		79.5	c–e B	▲	82.4	ab A	▲
C63	71.2	b–e A	▲□	43.0	ko B	▲□	53.7	c–h B	□
C103	56.5	d–j B		73.6	c–g B		84.7	ab A	
C107	55.4	d–j B		67.1	d–i A		46.3	e–i B	

Symbols mark significant differences from the C107 referent line (▲), or from the C103 referent line (□) according to LSD test; the small letters indicate significantly different means between the breeding lines within a treatment, and the capital letters indicate significantly different means between treatment levels within a breeding line according to Tukey-B test.

3.1.4. Changes in the SI of Root Number (SI_{RN})

PEG 6000 treatments strongly increased the number of roots in parental lines and in some of the breeding lines, and the best SI_{RN} values were obtained for referent genotypes in each treatment (Table 5, Figure S4a–c). However, 5% PEG 6000 in the medium resulted in decreased root number in a dozen breeding lines. Some lines responded with increased SI_{RN} values to raised (7.5%) PEG concentration. Increasing the PEG 6000 concentration to 10% resulted in significantly decreased SI_{RN} values in four breeding lines (C8, C26, C41 and C57) and yielded significantly increased SI_{RN} values in about a quarter of the lines (C2, C3, C11, C30, C35, C37 and C58) (Table 5, Figure S4c).

Table 5. Root number SI (SI_{RN}) values of potato genotypes under osmotic stress induced by PEG 6000 added to the medium at levels of 5.0%, 7.5% and 10%.

Breeding Line	SI_{RN} (%)								
	5% PEG 6000			7.5% PEG 6000			10% PEG 6000		
C2	108.1	c–f B	▲□	111.9	c–e A	▲□	135.1	c A	▲□
C3	65.6	jk A	▲□	58.6	j B	▲□	82.7	e–h A	▲□
C4	101.8	c–g A	▲□	95.3	d–h A	▲□	102.6	c–g A	▲□
C5	77.6	g–k A	▲□	76.9	f–j A	▲□	80.3	e–h A	▲□
C6	60.3	k A	▲□	55.9	j A	▲□	55.9	hi A	▲□
C8	98.2	d–h C	▲□	154.9	ab A		121.2	cd B	▲□
C9	113.8	c–e A	▲□	90.1	d–i B	▲□	82.4	e–h B	▲□
C10	75.9	h–k A	▲□	80.8	f–j A	▲□	78.8	e–h A	▲□
C11	110.2	c–f A	▲□	89.7	d–i B	▲□	113.9	c–e A	▲□
C12	101.9	c–g B	▲□	117.7	cd A	▲□	110.8	c–e A	▲□
C14	90.2	e–j A	▲□	87.9	e–i A	▲□	80.6	e–h A	▲□
C17	75.3	h–k A	▲□	70.2	g–j AB	▲□	40.3	i B	▲□
C19	115.8	b–d A	▲□	101.0	d–f B	▲□	107.9	c–f A	▲□
C20	90.4	e–j A	▲□	79.3	f–j A	▲□	78.9	e–h A	▲□
C21	88.9	e–j A	▲□	97.4	d–g A	▲□	94.9	d–g A	▲□
C22	96.6	d–i A	▲□	90.0	d–i A	▲□	82.5	e–h A	▲□
C26	75.4	h–k B	▲□	86.9	e–i A	▲□	70.2	g–i B	▲□
C28	81.5	g–k A	▲□	57.8	j B	▲□	57.7	hi B	▲□
C30	70.7	i–k B	▲□	67.2	h–j B	▲□	110.2	c–e A	▲□
C32	100.8	c–h B	▲□	134.7	bc A	▲□	122.7	cd A	▲□
C35	99.5	c–h A	▲□	75.5	f–j B	▲□	111.8	c–e A	▲□
C37	120.9	b–d AB	▲□	102.1	d–f B	▲□	129.0	cd A	▲□
C41	85.6	f–j A	▲□	77.9	f–j A	▲□	51.6	hi B	▲□
C42	95.9	d–i B	▲□	111.9	c–e A	▲□	120.9	cd A	▲□
C57	124.4	bc A	▲□	129.8	c A	▲□	106.3	c–g B	▲□
C58	84.8	f–k C	▲□	96.8	d–g B	▲□	112.4	c–e A	▲□
C63	78.8	g–k A	▲□	66.4	ij B	▲□	71.3	f–i AB	▲□
C103	149.7	a B		173.0	a A		180.3	a A	
C107	138.2	ab B		161.0	ab B		220.0	b A	

Symbols mark significant differences from the C107 referent line (▲), or from the C103 referent line (□) according to LSD test; the small letters indicate significantly different means between the breeding lines within a treatment, and the capital letters indicate significantly different means between treatment levels within a breeding line according to Tukey-B test.

3.2. Effect of Osmotic Stress Induced by PEG 600 on In Vitro Shoot Cultures

3.2.1. Changes in the SI of Survival Rate (SI_{SR})

The SI_{SR} values for survival rate decreased as PEG 600 level increased (Table 6, Figure S5a–c). At the lowest (2.5%) concentration of PEG 600, the SI_{SR} values of survival rates were usually higher than 70, and some breeding lines reached 100. Only explants of C9 and C32 lines survived in significantly lower rates than referent lines. The 5% level of PEG 600 resulted in significant decreases in the survival rates of several tested breeding lines and of the referent lines. Compared to the referent lines, higher values were obtained for the C57, C58 and C19 lines, although results differed significantly only when they were compared to the C107 referent line (its SI_{SR} was 69). Moreover, C19 and C20 breeding lines showed very good survival ability, too. Similarly, significantly higher survival rates were obtained for the C20 breeding line compared to both referent lines, although the highest concentration of PEG 600 (7.5%) significantly reduced the SI_{SR} in each genotype. In contrast, most breeding lines showed lower survival rates compared to the referent lines, and most frequently the differences were significant.

Table 6. Survival rate SI (SI_{SR}) values of potato genotypes under osmotic stress induced by PEG 600 added to the medium at level of 2.5%, 5.0% and 7.5%.

Breeding Line	SI_{SR} (%)								
	2.5% PEG 600			5.0% PEG 600			7.5% PEG 600		
C2	100	a A	▲	63	a–e B	□	24	b–f C	▲□
C3	90	a–c A		55	b–e B	□	3	ef C	▲□
C4	92	ab A	▲	32	ef B	▲□	0	C	
C5	92	ab A	▲	58	b–e B	□	36	b–e B	
C6	75	a–c A	□	60	a–e A	□	15	c–f B	▲□
C8	100	a A	▲	77	a–c B		34	b–f C	□
C9	46	d A	▲□	2	f B	▲□	1	f B	▲□
C10	88	a–c A		32	ef B	▲□	5	ef C	▲□
C11	100	a A	▲	47	c–e B	▲□	27	b–f C	▲□
C12	100	a A	▲	57	b–e AB	□	34	b–f B	□
C14	89	a–c A		50	c–e B	▲□	29	b–f C	□
C17	93	ab A	▲	35	d–f B	▲□	5	ef C	▲□
C19	75	a–c A	□	88	ab A	▲	21	b–f B	▲□
C20	98	ab A	▲	87	ab AB		70	a B	▲□
C21	88	a–c A		67,5	a–d A		34	b–f B	□
C22	92	ab A	▲	58	b–e B	□	20	c–f C	▲□
C26	67	b–d A	□	34	d–f B	▲□	7	d–f C	▲□
C28	60	cd A	□	14	f B	▲□	14	c–f B	▲□
C30	100	a A	▲	64	a–e B	□	13	c–f C	▲□
C32	40	d A	▲□	6	f B	▲□	6	ef B	▲□
C35	99	a A	▲	62	a–e B	□	3	ef C	▲□
C37	94	ab A	▲	69	a–d B		30	b–f C	□
C41	74	a–c A	□	13	f B	▲□	13	c–f B	▲□
C42	100	a A	▲	76	a B		7	d–f C	▲□

Table 6. *Cont.*

Breeding Line	SI$_{SR}$ (%)								
	2.5% PEG 600			5.0% PEG 600			7.5% PEG 600		
C57	100	a A	▲	94	a A	▲	6	ef B	▲□
C58	93	ab A	▲	95	a A	▲	40	b–d B	▲
C63	84	a–c A		78	a–c A		34	b–f B	□
C103	99	a A		87	ab B		53	bc C	
C107	74	a–c A		69	a–d A		45	b–d B	

Symbols mark significant differences from the C107 referent line (▲), or from the C103 referent line (□) according to LSD test; the small letters indicate significantly different means between the breeding lines within a treatment, and the capital letters indicate significantly different means between treatment levels within a breeding line according to Tukey-B test.

3.2.2. Changes in the SI of Shoot Length (SI$_{SL}$)

Each breeding line and both referent lines responded with significantly decreased shoot lengths to the presence of PEG 600 at a concentration of 2.5%, and inhibition of shoot growth increased with increasing PEG 600 concentration (Table 7, Figure S6a). Referent genotypes showed very good stress tolerance compared to breeding lines. SI$_{SL}$ values of C103 were higher than those of C107 at each level of PEG 600, and differences between SI$_{SL}$ values of referent lines increased as PEG 600 level increased. When 2.5% PEG 600 was applied, only the SI$_{SL}$ values of C6 and C42 were significantly higher than those of both referent lines. In addition to them, the C19 breeding line showed very good SI$_{SL}$ results in both treatments by 2.5% and 5.0% PEG 600 (SI$_{SL}$ were 75 and 53.9, respectively). At the highest level of PEG 600 treatment, the C103 referent line achieved the significantly best SI$_{SL}$ result, followed by the C107 referent line, but lagging far behind.

Table 7. Shoot length SI (SI$_{SL}$) values of potato genotypes under stress treatment induced by PEG 600 added to the medium at levels of 2.5%, 5.0% and 7.5%.

Breeding Line	SI$_{SL}$ (%)								
	2.5% PEG 600			5% PEG 600			7.5% PEG 600		
C2	57.9	c–g A	□	48.1	a–c B	▲	17.40	c–g C	▲□
C3	57.5	c–g A	□	36.0	b–f B	□	14.30	c–h C	▲□
C4	36.1	j A	▲□	22.3	g–l B	▲□	10.2	c–h C	▲□
C5	61.4	c–f A		28.6	e–i B	▲□	16.0	c–h C	▲□
C6	75.7	ab A	▲□	46.6	a–d B		21.3	c–f C	□
C8	68.7	a–c A		21.9	g–l B	▲□	5.0	gh C	▲□
C9	55.1	d–h A	▲□	27.5	e–j B	▲□	24.1	bc B	□
C10	41.8	h–j A	▲□	23.5	f–k B	▲□	9.7	d–h B	▲□
C11	61.4	c–f A		22.7	f–l B	▲□	14.0	c–h C	▲□
C12	24.2	k AB	▲□	29.3	e–h A	□	20.4	c–f B	▲□
C14	45.3	g–j A	▲□	28.7	e–i B	▲□	19.3	c–f C	▲□
C17	48.1	f–j A	▲□	28.1	e–i B	▲□	22.4	c–e B	□
C19	75.0	ab A	□	53.9	a B	▲	14.9	c–h C	▲□
C20	42.0	h–j A	▲□	22.5	f–l B	▲□	13.7	c–h C	▲□

Table 7. *Cont.*

Breeding Line	SI$_{SL}$ (%)								
	2.5% PEG 600			5% PEG 600			7.5% PEG 600		
C21	63.0	b–e A	□	21.5	h–l B	▲□	17.6	c–f B	▲□
C22	51.6	e–i A	▲□	14.1	j–n B	▲□	12.0	c–h B	▲□
C26	39.0	ij A	▲□	11.6	k–m B	▲□	12.0	c–h B	▲□
C28	53.5	e–h A	▲□	15.4	i–m B	▲□	12.3	c–h B	▲□
C30	42.0	h–j A	▲□	9.5	lm B	▲□	2.8	h C	▲□
C32	18.0	k A	▲□	5.9	m B	▲□	7.4	f–h B	▲□
C35	52.3	e–i A	▲□	18.6	h–m B	▲□	10.8	c–h B	▲□
C37	51.6	e–i A	□	38.1	b–e B	□	16.0	c–h C	▲□
C41	45.6	g–j A	▲□	48.5	a–c A	▲	25.5	b A	□
C42	79.5	a A	▲□	48.4	a–c B	▲	21.7	c–e C	□
C57	52.3	e–i A	▲□	46.1	a–d A	▲	9.1	e–h B	▲□
C58	67.6	a–d A		44.1	a–d B	▲□	23.5	b–d C	□
C63	54.9	d–h A	▲□	33.6	d–h B	□	12.5	c–h C	▲□
C103	70.4	a–c A		49.3	a–c B		43.5	a B	
C107	62.7	c–f A		35.6	b–f B		25.6	b C	

Symbols mark significant differences from the C107 referent line (▲), or from the C103 referent line (□) according to LSD test; the small letters indicate significantly different means between the breeding lines within a treatment, and the capital letters indicate significantly different means between treatment levels within a breeding line according to Tukey-B test.

3.2.3. Changes in the SI of Root Length (SI$_{RL}$)

In general, in vitro potato shoot cultures responded to PEG 600 treatment with decreased SI$_{RL}$ values for root length, and the degree of inhibition increased with increasing PEG 600 concentration (Table 8, Figure S7a–c). The use of 2.5% PEG 600 had a stimulatory effect on the root length of four lines (C5, C20, C57 and C63) compared to the referent lines, with the highest SI$_{RL}$ value obtained in the C63 line (131.1). Considering the SI$_{RL}$ values for root length, the C63 breeding line was the best at each PEG 600 level. According to its response, the C5 line tolerated well the stress induced by each PEG 600 treatment, while the C57 and C20 lines tolerated only the mild stress (2.5% PEG 600) according to their SI$_{RL}$ values for root length. However, the majority of breeding lines showed lower SI$_{RL}$ results than those of the referent lines for root length in each PEG treatment, especially when PEG 600 was applied in the highest (7.5%) concentration.

Table 8. Root length SI (SI$_{RL}$) values of potato genotypes under osmotic stress induced by PEG 600 added to the medium at levels of 2.5%, 5.0%, and 7.5%.

Breeding Line	SI$_{RL}$ (%)								
	2.5% PEG 600			5% PEG 600			7.5% PEG 600		
C2	57.7	h–j A	▲□	31.1	e–j B	▲□	8.5	de C	▲□
C3	54.0	h–k A	▲□	35.2	d–i B	▲□	13.9	c–e B	▲□
C4	64.7	g–i A	▲□	27.3	e–k B	▲□	0		
C5	99.5	bc A	▲	92.3	a A	▲	46.7	a–d B	□
C6	89.9	c–e A		47.9	b–g B	▲□	21.0	c–e C	▲□

Table 8. *Cont.*

Breeding Line	SI$_{RL}$ (%)								
	2.5% PEG 600			5% PEG 600			7.5% PEG 600		
C8	73.9	f–h A	▲□	26.0	f–k B	▲□	3.7	de C	▲□
C9	50.3	i–l A	▲□	23.1	h–k B	▲□	0		
C10	75.9	fg A	□	37.9	c–h B	▲□	10.6	de B	▲□
C11	93.4	b–d A		25.5	g–k B	▲□	14.5	c–e C	▲□
C12	51.8	h–l AB	▲□	66.1	b A	□	27.4	b–e B	▲□
C14	92.4	b–e A		25.4	g–k B		9.3	de B	▲□
C17	61.7	g–j A	▲□	26.2	f–k B	▲□	21.4	c–e B	▲□
C19	79.0	ef A	□	56.4	b–d B	□	9.7	de C	▲□
C20	105.6	b A	▲	47.3	b–g B	▲□	18.6	c–e C	▲□
C21	81.3	d–f A	□	26.9	f–k B	▲□	21.71	c–e B	▲□
C22	57.5	h–j A	▲□	17.5	h–k B	▲□	15.4	c–e B	▲□
C26	39.2	l A	▲□	4.8	k B	▲□	1.6	e B	▲□
C28	57.9	h–j A	▲□	13.5	i–k B	▲□	6.1	de B	▲□
C30	49.0	i–l A	▲□	8.9	jk B	▲□	2.5	e B	▲□
C32	23.6	m A	▲□	5.2	k B	▲□	2.8	e B	▲□
C35	74.8	f–h A	□	31.1	e–j B	▲□	7.1	de B	▲□
C37	64.9	g–i A	▲□	50.0	b–f B	□	11.5	de C	▲□
C41	41.1	kl A	▲□	38.0	c–h A	▲□	15.8	c–e B	▲□
C42	74.5	f–h A	□	37.8	c–h B	▲□	16.2	c–e C	▲□
C57	105.6	b A	▲□	50.9	b–e B	□	11.7	de C	▲□
C58	81.7	d–f A	□	59.0	bc B	□	44.1	a–e C	□
C63	131.1	a A	▲□	107.8	a B	▲□	73.4	a C	▲
C103	98.6	bc A		87.1	A		64.2	ab B	
C107	84.9	de A		59.7	B		54.2	a–c B	

Symbols mark significant differences from the C107 referent line (▲), or from the C103 referent line (□) according to LSD test; the small letters indicate significantly different means between the breeding lines within a treatment, and the capital letters indicate significantly different means between treatment levels within a breeding line according to Tukey-B test.

3.2.4. Changes in the SI of Root Number (SI$_{RN}$)

The number of roots decreased in the majority of breeding lines when media contained PEG 600, and the SI values decreased as PEG 600 level increased (Table 9, Figure S8a–c). However, SI values of higher than 100 were obtained for the C2 and C3 breeding lines in the 2.5% PEG 600 treatment, and their results were significantly higher compared to those of the referent lines and the majority of other breeding lines when treated with 2.5 and 5.0% PEG 600. In addition, the C42 breeding line also achieved significantly higher SI results with the 5.0% PEG treatment. When 7.5% PEG 600 was added to the medium, the C103 referent line showed the significantly best result (72.2), while the SI value of the other referent line (C107) was significantly lower (57.7) than that of C103, but significantly higher than those of most breeding lines. Moreover, the C2 and C3 breeding lines showed results very similar to those of the C107 referent line.

Table 9. Root number SI (SI_{RN}) values of potato genotypes under osmotic stress induced by PEG 600 added to the medium at levels of 2.5%, 5.0%, and 7.5%.

Breeding Line	SI_{SN} (%)								
	2.5% PEG 600			5% PEG 600			7.5% PEG 600		
C2	109.3	a A	▲□	103.0	a A	▲□	58.9	ab B	□
C3	108.0	a A	▲□	90.7	a–c B	▲□	62.2	ab B	
C4	82.6	b–e A	□	57.9	c–f B	▲□	0.0		
C5	74.2	c–g A	□	72.2	a–e A	□	56.7	ab B	□
C6	59.2	gh A	▲□	45.1	d–f B	▲□	24.2	ab C	▲□
C8	70.7	d–h A	▲□	47.4	d–f B	▲□	22.5	ab C	▲□
C9	84.8	b–d		0.0			0.0		
C10	74.6	c–h A	□	53.7	d–f B	□	22.7	ab C	▲□
C11	96.7	ab A	▲	48.9	d–f B	▲□	35.3	ab B	▲□
C12	50.7	h A	▲□	43.2	d–f A	▲□	28.4	ab B	▲□
C14	81.5	b–e A	□	59.8	c–f B	□	35.7	ab C	▲□
C17	93.9	a–c A		62.7	b–f B		0.0		
C19	76.2	c–g A	□	52.7	d–f B	□	27.7	ab C	▲□
C20	51.8	h A	▲□	41.9	d–f B	▲□	34.4	ab C	▲□
C21	70.9	d–h A	▲□	52.9	d–f B	□	38.4	ab C	▲□
C22	71.5	d–h A	▲□	47.8	d–f B	▲□	24.2	ab C	▲□
C26	78.3	c–g A	□	33.5	f B	▲□	29.9	ab B	▲□
C28	77.2	c–g A	□	37.2	ef B	▲□	34.5	ab B	▲□
C30	91.9	a–d A		32.4	f B	▲□	25.4	ab B	▲□
C32	62.1	f–h A	▲□	33.6	f A	▲□	38.4	ab A	□
C35	77.7	c–g A	□	41.7	d–f B	▲□	13.6	b B	▲□
C37	91.8	a–d A		52.5	d–f B	□	28.8	ab C	▲□
C41	71.6	d–h A	□	52.8	d–f A	□	29.5	ab B	▲□
C42	96.7	ab A	▲	95.9	ab A	▲□	31.3	ab B	▲□
C57	65.5	e–h B	▲□	75.5	a–d A	▲	47.3	ab B	▲□
C58	80.2	b–f A	□	73.3 B	a–e	▲	41.6	ab C	▲□
C63	84.8	b–d A		58.6	c–f B	□	28.8	ab C	▲□
C103	94.3	a–c A		73.0	a–d B		72.2	ab B	
C107	83.6	b–d A		61.4	c–f B		57.7	ab B	

Symbols mark significant differences from the C107 referent line (▲), or from the C103 referent line (□) according to LSD test; the small letters indicate significantly different means between the breeding lines within a treatment, and the capital letters indicate significantly different means between treatment levels within a breeding line according to Tukey-B test.

3.3. Effect of Osmotic Stress Induced by D-Mannitol in In Vitro Shoot Cultures

3.3.1. Changes in the SI of Survival Rate (SI_{SR})

The survival rates of the shoot cultures were not significantly affected by any concentration of D-mannitol in the case of about 41% of the breeding lines, and in fact, each explant survived in all treatments in five breeding lines (C2, C8, C20, C30 and C63) (Table 10, Figure S9a–c). Using D-mannitol at a concentration of 0.1 M resulted in significantly lower survival in C9, C21, and C58 and in referent line C107 than in referent line C103. A remarkable number of the breeding lines showed significantly higher SI_{SR} values than the C107 referent line. In the treatment with 0.2 M D-mannitol, significantly lower SI_{SR} values were observed in three breeding lines (C6, C22 and C35) and the C107 referent line compared to their respective values obtained with the 0.1 M and 0.2 M D-mannitol levels. When 0.3 M D-mannitol was applied, decreased SI_{SR} values were found in 12 breeding lines and in the C103 referent line. Moreover, all tested genotypes showed significantly higher SI_{SR} values compared to the C107 referent line.

Table 10. Survival rate SI (SI_{SR}) values of potato genotypes under stress treatment induced by D-mannitol added to the medium at levels of 0.1 M, 0.2 M and 0.3 M.

Breeding Line	SI_{SR} (%)								
	0.1 M D-Mannitol			0.2 M D-Mannitol			0.3M D-Mannitol		
C2	100.0	a A	▲	100.0	a A	▲	100.0	a A	▲□
C3	100.0	a A	▲	100.0	a A	▲	74.0	a–f B	▲
C4	96.0	a A		91.0	ab A		54.0	e–g B	
C5	100.0	a A	▲	95.0	ab AB		89.0	a–d B	▲□
C6	100.0	a A		67.0	c B	▲□	54.0	e–g B	
C8	100.0	a A	▲	100.0	a A	▲	100.0	a A	▲□
C9	90.0	a A	□	83.0	a–c A	□	73.0	a–f A	▲
C10	100.0	a A	▲	100.0	a A	▲	92.0	a–d B	▲□
C11	100.0	a A	▲	95.0	ab A		92.0	a–d A	▲□
C12	100.0	a A	▲	97.0	a A		74.0	a–f B	▲
C14	100.0	a A	▲	94.0	ab A		77.0	a–e B	▲
C17	100.0	a A	▲	90.0	a–c AB		75.0	a–f B	▲
C19	96.0	a A		94.0	ab A		89.0	a–d A	▲□
C20	100.0	a A	▲	100.0	a A	▲	100.0	a A	▲□
C21	91.3	a A	□	83.0	a–c A	□	51.0	e–g B	□
C22	96.0	a A		83.0	a–c B	□	56.0	e–g C	
C26	100.0	a A	▲	96.0	a A		76.0	a–f B	▲
C28	100.0	a A	▲	100.0	a A	▲	98.0	ab A	▲□
C30	100.0	a A	▲	100.0	a A	▲	100.0	a A	▲□
C32	99.0	a A	▲	100.0	a A	▲	92.0	a–d A	▲□
C35	100.0	a A	▲	88.0	a–c B		72.0	b–f C	▲
C37	100.0	a A	▲	94.0	ab A		96.0	a–c A	▲□
C41	100.0	a A	▲	85.0	a–c A	□	50.0	fg B	□
C42	100.0	a A	▲	92.0	ab A		68.0	d–g B	▲
C57	100.0	a A	▲	96.0	a A		93.0	a–d A	▲□
C58	87.0	a A	□	73.0	bc A	□	70.0	c–f A	▲
C63	100.0	a A	▲	100.0	a A	▲	100.0	a A	▲□
C103	100.0	a A		100.0	a A		68.0	d–g B	
C107	90.0	a A		85.0	a–c A		44.0	g B	

Symbols mark significant differences from the C107 referent line (▲), or from the C103 referent line (□) according to LSD test; the small letters indicate significantly different means between the breeding lines within a treatment, and the capital letters indicate significantly different means between treatment levels within a breeding line according to Tukey-B test.

3.3.2. Changes in the SI of Shoot Length (SI_{SL})

The presence of D-mannitol at a concentration of 0.1 M in medium inhibited the shoot growth of almost all genotypes, and in general, SI_{SL} values decreased significantly with increasing D-mannitol concentration (Table 11, Figure S10a–c). The SI_{SL} value of the C12 line was higher than 100 in the treatment with 0.1 M D-mannitol, and 13 breeding lines achieved significantly higher SI_{SL} values than both referent lines. Similarly, at 0.2 M and 0.3 M levels of D-mannitol, the SI_{SL} values of several breeding lines were significantly higher than those of the referent lines.

Table 11. Shoot length SI (SI_{SL}) values of potato genotypes under stress treatment induced by D-mannitol added to the medium at levels of 0.1 M, 0.2 M and 0.3 M.

Breeding Line	SI_{SL} (%)								
	0.1 M D-Mannitol			0.2 M D-Mannitol			0.3 M D-Mannitol		
C2	54.4	h–l A		34.6	g–j B	□	23.0	d–g C	▲
C3	67.4	e–g A	▲□	37.2	f–i B		21.1	e–i C	▲
C4	65.4	e–i A	▲	32.4	h–k B	□	11.7	k C	□
C5	65.4	g–l A		29.2	i–k B	▲□	24.5	c–f B	▲
C6	53.0	i–l A	□	33.6	h–j B	□	21.4	e–h C	▲
C8	93.2	b A	▲□	46.8	a–e B	▲□	24.2	c–f C	▲
C9	60.4	f–l A	▲	46.9	a–e B	▲□	29.5	bc C	▲□
C10	52.4	j–l A	□	27.3	jk B	▲□	20.8	e–i C	▲
C11	68.3	e–g A	▲□	36.1	g–i B	□	19.0	f–i C	▲□
C12	119.7	a A	▲□	51.8	ab B	▲□	36.1	a C	▲□
C14	89.0	bc A	▲□	52.6	ab B	▲□	40.9	a C	▲□
C17	74.1	de A	▲□	54.7	a B	▲□	38.0	a C	▲□
C19	87.4	bc A	▲□	39.4	e–i B		27.5	b–d C	▲□
C20	56.0	g–l A		24.4	k B	▲□	15.3	i–k C	□
C21	57.9	g–l A		28.5	i–k B	▲□	16.3	h–k C	□
C22	71.4	d–f A	▲□	37.0	g–i B		21.5	e–h C	▲
C26	70.8	d–f A	▲□	49.8	a–c B	▲□	25.5	b–e C	▲
C28	66.5	e–h A	▲	41.1	d–h B	▲	28.1	b–d C	▲□
C30	67.8	e–g A	▲□	35.0	g–j B	□	17.7	g–j C	▲□
C32	75.1	de A	▲□	48.1	a–d B	▲□	30.0	bc C	▲□
C35	59.2	f–l A	▲	32.5	h–k B	□	15.6	h–k C	□
C37	53.2	i–l A	□	34.3	g–j B	□	31.0	b B	▲□
C41	65.0	e–j A	▲	45.6	b–f B	▲	26.2	b–f C	▲
C42	51.7	kl A	□	30.2	i–k B	□	17.9	i–k C	▲□
C57	47.5	l A	□	37.2	f–i B		21.3	f–i C	
C58	81.2	cd A	▲□	46.5	a–e B	▲□	24.7	a–e C	▲
C63	89.3	bc A	▲□	42.5	c–g B	▲	25.1	c–g C	
C103	60.6	g–k A		41.1	d–h B		23.3	d–g C	
C107	51.0	kl A		34.9	g–j B		13.2	jk C	

Symbols mark significant difference from the C107 referent line (▲), or from the C103 referent lines (□) according to LSD test; the small letters indicate significantly different means between the breeding lines within a treatment, and the capital letters indicate significantly different means between treatment levels within a breeding line according to Tukey-B test.

3.3.3. Changes in the SI of Root Length (SI$_{RL}$)

About half of the breeding lines responded to 0.1 M D-mannitol with significant decreases in their root length, but the SI$_{RL}$ values of two breeding lines were significantly increased (Table 12, Figure S11a). One of them (C63) showed significantly a higher result than the C103 referent line, although the C103 referent line also responded with increased root length. However, as D-mannitol concentration increased, the SI$_{RL}$ values decreased, mostly significantly (Table 11, Figure S10a–c). When 0.2 M D-mannitol was applied, only the C103 referent line showed SI$_{RL}$ values higher than 100, but the SI$_{RL}$ values of C17 and C63 breeding lines did not differ significantly. Decreased SI$_{RL}$ values were found at this level of D-mannitol in almost all breeding lines. SI$_{RL}$ values of breeding lines C17 and C63 were significantly higher than those of the C107 referent line, and increased SI$_{RL}$ values were detected in more than half the of breeding lines. At the highest D-mannitol level (0.3 M), each genotype involved in the experiment showed decreased SI$_{RL}$ values, and we found significantly higher SI$_{RL}$ values in more than a third of the breeding lines compared to the C103 referent line.

Table 12. Root length SI (SI$_{RL}$) values of potato genotypes under stress treatment induced by D-mannitol added to the medium at levels of 0.1 M, 0.2 M and 0.3 M.

Breeding Line	SI$_{RL}$ (%)								
	0.1 M D-Mannitol			0.2 M D-Mannitol			0.3 M D-Mannitol		
C2	97.9	b–f A	▲□	61.0	e–h B	▲□	37.6	j–m C	▲
C3	86.8	e–i A	□	59.0	e–h B	▲□	37.6	j–m C	▲
C4	74.0	i–k A	▲□	48.8	gh B	▲□	18.4	o C	▲□
C5	86.5	e–i A	□	73.3	c–e B	□	78.5	a B	□
C6	59.2	k A	▲□	50.5	gh B	▲□	34.5	k–n C	▲□
C8	101.9	b–e A	▲	84.6	bc B	□	70.0	a–c C	□
C9	81.0	f–j A	□	77.8	cd A	□	39.6	i B	▲
C10	105.1	a–d A	▲	70.0	c–e B	□	48.5	f–k C	▲
C11	90.0	d–i A	□	79.3	cd B	□	60.3	b–f C	□
C12	94.5	c–g A	▲□	69.3	c–e B	▲□	56.5	c–h C	▲□
C14	93.7	c–g A	□	50.6	gh B	▲□	21.7	no C	▲□
C17	95.2	c–g A	▲□	97.7	ab A	▲	72.6	ab B	□
C19	92.4	c–g A	□	60.3	e–h B	▲□	44.3	f–l C	▲
C20	99.1	b–e A	▲□	74.3	c–e B	□	55.9	c–i C	▲□
C21	67.2	jk A	▲□	48.6	gh B	▲□	30.3	l–o C	▲□
C22	67.8	jk A	▲□	46.0	h B	▲□	37.8	j–m B	▲
C26	91.4	d–h A	□	67.2	d–f B	▲□	27.5	m–o C	▲□
C28	84.7	e–i A	□	52.5	f–h B	▲□	43.7	f–h C	▲
C30	74.7	h–k A	▲□	68.9	c–e A	▲□	39.8	i–m B	▲
C32	91.3	d–h A	□	70.2	c–e B	□	57.9	b–g C	□
C35	78.1	g–j A	□	49.4	gh B	▲□	37.2	j–m C	▲
C37	96.6	b–f A	▲□	68.6	c–e B	▲□	47.1	f–k C	▲
C41	112.7	ab A	▲	63.9	d–h B	▲□	41.4	h–m C	▲
C42	86.2	e–i A	□	69.1	c–e B	□	48.3	f–k C	▲

Table 12. *Cont.*

Breeding Line	SI$_{RL}$ (%)								
	0.1 M D-Mannitol			0.2 M D-Mannitol			0.3 M D-Mannitol		
C57	104.8	a–d A	▲	84.2	bc B	□	64.3	a–e C	□
C58	88.6	d–i A	□	78.9	cd B	□	52.0	e–j C	▲□
C63	119.9	a A	▲□	96.5	ab B	▲	53.0	d–j C	▲□
C103	108.8	ab A		103.7	ab A		44.2	g–l B	
C107	85.4	e–i A		77.9	cd A		68.4	a–c A	

Symbols mark significant differences from the C107 referent line (▲), or from the C103 referent line (□) according to LSD test; the small letters indicate significantly different means between the breeding lines within a treatment, and the capital letters indicate significantly different means between treatment levels within a breeding line according to Tukey-B test.

3.3.4. Changes in the SI of Root Number (SI$_{RN}$)

About half of the breeding lines responded with decreased SI$_{RN}$ values to 0.1 M D-mannitol treatment (Table 13, Figure S12a). Some breeding lines showed significantly increased SI$_{RN}$ values compared to both referent lines. In general, decreasing tendencies could be observed in SI$_{RN}$ values as D-mannitol level increased (0.2 M), and significant reductions were verified in the majority of breeding lines. SI$_{RN}$ values increased in some breeding lines. There were significant differences between C103 and C107 when 0.1 M and 0.2 M D-mannitol concentrations were applied to the medium. In addition, four, eleven and nine breeding lines showed significantly higher SI$_{RN}$ values compared to both referent lines at the levels of 0.1, 0.2 and 0.3 M D-mannitol, respectively (Table 13, Figure S12a–c).

Table 13. Root number SI (SI$_{RN}$) values of potato genotypes under stress treatment induced by D-mannitol added to the medium at levels of 0.1 M, 0.2 M and 0.3 M.

Breeding Line	SI$_{RN}$ (%)								
	0.1 M D-Mannitol			0.2 M D-Mannitol			0.3 M D-Mannitol		
C2	91.4	d–j A	□	74.4	g–j B		67.6	f–i B	
C3	91.2	d–k A	□	75.6	f–j B		43.2	kl C	▲□
C4	88.3	e–k A	□	75.7	f–j B		50.7	i–l C	▲□
C5	113.2	b A	▲□	93.1	c–h B	▲	97.5	bc B	▲□
C6	79.6	i–l A	▲□	65.4	ij B	▲□	77.7	c–g AB	
C8	108.0	b–d A	▲	100.9	cd A	▲□	77.0	d–h B	
C9	87.2	e–k A	□	85.5	d–i A		64.4	g–j B	
C10	97.2	b–i A		79.4	e–i B		63.0	g–k C	
C11	80.5	h–l B	▲□	121.2	b A	▲□	121.7	a A	▲□
C12	80.5	h–l A	▲□	67.7	ij B	□	45.0	j–l C	▲□
C14	108.5	b–d A	▲	94.7	c–f B	▲□	70.6	f–i C	
C17	111.1	bc A	▲□	104.9	bc A	▲□	98.4	b A	▲□
C19	114.6	b A	▲□	82.8	d–i B		83.8	b–g B	▲□
C20	68.0	l A	▲□	73.3	h–j A		55.7	h–j B	□
C21	72.9	kl A	▲□	58.7	j B	▲□	55.9	h–k B	□
C22	99.2	b–h A		94.6	c–g A	▲□	49.9	i–l B	▲□

Table 13. Cont.

Breeding Line	SI$_{RN}$ (%)								
	0.1 M D-Mannitol			0.2 M D-Mannitol			0.3 M D-Mannitol		
C26	98.5	b–h A		95.3	c–f A	▲□	69.6	f–i B	
C28	94.3	c–j A		93.2	c–h A	▲□	74.1	eh B	
C30	105.3	b–e B	▲	119.4	b A	▲□	96.5	b–d B	▲□
C32	134.0	a B	▲□	173.2	a A	▲□	135.2	a B	▲□
C35	82.0	g–l A	□	96.8	c–e A	▲□	64.6	g–j B	
C37	88.1	e–k A	□	75.4	f–j B		66.7	f–i B	
C41	76.3	j–l A	▲□	59.0	j B	▲□	35.2	l C	▲□
C42	78.8	i–l A	▲□	73.7	h–j A		56.3	h–k B	□
C57	82.3	g–l A	□	85.6	c–i A		85.9	b–f A	▲□
C58	105.1	b–e A	▲	104.8	bc A	▲□	87.7	b–f B	▲□
C63	84.2	f–l B	□	83.3	d–i B		91.4	b–e A	▲□
C103	100.9	b–g A		82.4	d–i B		68.8	f–i C	
C107	90.5	d–j A		76.6	f–i B		67.0	f–i B	

Symbols mark significant differences from the C107 referent line (▲), or from the C103 referent line (□) according to LSD test; the small letters indicate significantly different means between the breeding lines within a treatment, and the capital letters indicate significantly different means between treatment levels within a breeding line according to Tukey-B test.

4. Discussion

In conventional breeding work including field experiments, the test for stress tolerance of breeding lines is expensive, takes a long time [46], and the results obtained in different experiments can be variable because the reaction of the plants is strongly influenced by environmental conditions [47]. That is why several new approaches are applied to evaluate traits that can be associated with drought tolerance, including laboratory mo-dels and biotechnological tools. Responses of several potato varieties to osmotic stress under in vitro conditions matched well with the drought tolerance results obtained in the field experiments [33,34]. In addition, if we take into account the very large number of offspring produced by crosses to be screened for the desired traits, the importance of alternative experiments cannot be ignored.

In field and greenhouse and also pot experiments, drought affected all physiological and agronomic traits of the potato cultivars studied [35,36]. Osmotic stress applied to model drought stress under in vitro conditions also resulted in significant changes in all studied traits in potato, e.g., survival rate, shoot length, fresh and dry weight, and root number and length [37,38]. However, no difference could be revealed between genotypes if the evaluation was based only on the measured parameters, even though significant inhibition of the growth of in vitro shoots and roots as an effect of osmotic stress in *Solanum* species was observed [32,39]. The development of a multi-parameter stress index (SI), when the results were expressed as a percentage of the values for the controls, allowed a differentiation of genotypes according to their stress tolerance [32,40]. Accordingly, a stress index was formed from our results for each treatment examined, and the examined genotypes were evaluated based on their SI values.

In our experiments, the responses of potato breeding lines were compared both to each other and to two referent lines. We observed that the osmotic stress resulted in explant deaths in varying proportions in the breeding lines and the referent lines, most often without shoot development. High survival rates were found in C2, C8, C20, C30, and C63 in the treatments with D-mannitol, in C8 and C30 in the treatments with PEG 6000, and in C8, C12, C20, C57, C58 and C63 in the treatments with PEG 600. After ranking the breeding lines based on their SI values, nine genotypes (in descending order: C8, C63, C14,

C2, C5, C58, C12, C11 and C30) were found to be valuable breeding material (Tables S1–S4). However, the C103 referent line was the 1st in the ranking, while C107 was the 11th.

Considering the responses of referent genotypes to osmotic stress, they showed (sometimes significantly) different SI results. In general, SI results of C103 were higher than those of C107 in almost all treatments and for almost all traits studied.

It could be supposed from these results that drought tolerance of referent lines could be based on a different mechanism. Although osmotic adjustment ability in potato was found to be restricted to 0.16 MPa [48], its role in drought tolerance was reported for several crops [49]. Osmotic stress tolerance expressed at the tissue level may play a greater role in referent genotype C103, while other factors should be considered in the case of the C107 genotype. As drought tolerance is rather complex in nature, other factors, for example, regulation of stoma closure [16,50] and/or phenological properties (especially at time of maturity) could be relevant [14,51,52]. However, considering our results, morphological factors including root developmental characters formed under stress conditions can be of great importance [53]. All of these traits can help prevent dehydration of tissues either via reduction of water loss or increase in water uptake [14,52].

Simple morpho-physiological traits were proven to be suitable parameters for distinguishing genotypes according to their osmotic stress tolerance [32,35,41–43]. Although several biochemical markers [46,53,54] and QTLs [42] were found to be exact tools for the selection of drought-tolerant genotypes, when laboratory infrastructure and/or budget are limited, researchers are forced to use simple parameters. Therefore, we observed shoot and root length, number of roots, and survival rate and found that morpho-physiological responses of potato shoot cultures to osmotic stress included both the adaptive and—most frequently—the damage responses. Phenomena of reduced growth of shoot and root were mild physiological damage responses, while strong osmotic stress often resulted in explant death. However, several breeding lines showed increased root length and/or root number, maybe as an adaptive response that could play an important role in drought to-lerance and lead to plant escape from water stress [14].

The underground part of the plant (the root system), in addition to providing a site for fixation and water and nutrient uptake, plays a significant role in evolving abiotic stress responses [17,53]. Prolonged drought can induce adaptive responses in plants, and it can be manifested by modified structure and function of roots [53], which can play an important role in coping with water stress [19,35,55].

Although the morphology, function and even histology of roots can be highly varied in plants developed in the field, in growing containers, in hydroculture systems or in vitro [53,56], the separation of genotypes based on their rooting characters in laboratory tests can be applied in breeding work [57]. The relationship between root growth and water uptake was demonstrated by Iwama [58] under both in vitro and field conditions.

In our experiments many breeding lines responded to osmotic stress with longer and/or more root development, while others showed varying degrees of inhibited growth. In general, the lower concentrations of osmotic agent resulted in longer roots and, most frequently, more roots developed on shoots compared to the control culture. The root length SI results for the C103 referent line were very high under several osmotic treatments. Both referent lines developed the most roots under osmotic stress conditions at each level of PEG 6000, and very few breeding lines were able to outperform the referent genotypes in the PEG 600 treatments. However, SI results for root number obtained from several breeding lines were higher than those of the referent genotypes in the D-mannitol treatments.

Changes in rooting parameters under stress conditions can vary; for example, each observed rooting trait (root number and root length) decreased as D-mannitol concentration increased [32], but the root number of cv. Boró (a highly drought-tolerant variety) increased at the 0.2 M D-mannitol level. Similarly, the root dry mass increased by 25% on average in 43 potato genotypes [34]. Moreover, when Zaki and Radwan [33] tested the osmotic stress tolerance of 21 potato cultivars on media containing sorbitol at three concentration levels (0.1, 0.2 and 0.3 M), they also found that tolerant genotypes developed greater or slightly decreased root mass under stress compared to the sensitive cultivars, which suffered from strong inhibition in their root development. They also observed that stimulated root growth occurred frequently at the lowest level of stress, and sometimes it could be detected at the mid-level of stress.

Opposite results observed and published in terms of root parameters such as root length, root dry mass, and root number [19,20] may be attributed to the fact that some—tolerant—varieties can respond to drought stress with increased root length, while root length does not change or decrease in more sensitive varieties compared to referent lines [36]. In addition, different experimental settings and interactions between genotypes and environment can also contribute to the variability of results [19].

In general, drought had a greater effect on the above-ground than under-ground growth in potato crop [36,55], and in the case of other crops [59–61]. In vitro experiments with potatoes yielded similar results in osmotic stress-induced changes in shoots and root systems [32,34]. In our experiments, shoot growth was also more inhibited than the growth of the root system. SI values for all treatments and all breeding lines averaged 40.06, 60.79, 76.02 and 82.61 for SL, RL, SR and RN, respectively (Table 1).

Despite a significant reduction detected in shoot length, this alone did not appear to be an appropriate trait for grouping genotypes by their osmotic stress tolerance, because high SI values of shoot lengths were not regularly accompanied by higher survival rates in breeding lines. A similar conclusion was reached by other researchers who did not recommend the use of changes in shoot length as a suitable parameter for selection, for example, in wheat [62] or in potato [32].

However, potato clones did not show the same reactions to different osmotic agents. Responses observed on media supplemented with PEG 6000 and D-mannitol were sometimes similar, but most frequently the performance of potato shoot cultures varied on different media considering the type of osmotic material as well as its concentration. In addition, the evaluation of potato genotypes on media with PEG 600 seemed to be difficult because of the very high levels of inhibition detected in breeding lines. Gangopadhyay et al. [63] also found that considering the growth, viability and proline content of tobacco (*Nicotiana tabacum* L. *var.* Jayasri), the responses of callus lines depended on physicochemical characters of the used osmotic agents (PEG 6000, D-mannitol and NaCl).

The various responses of genotypes may be due to different effects of the three osmotic agents (PEG 6000, D-mannitol and PEG 600) on the growth and developmental characters of in vitro shoot cultures. In fact, we used PEG 6000, which is a non-ionic, non-penetrating osmotic substance due to its high molecular weight [63]. In contrast, D-mannitol is a sugar alcohol that is also a non-ionic but penetrating osmotic agent [31]. PEG 600, being less than 1000 in molecular weight, is a non-ionic osmotic, but due to its lower molecular weight, it is more likely to be absorbed by plants and may have toxic effects [64].

As we observed, they all inhibited shoot growth, and the already quite strong inhibition further increased with increasing concentrations of these osmotic agents. In contrast, the root length was inhibited and also stimulated when explants were grown on media supplemented with the lowest levels of PEG 6000 or D-mannitol, whereas an inhibitory effect was observed in the case of PEG 600. Moreover, the number of roots were often significantly increased by PEG 6000 and D-mannitol treatments, but this was not true for treatments with PEG 600. Survival rates were also decreased by each treatment.

Even though PEG can result in serious stress to plants, it is used frequently as an osmotically active agent [51]. In our experiments the significantly strongest inhibition on studied characters was observed on shoot cultures grown on media with PEG 600. PEG 6000 and D-mannitol had a broadly similar inhibitory effect on plantlets, except for the number of roots, where the presence of D-mannitol resulted in a significantly lower SI value compared to PEG 6000 in the mean of all concentrations and breeding lines.

According to results reported by Thimann et al. [65], a very small amount of exogenous D-mannitol was able to enter potato disc tissues. In contrast, Trip et al. [31] found that potato leaf discs absorbed D-mannitol in a very large proportion (99%), although only 1.3%

was metabolized. In potato tissue cultures, Lipavská and Vreugdenhil [66] revealed that in vitro potato shoots can readily absorb D-mannitol from the medium, and its transport to shoots was unobstructed as well. In spite of these results, in experiments involving potatoes, D-mannitol was also used to induce osmotic stress. The level that could be applied to distinguish genotypes was higher (0.8 M) for calli culture than that for shoot culture (0.2–0.4 M) [67,68]. The researchers detected strong inhibition in growth and survival with 0.4 M D-mannitol, even in the case of tolerant genotypes. Thus, we tested the effect of D-mannitol at concentrations of 0.1, 0.2 and 0.3 M, the last of which also led to strong inhibition of shoot length but had a weaker inhibitory effect on survival. Evers et al. [39] found the same tendencies when 0.2 and 0.3 M D-mannitol were applied in experiments with *Solanum phureja* and *S. tuberosum* clones.

In addition to variations in the tolerance of breeding lines to osmotic stress, there might be some differences in their ability to absorb D-mannitol, and/or tolerate the toxic effect of PEG 600; thus, interactions between genotypes and osmotic agents added to medium could lead to various responses. Significant interactions between genotypes and the degree of osmotic pressure also should be considered [40].

5. Conclusions

All tested parameters were affected by each osmotic material used in our experiments. In general, shoot length was the most inhibited, while changes in rooting parameters both stimulated and reduced growth, depending on the genotype. Usually, the survival rates decreased significantly in treatments with strong osmotic pressure. Responses of genotypes were affected by the type and concentration of osmotic agent. Comparing genotypes, we can conclude that C103 tolerated more osmotic stress than C107, although both of them are drought-tolerant. Out of the 27 total breeding lines examined, nine genotypes (C8, C63, C14, C2, C5, C58, C12, C11 and C30) were shown to be worthy of further investigation.

Besides their high survival rate SI values, their rooting parameters were stimulated or hardly inhibited comparing to other genotypes, and their shoot length SI values were also high in several cases. We found significant interactions between genotypes and osmotic agents and their concentrations. In fact, breeding lines with high survival rates showed high SI values in the PEG 6000 or D-mannitol treatment or in both, but not in treatments with PEG 600. In general, the PEG 6000 10.0% and D-mannitol 0.2 M treatments proved to be the most suitable for differentiating genotypes according to their osmotic stress tolerance. The PEG 6000 and D-mannitol-containing media were tested at different concentrations, but they resulted in very similar inhibitory effects in terms of shoot length, root length, and survival rate (Table 1). In fact, the rate of absorbed and metabolized D-mannitol is not known. The very strong inhibitions observed in the PEG 600 treatments may be attributable mainly to its toxicity, because their resulting osmolality values were higher than those of PEG 6000 treatments but lower than those of media with D-mannitol (Table S5). However, a significant root growth stimulating effect was also observed in the referent lines, which is probably due to the adaptation response to osmotic stress. There are ongoing in vivo greenhouse and field experiments with selected breeding lines to confirm their drought tolerance.

Supplementary Materials: The following supporting information can be downloaded at: https://www.mdpi.com/article/10.3390/horticulturae8070591/s1, Figure S1: Survival rate SI (SI_SR) values of potato genotypes under osmotic stress induced by PEG 6000 added to the medium at levels of 5.0% (c), 7.5% (b) and 10.0% (c). Figure S2: Shoot length SI values (SI_SL) of potato genotypes under osmotic stress induced by PEG 6000 added to the medium at levels of 5.0% (a), 7.5% (b) and 10.0% (c); Figure S3: Root length SI (SI_RL) values of potato genotypes under osmotic stress induced by PEG 6000 added to the medium at levels of 5.0% (a), 7.5% (b) and 10% (c); Figure S4: Root number SI (SI_RN) values of potato genotypes under osmotic stress induced by PEG 6000 added to the medium at levels of 5.0% (a), 7.5% (b) and 10% (c).; Figure S5: Survival rate SI (SI_SR) values of potato genotypes under osmotic stress induced by PEG 600 added to the medium at levels of 2.5% (a), 5.0% (b) and 7.5% (c); Figure S6: Shoot length SI (SI_SL) values of potato genotypes under stress treatment induced by PEG 600 added to the medium at levels of 2.5% (a), 5.0% (b) and 7.5% (c).; Figure S7: Root lengths SI (SI_{RL}) values of potato genotypes under osmotic stress induced by PEG 600 added to the medium at levels of 2.5% (a), 5.0% (b), 7.5% (c); Figure S8: Root number SI (SI_{RN}) values of potato genotypes under osmotic stress induced by PEG 600 added to the medium at levels of 2.5% (a), 5.0% (b), 7.5% (c); Figure S9: Survivor rate SI (SI_{SR}) values of potato genotypes under stress treatment induced by D-mannitol added to the medium at levels of 0.1 M (a), 0.2 M (b) and 0.3 M (c); Figure S10: Shoot length SI (SI_{SL}) values of potato genotypes under stress treatment induced by D-mannitol added to the medium at levels of 0.1 M (a), 0.2 M (b) and 0.3 M (c); Figure S11: Root length SI (SI_{RL}) values of potato genotypes under stress treatment induced by D-mannitol added to the medium at levels of 0.1 M (a), 0.2 M (b) and 0.3 M (c); Figure S12: Root number SI (SI_{RN}) values of potato genotypes under stress treatment induced by D-mannitol added to the medium at levels of 0.1 M (a), 0.2 M (b) and 0.3 M (c).; Table S1: Ranking of breeding lines cultured on medium supplemented with PEG 6000 including each SI value; Table S2: Ranking of breeding lines cultured on medium supplemented with PEG 600 including each SI value; Table S3: Ranking of breeding lines cultured on medium supplemented with D-mannitol including each SI value; Table S4: Ranking of breeding lines cultured on medium supplemented with PEG 6000, PEG 600 or D-mannitol including each SI value; Table S5: Osmolality values of MS media supplemented with different osmotic agents used in experiments.

Author Contributions: Writing—original draft preparation, A.H., N.M.-D. and K.M.-T.; writing—review and editing, A.H., N.M.-D., K.M.-T., J.D. and L.Z.; visualization, N.M.-D. and A.H.; supervision, K.M.-T. All authors have read and agreed to the published version of the manuscript.

Funding: The research was supported by the Hungarian Government; project number: AGR_PIAC_13 _1_2013_0006 ("Research on disease resistance of potato for decreasing the effect of climate change").

Institutional Review Board Statement: Not applicable.

Informed Consent Statement: Not applicable.

Data Availability Statement: Data are contained within the article.

Acknowledgments: The research was supported by the Thematic Excellence Programme (TKP2021-EGA-20) of the Ministry for Innovation and Technology in Hungary, within the framework of the Biotechnology thematic programme of the University of Debrecen.

Conflicts of Interest: The authors declare no conflict of interest.

References

1. Anithakumari, A.M.; Dolstra, O.; Vosman, B.; Visser, R.G.F.C.; van der Linden, G. In vitro screening and QTL analysis for drought tolerance in diploid potato. *Euphytica* **2011**, *181*, 357–369. [CrossRef]
2. Thiele, G.; Theisen, K.; Bonierbale, M.; Walker, T. Targeting the poor and hungry with potato science. *Potato J.* **2010**, *37*, 75–86.
3. Boyer, J.S. Plant productivity and environment. *Science* **1982**, *218*, 443–448. [CrossRef] [PubMed]
4. Seleiman, M.F.; Al-Suhaibani, N.; Ali, N.; Akmal, M.; Alotaibi, M.; Refay, Y.; Dindaroglu, T.; Abdul-Wajid, H.H.; Battaglia, M.L. Drought stress impacts on plants and different approaches to alleviate its adverse effects. *Plants* **2021**, *10*, 259. [CrossRef] [PubMed]
5. Verma, A.; Deepti, S. Abiotic stress and crop improvement: Current scenario. *APAR* **2016**, *4*, 00149.
6. Gopal, J.; Iwama, K. In vitro screening of potato against water-stress mediated through sorbitol and polyethylene glycol. *Plant Cell Rep.* **2007**, *26*, 693–700. [CrossRef]
7. Impa, S.M.; Nadaradjan, S.; Jagadish, S.V.K. *Drought Stress Induced Reactive Oxygen Species and Anti-Oxidants in Plants. Abiotic Stress Responses in Plants: Metabolism, Productivity and Sustainability*, 7th ed.; Ahmad, P., Prasad, M.N.V., Eds.; Springer Science & Business Media: Berlin/Heidelberg, Germany, 2012; pp. 131–147.
8. Zhang, H.; Xu, F.; Wu, Y.; Hu, H.-h.; Dai, X.-f. Progress of potato staple food research and industry development in China. *J. Integr. Agric.* **2017**, *16*, 2924–2932. [CrossRef]
9. Harris, P.M. Water. In *The Potato Crop*, 1st ed.; Harris, P.M., Ed.; Chapman & Hall: London, UK, 1978; pp. 244–277. [CrossRef]
10. Monneveux, P.; Ramírez, D.A.; Pino, M.T. Drought tolerance in potato (*S. tuberosum* L.). Can we learn from drought tolerance research in cereals? *Plant Sci.* **2013**, *205–206*, 76–86. [CrossRef]
11. Burton, W.G. Challenges for stress physiology in potato. *Am. Potato J.* **1981**, *58*, 3–14. [CrossRef]
12. Vos, J.; Haverkort, A.J. Water availability and potato crop performance. In *Potato Biology and Biotechnology: Advances and Perspectives*; Vreugdenhil, D., Ed.; Elsevier: Amsterdam, The Netherlands, 2007; pp. 333–351. [CrossRef]
13. Wishart, J.; George, T.S.; Brown, L.K.; White, P.J.; Ramsay, G.; Jones, H.; Gregory, P.J. Field phenotyping of potato to assess root and shoot characteristics associated with drought tolerance. *Plant Soil* **2014**, *378*, 351–363. [CrossRef]

14. Obidiegwu, J.E.; Bryan, G.J.; Jones, H.G.; Prashar, A. Coping with drought: Stress and adaptive responses in potato and perspectives for improvement. *Front. Plant Sci.* **2015**, *6*, 542. [CrossRef] [PubMed]
15. Paoletti, C.; Bagatta, M.; DiCandilo, M.; Bassi, F.; Ranalli, P. The morpho-physiological response to prolonged water-stress in potato: Variation among clones. In Proceedings of the 14th Triennial Conference of EAPR, Sorrento, Italy, 2–7 May 1999; pp. 182–183.
16. Gervais, T.; Creelman, A.; Li, X.-Q.; Bizimungu, B.; De Koeyer, D.; Dahal, K. Potato response to drought stress: Physiological and growth basis. *Front. Plant Sci.* **2021**, *12*, 698060. [CrossRef] [PubMed]
17. Bandurska, H. Drought stress responses: Coping strategy and resistance. *Plants* **2022**, *11*, 922. [CrossRef] [PubMed]
18. Spitters, C.J.T.; Schapendonk, A.H.C.M. Evaluation of breeding strategies for drought tolerance in potato by means of crop growth simulation. *Plant Soil* **1990**, *123*, 193–203. [CrossRef]
19. Hill, D.; Nelson, D.; Hammond, J.; Bell, L. Morphophysiology of potato (*Solanum tuberosum*) in response to drought stress: Paving the way forward. *Front. Plant Sci.* **2021**, *11*, 597554. [CrossRef]
20. Nasir, M.W.; Tóth, Z. Effect of drought stress on potato production: A review. *Agronomy* **2022**, *12*, 635. [CrossRef]
21. Hura, T.; Hura, K.; Grzesiak, S.; Banaszak, Z. Simulation of osmotic stress during the early stages of triticale development as a promising laboratory test for screening drought resistance. *Cereal Res. Commun.* **2010**, *38*, 327–334. [CrossRef]
22. Sakthivelu, G.; Akitha Devi, M.K.; Giridhar, P.; Rajasekaran, T.; Ravishankar, G.A.; Nedev, T.; Kosturkova, G. Drought-induced alterations in growth, osmotic potential and in vitro regeneration of soybean cultivars. *Gen. Appl. Plant Physiol.* **2008**, *34*, 103–112.
23. Bajji, M.; Lutts, S.; Kinet, J.M. Physiological changes after exposure to and recovery from polyethylene glycol-induced water deficit in callus culture issued from durum wheat (*Triticum durum* Desf.) cultivars differing in drought resistance. *J. Plant Physiol.* **2000**, *156*, 75–83. [CrossRef]
24. Badoni, A.; Chauhan, J.S. Effect of growth regulators on meristemtip development and in-vitro multiplication of potato cultivar 'Kufri Himalini'. *Nat. Sci.* **2009**, *7*, 31–34.
25. Pérez-Clemente, R.M.; Gómez-Cadenas, A. In vitro tissue culture, a tool for the study and breeding of plants subjected to abiotic stress conditions. In *Recent Advances in Plant In Vitro Culture*; Leva, A., Rinaldi, L.M.R., Eds.; IntechOpen: London, UK, 2012. [CrossRef]
26. Vreugdenhil, D.; Boogaard, Y.; Visser, R.G.F.; de Bruijn, S.M. Comparison of tuber and shoot formation from in vitro cultured potato explants. *Plant Cell Tissue Organ Cult.* **1998**, *53*, 197–204. [CrossRef]
27. Zgallai, H.; Steppe, K.; Lemeur, R. Photosynthetic, physiological and biochemical responses of tomato plants to polyethylene glycol- induced water deficit. *J. Integr. Plant Biol.* **2005**, *47*, 1470–1478. [CrossRef]
28. Kaydan, D.; Yagmur, M. Germination, seedling growth and relative water content of shoot in different seed sizes of triticale under osmotic stress of water and NaCl. *Afr. J. Biotechnol.* **2008**, *7*, 2862–2868.
29. Carpita, N.; Sabularse, D.; Montezinos, D.; Delmer, D.P. Determination of the pore size of cell walls of living plant cells. *Science* **1979**, *205*, 1144–1147. [CrossRef]
30. Verslues, P.E.; Ober, E.S.; Sharp, R.E. Root growth and oxygen relations at low water potentials. Impact of oxygen availability in polyethylene glycol solutions. *Plant Physiol.* **1998**, *116*, 1403–1412. [CrossRef]
31. Trip, P.; Krotkov, G.; Nelson, C.D. Metabolism of mannitol in higher plants. *Am. J. Bot.* **1964**, *51*, 828–835. [CrossRef]
32. Dobránszki, J.; Magyar-Tábori, K.; Takács-Hudák, Á. Growth and developmental responses of potato to osmotic stress under *in vitro* conditions. *Acta Biol. Hung.* **2003**, *54*, 365–372. [CrossRef]
33. Zaki, H.E.M.; Radwan, K.S.A. Response of potato (*Solanum tuberosum* L.) cultivars to drought stress under in vitro and field conditions. *Chem. Biol. Technol. Agric.* **2022**, *9*, 1. [CrossRef]
34. Gelmesa, D.; Dechassa, N.; Mohammed, W.; Gebre, E.; Monneveux, P.; Bündig, C.; Winkelmann, T. In vitro screening of potato genotypes for osmotic stress tolerance. *Open Agric.* **2017**, *2*, 308–316. [CrossRef]
35. Lahlou, O.; Ledent, J.-F. Root mass and depth, stolons and roots formed on stolons in four cultivars of potato under water stress. *Eur. J. Agron.* **2005**, *22*, 159–173. [CrossRef]
36. Boguszewska-Mańkowska, D.; Zarzyńska, K.; Nosalewicz, A. Drought differentially affects root system size and architecture of potato cultivars with differing drought tolerance. *Am. J. Potato Res.* **2020**, *97*, 54–62. [CrossRef]
37. Albiski, F.; Najla, S.; Sanoubar, R.; Alkabani, N.; Murshed, R. In vitro screening of potato lines for drought tolerance. *Physiol. Mol. Biol. Plants* **2012**, *18*, 315–321. [CrossRef] [PubMed]
38. Sirait, B.A.; Charloq, R. In vitro study of potato (*Solanum tuberosum* L.) tolerant to the drought stress. NRLS Conference Proceedings. *KnE Life Sci.* **2017**, 188–192. [CrossRef]
39. Evers, D.; Bonnechère, S.; Hoffmann, L.; Hausman, J.F. Physiological aspects of abiotic stress response in potato. *Belg. J. Bot.* **2007**, *140*, 141–150.
40. Hassanpanah, D. Evaluation of potato advanced cultivars against water deficit stress under in vitro and in vivo condition. *Biotechnol. J.* **2010**, *9*, 164–169. [CrossRef]
41. Steckel, J.R.A.; Gray, D. Drought tolerance in potatoes. *J. Agric. Sci.* **1979**, *92*, 375–381. [CrossRef]
42. Gorji, A.M.; Mátyás, K.K.; Dublecz, Z.; Decsi, K.; Cernák, I.; Hoffmann, B.; Taller, J.; Polgár, Z. In vitro osmotic stress tolerance in potato and identification of major QTLs. *Am. J. Pot. Res.* **2012**, *89*, 453–464. [CrossRef]

43. Laisina, J.K.J.; Maharijaya, A.; Sobir, A.; Purwito, A. Penapisan Secara In vitro kentang (*Solanum tuberosum* L.) toleran kekeringan koleksi IPB. (In vitro screening of drought tolerant potatoes (*Solanum tuberosum* L.) of IPB collections). *JIPI* **2021**, *26*, 235–242. [CrossRef]

44. Murashige, T.; Skoog, F. A revised medium for rapid growth and bioassays with tobacco tisssue cultures. *Phys. Plant* **1962**, *15*, 473–497. [CrossRef]

45. Kpoghomou, B.K.; Sapra, V.T.; Beyl, C.A. Sensitivity to drought stress of three soybean cultivars during different growth stages. *J. Agron. Crop Sci.* **1990**, *164*, 104–109. [CrossRef]

46. Ramakrishnan, A.P.; Ritland, C.E.; Blas Sevillano, R.H.; Riseman, A. Review of potato molecular markers to enhance trait selection. *Am. J. Potato Res.* **2015**, *92*, 455–472. [CrossRef]

47. Schafleitner, R.; Gutierrez, R.; Espino, R.; Gaudin, A.; Pérez, J.; Martínez, M.; Domínguez, A.; Tincopa, L.; Alvarado, C.; Numberto, G. Field screening for variation of drought tolerance in *Solanum tuberosum* L. by agronomical, physiological and genetic analysis. *Pot. Res.* **2007**, *50*, 71–85. [CrossRef]

48. Jefferies, R.A. Responses of potato genotypes to drought. I. Expansion of individual leaves and osmotic adjustment. *Ann. Appl. Biol.* **1993**, *122*, 93–104. [CrossRef]

49. Blum, A. Osmotic adjustment is a prime drought stress adaptive engine in support of plant production. *Plant Cell Environ.* **2017**, *40*, 4–10. [CrossRef]

50. Banik, P.; Zeng, W.; Tai, H.; Bizimungu, B.; Tanino, K. Effects of drought acclimation on drought stress resistance in potato (*Solanum tuberosum* L.) genotypes. *Environ. Exp. Bot.* **2016**, *126*, 76–89. [CrossRef]

51. Dorneles, A.O.S.; Pereira, A.S.; da Silva, T.B.; Taniguchi, M.; Bortolin, G.S.; Castro, C.M.; da Silva Pereira, A.; Júnior, C.R.; do Amarante, L.; Haerter, J.A.; et al. Responses of *Solanum tuberosum* L. to water deficit by matric or osmotic induction. *Pot. Res.* **2021**, *64*, 515–534. [CrossRef]

52. Aliche, E.B.; Oortwijn, M.; Theeuwen, T.P.J.M.; Bachem, C.W.B.; Visser, R.G.F.; van der Linden, C.G. Drought response in field grown potatoes and the interactions between canopy growth and yield. *Agric. Water Manag.* **2018**, *206*, 20–30. [CrossRef]

53. Ryan, P.R.; Delhaize, E.; Watt, M.; Richardson, A.E. Plant roots: Understanding structure and function in an ocean of complexity. Preface: Part of a Special Issue on Root Biology. *Ann. Bot.* **2016**, *118*, 555–559. [CrossRef]

54. Qaderi, M.; Martel, A.; Dixon, S. Environmental factors influence plant vascular system and water regulation. *Plants* **2019**, *8*, 65. [CrossRef]

55. Nasir, M.W.; Toth, Z. Response of different potato genotypes to drought stress. *Agriculture* **2021**, *11*, 763. [CrossRef]

56. Jámbor-Benczúr, E.; Kissimon, J.; Fábián, M.; Mészáros, A.; Sinkó, Z.; Gazdag, G.Y.; Nagy, T. In vitro rooting and anatomical study of leaves and roots of in vitro and ex vitro plants of *Prunus x davidopersica* "Piroska". *Int. J. Hortic. Sci.* **2001**, *7*, 42–46. [CrossRef]

57. Rao, I.M.; Miles, J.W.; Beebe, S.E.; Horst, W.J. Root adaptations to soils with low fertility and aluminium toxicity. Review: Part of a special issue on root biology. *Ann. Bot.* **2016**, *118*, 593–605. [CrossRef] [PubMed]

58. Iwama, K. Physiology of the potato: New insights into root system and repercussions for crop management. *Pot. Res.* **2008**, *51*, 333–353. [CrossRef]

59. Boudiar, R.; Casas, A.M.; Gioia, T.; Fiorani, F.; Nagel, K.A.; Igartua, E. Effects of low water availability on root placement and shoot development in landraces and modern barley cultivars. *Agronomy* **2020**, *10*, 134. [CrossRef]

60. Xu, W.; Cui, K.; Aihui Xu, A.; Nie, L.; Huang, J.; Peng, S. Drought stress condition increases root to shoot ratio via alteration of carbohydrate partitioning and enzymatic activity in rice seedlings. *Acta Physiol. Plant* **2015**, *37*, 9. [CrossRef]

61. Khan, A.S.; Allah, S.U.; Sadique, S. Genetic variability and correlation among seedling traits of wheat (*Triticum aestivum*) under water stress. *Int. J. Agric. Biol.* **2010**, *12*, 247–250.

62. Ahmed, H.G.M.-D.; Sajjad, M.; Li, M.; Azmat, M.A.; Rizwan, M.; Maqsood, R.H.; Khan, S.H. Selection criteria for drought-tolerant bread wheat genotypes at seedling stage. *Sustainability* **2019**, *11*, 2584. [CrossRef]

63. Gangopadhyay, G.; Basu, S.; Mukherjee, B.B.; Gupta, S. Effects of salt and osmotic shocks on unadapted and adapted callus lines of tobacco. *Plant Cell Tissue Organ Cult.* **1997**, *49*, 45–52. [CrossRef]

64. Lawlor, D.W. Absorption of polyethylene glycols by plants and their effects on plant growth. *New Phytol.* **1970**, *69*, 501–513. [CrossRef]

65. Thimann, K.V.; Loos, G.M.; Samuel, E.W. Penetration of D-mannitol into potato discs. *Plant Physiol.* **1960**, *35*, 848–853. [CrossRef]

66. Lipavská, H.; Vreugdenhil, D. Uptake of mannitol from the media by in vitro grown plants. *Plant Cell Tissue Organ Cult.* **1996**, *45*, 103–107. [CrossRef]

67. Hudák, I.; Dobránszki, J.; Stefanovits-Bányai, É.; Hevesi, M.; Magyar-Tábori, K. Influence of osmotic stress on biochemical properties in potato. *Acta Hortic.* **2009**, *812*, 237–240. [CrossRef]

68. Hudák, I.; Dobránszki, J.; Sárdi, É.; Hevesi, M. Changes in carbohydrate content of potato calli during osmotic stress induced by mannitol. *Acta Biol. Hung.* **2010**, *61*, 234–236. [CrossRef] [PubMed]

Article

Gamma Radiation (^{60}Co) Induces Mutation during In Vitro Multiplication of Vanilla (*Vanilla planifolia* Jacks. ex Andrews)

María Karen Serrano-Fuentes [1], Fernando Carlos Gómez-Merino [1], Serafín Cruz-Izquierdo [2], José Luis Spinoso-Castillo [1] and Jericó Jabín Bello-Bello [3,*]

[1]   Colegio de Postgraduados, Campus Córdoba, Carretera Córdoba Veracruz, Amatlán de los Reyes Km 348, Veracruz 94946, Mexico; ame_karen15@hotmail.com (M.K.S.-F.); fernandg@colpos.mx (F.C.G.-M.); jlspinoso@gmail.com (J.L.S.-C.)
[2]   Colegio de Postgraduados, Campus Montecillo, Carretera México-Texcoco, Montecillo, Texcoco Km 36.5, Texcoco 56230, Mexico; sercruz@colpos.mx
[3]   CONACYT—Colegio de Postgraduados, Campus Córdoba, Carretera Córdoba Veracruz, Amatlán de los Reyes Km 348, Veracruz 94946, Mexico
*   Correspondence: jericobello@gmail.com; Tel.: +52-228-753-3476

Abstract: In vitro mutagenesis is an alternative to induce genetic variation in vanilla (*Vanilla planifolia* Jacks. ex Andrews), which is characterized by low genetic diversity. The objective of this study was to induce somaclonal variation in *V. planifolia* by gamma radiation and detect it using inter-simple sequence repeat (ISSR) molecular markers. Shoots previously established in vitro were multiplied in Murashige and Skoog culture medium supplemented with 2 mg·L^{-1} BAP (6-benzylaminopurine). Explants were irradiated with different doses (0, 20, 40, 60, 80 and 100 Gy) of ^{60}Co gamma rays. Survival percentage, number of shoots per explant, shoot length, number of leaves per shoot, and lethal dose (LD$_{50}$) were recorded after 60 d of culture. For molecular analysis, ten shoots were used for each dose and the donor plant as a control. Eight ISSR primers were selected, and 43 fragments were obtained. The percentage of polymorphism (% P) was estimated. A dendrogram based on Jaccard's coefficient and the neighbor joining clustering method was obtained. Results showed a hormetic effect on the explants, promoting development at low dose (20 Gy) and inhibition and death at high doses (60–100 Gy). The LD$_{50}$ was observed at the 60 Gy. Primers UBC-808, UBC-836 and UBC-840 showed the highest % P, with 42.6%, 34.7% and 28.7%, respectively. Genetic distance analysis showed that treatments without irradiation and with irradiation presented somaclonal variation. The use of gamma rays during in vitro culture is an alternative to broaden genetic diversity for vanilla breeding.

Keywords: gamma rays; hormetic effect; ISSR markers; polymorphism; somaclonal variation

Citation: Serrano-Fuentes, M.K.; Gómez-Merino, F.C.; Cruz-Izquierdo, S.; Spinoso-Castillo, J.L.; Bello-Bello, J.J. Gamma Radiation (^{60}Co) Induces Mutation during In Vitro Multiplication of Vanilla (*Vanilla planifolia* Jacks. ex Andrews). *Horticulturae* **2022**, *8*, 503. https://doi.org/10.3390/horticulturae8060503

Academic Editor: Sergio Ruffo Roberto

Received: 11 May 2022
Accepted: 2 June 2022
Published: 5 June 2022

Publisher's Note: MDPI stays neutral with regard to jurisdictional claims in published maps and institutional affiliations.

1. Introduction

Vanilla (*Vanilla planifolia* Jacks. ex Andrews), of the family Orquidaceae, is cultivated for its fruits for the extraction of vanillin, one of the most valuable spices in the food, cosmetic and pharmaceutical industries [1,2]. Despite its economic importance, *V. planifolia* is classified in the category B2ab (iii, v) "Endangered" in The International Union for Conservation of Nature (IUCN) Red List (http://www.iucnredlist.org accessed on 22 June 2021) version 3.1. In Mexico, *V. planifolia* is a classified species with special protection due to severe fragmentation of its habitat [3].

Vanilla is propagated asexually by cuttings and is manually pollinated, obtaining pods that contain seeds with low or no germination [1,4]. Commercial propagation by cuttings limits the genetic diversity of this species, causing susceptibility to pests and diseases and loss of tolerance to abiotic factors [5], leading to premature fruit drop. Therefore, expanding the genetic diversity of *V. planifolia* is important in breeding programs [6].

An alternative to induce genetic variability is by in vitro mutagenesis techniques [7–9]. Genetic variations obtained from in vitro culture are also called somaclonal variations [10].

Somaclonal variations obtained by in vitro mutagenesis with gamma radiation are free of regulatory constraints and allow the regeneration of genetic variations in a short period of time at low cost, added to which this system provides easy manipulation of explants in confined and controlled spaces under aseptic conditions [11–13]. Mutagenesis with cobalt 60 (^{60}Co) has high penetration potential, poses no risk to the environment and can be used to irradiate cells, tissues, organs and whole plants [14,15]. In vitro mutagenesis with ^{60}Co has been used in the breeding of San Francisco lily (*Laelia autumnalis*) [16], rice (*Oryza sativa* L.) [17], ginger (*Zingiber officinale* Rosc.) [8], potato (*Solanum tuberosum* L.) [18], and tomato (*Lycopersicon esculentum* L.) [19].

Molecular marker analysis is an important tool to estimate somaclonal variation [20–22]. Among molecular markers, inter simple sequence repeats (ISSRs) are characterized by being dominant, reproducible, inexpensive and they do not require prior knowledge of the genome [23]. Recently, ISSRs have been studied to analyze somaclonal variation in broadleaf plantain (*Plantago major*) [24], sugarcane (*Saccharum* spp.) [25], tulip (*Tulipa suaveolens*) [26], arracacha (*Arracacia xanthorrhiza*) [27] and *Disanthus cercidifolius* Maxim., an ornamental shrub [28]. In *V. planifolia*, ISSRs have been previously used by Bello-Bello et al. [29], Ramírez-Mosqueda et al. [30] and Pastelín-Solano et al. [31]. The aim of this study was to induce somaclonal variation in *V. planifolia* by means of gamma irradiation with ^{60}Co and detect it using ISSR molecular markers.

2. Materials and Methods

2.1. Plant Material

The mother plant was collected from a commercial plantation in Veracruz, Mexico (19.39°81′00″ N, −96.76°22′62″ W). The accession was identified by the National Agro-Alimentary Health, Safety and Quality Service (SENASICA, certificate No. LAB 30044001/2018) and handling authorization was approved by the Mexico program of the Ministry of the Environment and Natural Resources (permission No. SGPA/DGVS/2868/19). The accession has been deposited in a public in vitro germplasm bank at the Plant Tissue Culture Laboratory of Colegio de Postgraduados, Veracruz, Mexico. In this study, all procedures performed were in compliance with the IUCN Policy Statement on Research Involving Species at Risk of Extinction and the Convention on the Trade in Endangered Species of Wild Fauna and Flora.

2.2. In Vitro Establishment and Multiplication

Young, 20–30 cm long vanilla mother seedling stems containing three buds were used for in vitro establishment under greenhouse conditions. Leaves were removed from the stems and 2 cm long nodal segments containing one bud were cut. The nodes were used as explants and washed with running water and two drops of Tween 20 (Sigma-Aldrich, Chemical Company, St. Louis, MO, USA) for 30 min. Subsequently, the explants were immersed in a 0.1 mg·L^{-1} fungicide solution of 50 WP Captan-ultra (Arysta Life Science México S.A. de C.V. Coah., MX) for 30 min. The explants were submerged in a 5% (w/v) solution of NaClO (Cloralex, Industrias Alen, S.A. de C.V, NL, MX) (6% a.i.). Finally, they were immersed in a 1.3% (w/v) mercuric chloride (HgCl$_2$) solution for 15 min and rinsed five times with sterile distilled water. The explants were cultured individually in 22 × 150 mm test tubes containing 10 mL MS [32] medium, supplemented with 2 mg·L^{-1} 6-benzylaminopurine (BAP) (Sigma-Aldrich, St. Louis, MO, USA) and 30 g·L^{-1} sucrose. The Murashige and Skoog (MS) medium was adjusted to a pH of ±5.8 with 1 N sodium hydroxide (NaOH), and 0.25% (w/v) phytagel (Sigma-Aldrich) was used. The material was sterilized in an autoclave at 120 °C for 20 min. Cultures were incubated at 24 °C and the photoperiod was 16 h with a white LED light (460 and 560 nm) at an irradiance of 45 ± 5 µmol m^{-2} s^{-1}. After two weeks of culture, the explants were transferred to 500 mL jars with 30 mL MS multiplication medium supplemented with 2 mg·L^{-1} BAP under the aforementioned light and temperature conditions. Three subcultures were performed in 60-day periods prior to the ^{60}Co irradiation treatments.

2.3. Gamma Irradiation of In Vitro Explants

In vitro regenerated shoots were obtained, at the differentiation during multiplication stage, of approximately 2 cm in length and were irradiated at the National Institute for Nuclear Research, located in the city of Toluca, Mexico, with a Transelektro irradiator (LG1-01, Budapest, HU) using gamma rays with ^{60}Co. The explants were treated with doses of 0, 20, 40, 40, 60, 80 and 100 Gy, with an exposure time of 2.1, 4.2, 6.3, 8.3, and 10.4 min, respectively. Shoot irradiation was carried out in glass Petri dishes containing MS medium without growth regulators. Twelve explants were used, with four explants per petri dish. Subsequently, the irradiated explants were immediately transferred to MS multiplication medium to avoid the possible effects of denaturation of the medium components. The explants were cultured for 60 d in the multiplication medium described above, under the aforementioned light and temperature conditions. The variables to be measured were survival percentage, number of shoots per explant, shoot length and number of leaves per shoot. The lethal dose (LD_{50}) was calculated by observing survival percentage of explants at 60 days.

2.4. Data and Molecular Analysis

A completely random design was used for the gamma irradiation and in vitro multiplication. All treatments were performed in triplicate. Data were subjected to an analysis of variance and Tukey test ($p < 0.05$) using IBM SPSS statistical software v21. Percentage data were transformed with the formula $Y = \text{arcsine} (\sqrt{(\times/100)})$, where $\times$ is the percentage value.

A binary matrix was made with the ISSR fragment bands and recorded as present (1) or absent (0). For each primer, the percentage of polymorphism was calculated. In addition, a cluster analysis was performed using Jaccard's similarity coefficient based on similarity between sets of samples relative to matches, where 1 is similarity and 0 is divergence, and the neighbor joining (NJ) agglomeration model based on minimum evolution [33]. The donor plant was set as an outgroup. Neighbor joining clustering was performed after 1000 Bootstrap replicates. The resulting cluster was expressed as a dendrogram using the PAST software v3.04.

2.5. DNA Extraction and ISSR Analysis

The donor plant and non-irradiated explants were used as controls. Leaf samples from the donor plant and ten randomly selected shoots per irradiation dose were used in all experiments. For DNA isolation, the Stewart and Via [34] method was performed. The amount and purity of the DNA was determined by a spectrophotometer (Nanodrop 2000, Thermo Scientific, Waltham, MA, USA). The DNA samples used were those obtained with an A260/230 ratio of 1.8–2.2. The DNA was verified on a 1.5% (w/v) agarose gel stained with 3 mg·mL^{-1} ethidium bromide (Sigma-Aldrich) at a concentration of 0.1 mg·mL^{-1} using TAE 1X as buffer. To detect the polymorphism of the *V. planifolia* genomic DNA, 14 ISSR primers were evaluated. Eight primers were selected based on its reproducibility of the banding patterns (Table 1).

The PCR reactions were brought to a volume of 25 μL, with 30 ng template DNA, 1.5 U GoTaq DNA Polymerase (Promega, Madison, WI, USA), 1X Buffer (10 mM Tris-HCL and 50 mM KCL), 2.5 mM $MgCl_2$, 0.2 mM dNTPs and 0.5 μM primer. The products were amplified in an Engine System thermal cycler (PTC-200, BIO-RAD, Watertown, MA). A program starting with one cycle for 4 min at 94 °C was used, followed by 35 cycles for 50 s at 94 °C, 45–53 °C according to the primer for 50 s and 72 °C for 90 s. Finally, an extension at 72 °C for 10 min.

Table 1. Inter simple sequence repeat (ISSR) molecular markers selected to evaluate the somaclonal variation of vanilla (*Vanilla planifolia* Jacks. ex Andrews) at different doses of gamma irradiation with ^{60}Co.

Primers	Sequence (5′–3′)	Annealing Temperature (°C)	Bands	Range (bp)	Polymorphism (%)
UBC-809	AGAGAGAGAGAGAGAGG	45	7	250–1500	4
T06	AGAGAGAGAGAGAGAGT	50	4	400–750	3
UBC-840	GAGAGAGAGAGAGAGAYT	52	6	500–1500	28.7
UBC-836	AGAGAGAGAGAGAGAGYA	50	6	400–1500	34.7
UBC-812	GAGAGAGAGAGAGAGAA	50	4	500–1500	0
UBC-825	ACACACACACACACACT	50	6	500–2000	14.9
UBC-808	AGAGAGAGAGAGAGAGC	53	9	400–1000	42.6
T05	CGTTGTGTGTGTGTGTGT	53	1	750	0

bp = base pair; Y = C or T.

The amplification fragments were separated by electrophoresis on 2.5% (*w/v*) agarose gels in a 1X TAE buffer solution at 90 V for 1.5 h. The gels were stained with 3 mg·mL^{-1} ethidium bromide. A 1 kb Plus DNA ladder (Promega, Madison, WI, USA) was used. Finally, the gels were photographed using a gel documentation system (ChemiDocXRS, Bio-Rad, Hercules, CA, USA).

3. Results

3.1. Effect of Gamma Radiation on In Vitro Survival and Development

The different doses of gamma irradiation showed a significant effect on survival percentage, number of shoots per explant, and length and number of leaves per shoot (Figure 1). The LD$_{50}$ was observed at the 60 Gy, this dose reduced survival of explants by 52%. The highest survival percentage was observed at the 0 and 20 Gy doses, with 100% survival, whereas the lowest survival percentages were obtained at the 80 and 100 Gy doses, with 31 and 24% survival, respectively (Figure 1a). Regarding the number of shoots per explant, the highest number of new shoots was obtained at the 20 and 40 Gy doses, with 6 and 5.3 shoots per explant, respectively, whereas the lowest number of shoots was observed at the 60, 80 and 100 Gy doses, obtaining only 3 shoots per explant (Figure 1b). The longest shoot length was obtained at the 0 and 20 Gy doses, with 3.1 and 3 cm in length, respectively. The shortest length was observed at the 60, 80 and 100 Gy doses, with 1.6 to 1.8 cm in height (Figure 1c). The highest number of leaves was observed at the 0 and 20 Gy doses, with 3.3 and 3 leaves per shoot, whereas the lowest number of leaves per shoot was observed at 60, 80 and 100 Gy, with 1.6 to 1.7 leaves per shoot on average (Figure 1d). In addition, the administration of different doses of gamma irradiation had an effect on the multiplication rate in vanilla shoot development (Figure 2).

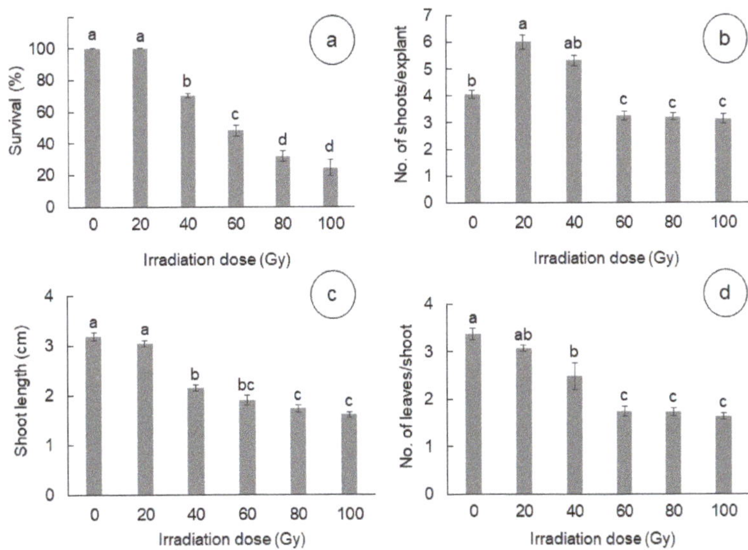

Figure 1. Effect of gamma radiation on in vitro survival and development of vanilla (*Vanilla planifolia* Jacks. ex Andrews). (**a**) Survival percentage, (**b**) shoots per explant, (**c**) shoot length and (**d**) leaves per shoot at 60 d of culture. Values represent the mean ± standard error. Means with different letters are significantly different (Tukey, $p < 0.05$).

Figure 2. Effect of different doses of gamma irradiation on in vitro shoot development of vanilla (*Vanilla planifolia* Jacks. ex Andrews) at 60 days of culture; (**a–f**) 0, 20, 40, 60, 80 and 100 Gy, respectively. Black bar = 1 cm.

3.2. DNA Polymorphism of Gamma Radiation on Somaclonal Variation

The ISSR analysis revealed the presence of monomorphic and polymorphic bands between the control and the different irradiation doses with respect to the donor plant (Figure 3 and Supplementary Figure S1). The eight selected ISSR primers amplified a total of 43 bands ranging from 250 to 2000 bp. The primers that showed the highest percentage of polymorphism were UBC-808 with 42.6% (9 bands), followed by UBC-836 with 34.7% (6 bands) and UBC-840 with 28.7% (6 bands). On the other hand, the primers that showed less than 15% polymorphism were UBC-825 (6 bands), T06 (4 bands) and UBC-809 (7 bands). Primers T05 (1 band) and UBC-812 (4 bands) revealed the presence of monomorphic fragments (Table 1).

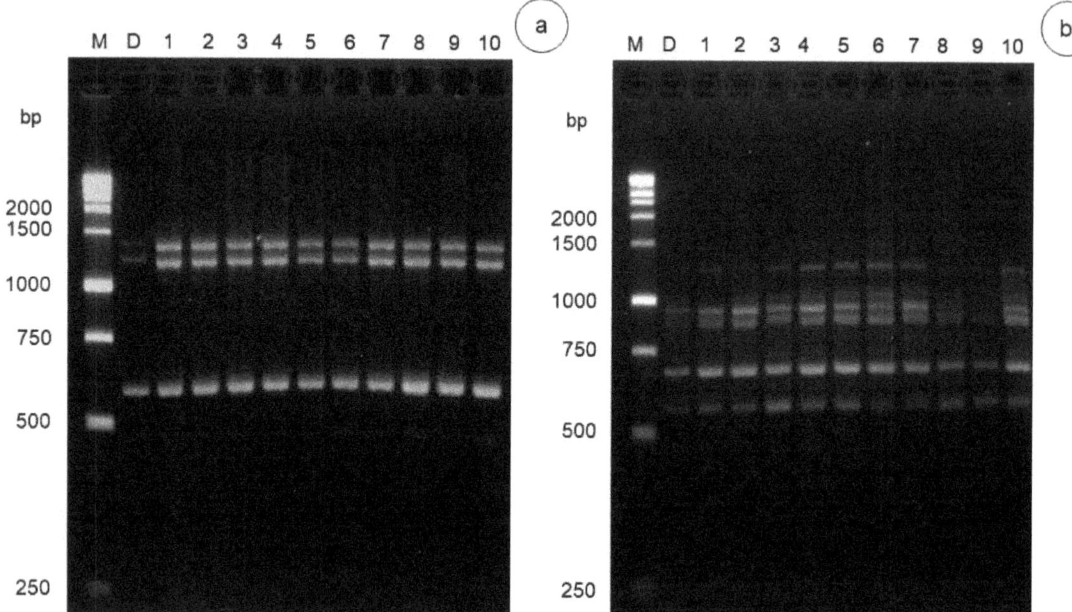

Figure 3. Electrophoresis pattern with ISSR markers of ten individuals (1–10) of vanilla (*Vanilla planifolia* Jacks. ex Andrews) exposed to gamma radiation with respect to donor plant (D). (**a**) Dose 60 Gy, primer UBC-812 (**b**) dose 20 Gy, primer UBC-840. M = molecular weight marker and bp = base pairs.

The dendrogram based on the neighbor joining model showed no similarity in the irradiation doses evaluated with respect to the donor plant. According to Bootstrap probabilities, similarity distances and branch lengths, the first group was considered to be the donor plant, whereas the second group comprised the doses of 0, 20, 40, 60, 80 and 100 Gy. The second group was divided into six subgroups, where subgroup six had the greatest similarity distance (0.76), formed by nine individuals of 100 Gy doses. The shortest distance was observed in the first and second subgroups with a distance of 0.88 and 0.86, respectively, with one individual of 0 and 20 Gy, respectively. For the rest of the subgroups, no clustering trend was observed with respect to the gamma radiation doses evaluated (Figure 4).

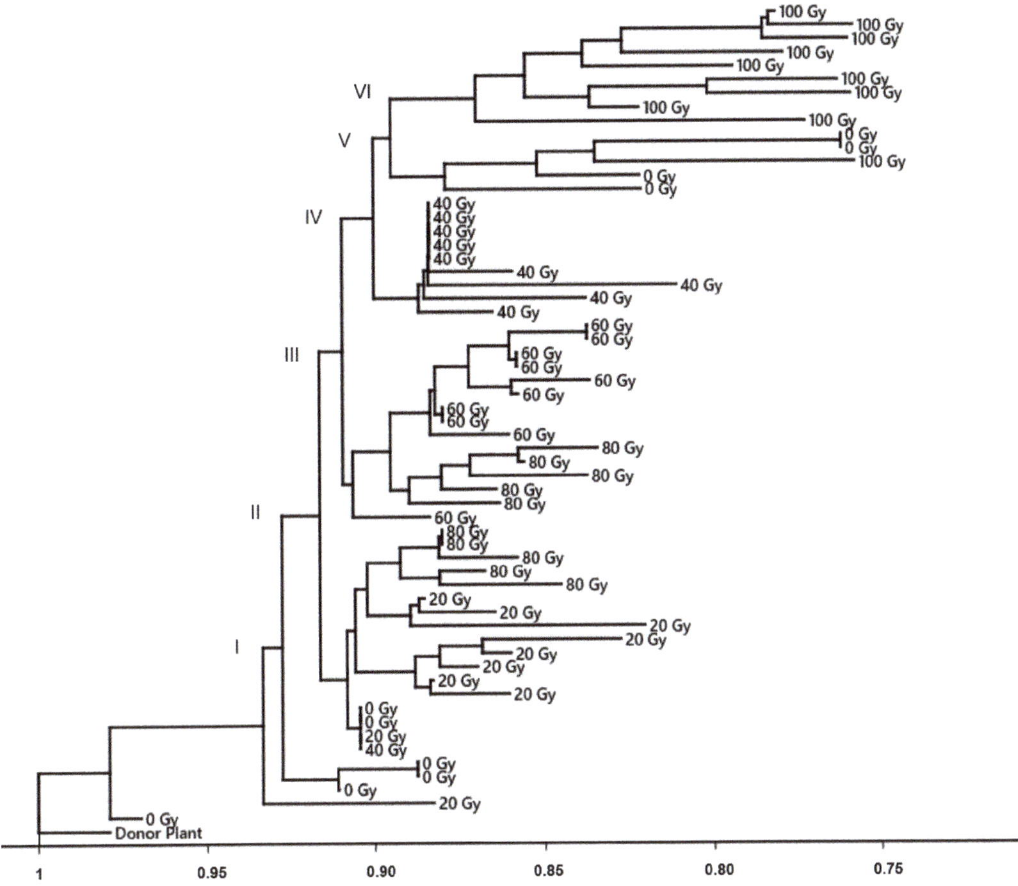

Figure 4. Neighbor joining dendrogram based on Jaccard's coefficient, calculated for each gamma irradiation dose with respect to the donor plant of vanilla (*Vanilla planifolia* Jacks. ex Andrews). The values under line represent: 1 similarity, and 0 divergence. I-VI: subgroups with different similarity distances.

4. Discussion

This study showed that gamma irradiation has an effect on survival and development during in vitro multiplication of *V. planifolia*. The mortality of the explants at doses higher than 60 Gy (DL_{50}) could be explained by the radiosensitivity of explants exposed to gamma rays. The sensitivity of explants to gamma radiation depends on tissue type, size, degree of development and water content [35]. Ionizing radiation interacts with atoms and molecules to produce free radicals in cells. These radicals can damage the structure of biomolecules such as carbohydrates, lipids, proteins, enzymes and nucleic acids, affecting the primary metabolism of plants [36,37]. In addition, irradiation can affect biochemical processes such as photosynthesis, respiration, krebs cycle and the metabolism of biomolecules. According to Hasbullah et al. [38] and Hernández-Muñoz et al. [35], irradiation can also affect cell division and cause damage to chromosomes and DNA.

The increase in the number of shoots at a dose of 20 Gy could be associated with a hormetic effect. According to Calabrese [39], the hormetic effect is characterized by beneficial or stimulation of development at low doses; and toxicity, inhibition or death at high doses. According to Jalal et al. [40], reactive oxygen species (ROS) are associated with hormesis because they are signaling molecules that trigger different physiological, biochemical and molecular processes in plant development. In this study, doses higher than 60 Gy caused the death of the explants and a reduction in the number of shoots per explant, shoot length and number of leaves per shoot. The reduced development and increased mortality rate at high doses could be associated with a longer time exposure to ^{60}Co. The high dosage of gamma ray causes production and accumulation of ROS, which are toxic to plant tissues [41,42]".

In this regard, Oliveira et al. [41] stated that the excess accumulation of free radicals resulting from water radiolysis produces negative effects on structural and functional biomolecules causing alterations in cellular metabolism, while Liu et al. [42] noted that high doses of gamma radiation can induce oxidative stress; this stress generates the formation of ROS that affects cell division and leads to apoptosis.

The mechanisms of the hormetic effect have yet to be fully elucidated; however, Iavicoli et al. [43] stated that hormesis is an adaptive preconditioning response to a stress of greater magnitude based on an evolutionary event. The hormetic effect has been observed in other in vitro mutagenesis studies with ^{60}Co in golden-flowered vetchling (*Lathyrus chrysanthus* Boiss) [44], San Francisco lily (*Laelia autumnalis*) [16], shoreline purslane (*Sesuvium portulacastrum*) [45] and more recently in rice (*Oryza sativa* L.) [17]. These results suggest that gamma irradiation with ^{60}Co could promote in vitro morphogenesis in explants in a dormant state in recalcitrant species.

The ISSR markers were able to detect somaclonal variations between individuals and the different doses of gamma irradiation evaluated. In this regard, Khan et al. [46] stated that ISSR markers produce multiple bands at the same locus, are highly reproducible and do not need prior information from the plant genome. In this study, primers UBC-808, UBC-836 and UBC-840 revealed the highest percentage of polymorphism and can be used for future analysis of somaclonal variation or genetic diversity in *V. planifolia*.

In general, individuals irradiated with doses of 100 Gy showed the least genetic similarity; however, for the rest of the doses evaluated, no clustering trend was observed. In vanilla, other studies involving somaclonal variation analysis using ISSR markers have observed that this species tends to be genetically unstable upon in vitro regeneration [29,31,47]. Ramírez-Mosqueda and Iglesias-Andreu [47] reported somaclonal variation during indirect organogenesis, with 71.66% polymorphism. Bello-Bello et al. [29] found an increase in the percentage of polymorphism with increasing concentrations of plant nanoparticles (AgNPs) in the culture medium during the growth of *V. planifolia*, with 25% polymorphism at a concentration of 200 mg·L^{-1} AgNPs. Pastelín-Solano et al. [31] demonstrated that the number of subcultures during direct organogenesis is an important factor in the increase in somaclonal variation, obtaining % P greater than 15% from subculture number six. On the other hand, other studies found no somaclonal variation in *V. planifolia* [30,48–50]. Sreedhar et al. [49] did not observe somaclonal variation during long-term growth using ISSR and RAPD markers. Gantait et al. [48] did not observe somaclonal variation during direct organogenesis using ISSR markers. Ramírez-Mosqueda et al. [30] in variegated plants obtained during direct organogenesis in temporary immersion obtained 0% polymorphism using ISSR markers. Recently, Manokari et al. [50] through direct organogenesis demonstrated 0% polymorphism using markers based on start codon targeted (SCoT) polymorphism.

Somaclonal variation during in vitro culture can originate through various aspects such as: explant type, regeneration pathway, subculture number, culture duration, growth regulator type, genotype and ploidy level [22–51]. However, somaclonal variation can be induced by chemical and physical mutagenic agents. Gamma irradiation using ^{60}Co can generate different types of mutations, namely deletions and insertions, translocations and base substitutions [52,53]. According to Jain [54], mutations produced by somaclonal variability are very similar to those produced spontaneously or by mutagenesis methods.

The somaclonal variation obtained in non-irradiated explants could be explained by the genetic nature of *V. planifolia*. In this regard, Nair and Ravindra [55], and Bory et al. [56]

observed in vanilla somatic associations and anomalies in the number of chromosomes, being lower than the reported $2n = 32$. This could explain why, during in vitro regeneration of vanilla, higher somaclonal variation is expected compared to other species that do not show somatic association or anomalies in the ploidy level. The somaclonal variation found in irradiated treatments, in addition to the genetic nature of the species, could be due to the high penetrating potential of gamma rays and mainly to the breaking of the chemical bonds in the DNA double strand, eliminating nucleotides or replacing them with new ones [14]. In this study, DNA mutations can probably affect homeotic genes with effects on the ability to regenerate new shoots.

Predieri [57] and Bairu et al. [51] state that in vitro culture increases the efficiency of mutagenic treatments due to the manipulation of explants in constant cell division under controlled conditions without biotic or abiotic factors that interfere with the mutagenic treatment. The effect of in vitro mutagenesis using ^{60}Co to broaden genetic variation for breeding purposes has been studied in San Francisco lily (*Laelia autumnalis*) [16], rice (*Oryza sativa* L.) [17], ginger (*Zingiber officinale* Rosc.) [8], potato (*Solanum tuberosum* L.) [18], and tomato (*Lycopersicon esculentum* L.) [19].

The species *V. planifolia* has low genetic diversity due to the cuttings-based asexual reproduction [58]. This commercial propagation method limits the diversity of the species. In vanilla, somaclonal variation is an alternative to broaden the genetic base of this species and generate new alleles [31] that can address the inbreeding depression of this species. The increase in the genetic diversity of vanilla is an important factor contributing to tolerance to abiotic and resistance to biotic factors caused by different climate change scenarios to avoid its extinction. Gamma radiation is a very useful mutagenesis method to generate genetic variations for the improvement of this species. In addition, future studies are required to analyze flowering stage and ripe fruits with morphological and biochemical markers to find possible phenotype variation.

5. Conclusions

In this study, it was observed that gamma radiation has a hormetic effect on explants, promoting the formation of new shoots at low dose (20 Gy) and inhibition of sprouting and death at high doses (60–100 Gy). Furthermore, in vitro regeneration via direct organogenesis and the different doses of gamma irradiation evaluated with ^{60}Co were shown to have an effect on somaclonal variation. The analysis of NJ clustering and Jaccard's genetic distance showed that the treatment without irradiation and the treatments with irradiation present genetic divergence from the donor plant. ISSR markers were shown to be efficient in detecting somaclonal variation. These results support the possibility of using gamma rays during in vitro culture to increase genetic diversity and undertake a vanilla breeding program.

Supplementary Materials: The following supporting information can be downloaded at: https://www.mdpi.com/article/10.3390/horticulturae8060503/s1, Figure S1: Uncropped blot of Figure 3a, Figure S2: Uncropped blot of Figure 3b.

Author Contributions: Conceptualization, M.K.S.-F. and J.J.B.-B.; designed the experiments, analyzed the data, conducted data interpretation and drafted the manuscript, M.K.S.-F.; conducted all the experimental work, F.C.G.-M., S.C.-I., J.L.S.-C. and J.J.B.-B.; contributed to the conceptualization of the experiment and revising the manuscript. All authors have read and agreed to the published version of the manuscript.

Funding: This research received no external funding.

Institutional Review Board Statement: Not applicable.

Informed Consent Statement: Not applicable.

Data Availability Statement: All data in this study can be found in the manuscript or in the Supplementary Materials.

Conflicts of Interest: The authors declare no conflict of interests.

References

1. Gantait, S.; Kundu, S. In vitro biotechnological approaches on *Vanilla planifolia* Andrews: Advancements and opportunities. *Acta Physiol. Plant.* **2017**, *39*, 196. [CrossRef]
2. Arya, S.S.; Rookes, J.E.; Cahill, D.M.; Lenka, S.K. Vanillin: A review on the therapeutic prospects of a popular flavouring molecule. *Adv. Tradit. Med.* **2021**, *21*, 1–17. [CrossRef]
3. SEMARNAT (Secretaría de Medio Ambiente y Recursos Naturales). *Norma Oficial Mexicana (NOM-059-ECOL-2001) de Protección Especial de Especies Nativas de México de Flora y Fauna Silvestres*; Diario Oficial de la Federación: Mexico City, Mexico, 2010; pp. 2–56.
4. Wan Abdullah, W.M.A.N.; Low, L.; Mumaiyizah, S.B.; Chai, Q.Y.; Loh, J.Y.; Ong-Abdullah, J.; Lai, K.S. Effect of lignosulphonates on *Vanilla planifolia* shoot multiplication, regeneration and metabolism. *Acta Physiol. Plant.* **2020**, *42*, 107. [CrossRef]
5. Li, J.; Demesyeux, L.; Brym, M.; Chambers, A.H. Development of species-specific molecular markers in Vanilla for seedling selection of hybrids. *Mol. Biol. Rep.* **2020**, *47*, 1905–1920. [CrossRef]
6. Hu, Y.; Resende, M.F.R.; Bombarely, A.; Brym, M.; Bassil, E.; Chambers, A.H. Genomics-based diversity analysis of Vanilla species using a *Vanilla planifolia* draft genome and Genotyping-By-Sequencing. *Sci. Rep.* **2019**, *9*, 3416. [CrossRef]
7. Li, F.; Shimizu, A.; Nishio, T.; Tsutsumi, N.; Kato, H. Comparison and Characterization of Mutations Induced by Gamma-Ray and Carbon-Ion Irradiation in Rice (*Oryza sativa* L.) Using Whole-Genome Resequencing. *G3 Genes Genom. Genet.* **2019**, *9*, 3743–3751. [CrossRef]
8. Sharma, V.; Thakur, M.; Tomar, M. In vitro selection of gamma irradiated shoots of ginger (*Zingiber officinale* Rosc.) against *Fusarium oxysporum* f.sp. zingiberi and molecular analysis of the resistant plants. *Plant Cell Tissue Organ Cult.* **2020**, *143*, 319–330. [CrossRef]
9. IAEA. Organismo Internacional de Energía Atómica. Inducción de Mutaciones. Available online: https://www.iaea.org/es/temas/induccion-de-mutaciones (accessed on 14 June 2021).
10. Larkin, P.J.; Scowcroft, W.R. Somaclonal variation—A novel source of variability from cell cultures for plant improvement. *Theor. Appl. Genet.* **1981**, *60*, 197–214. [CrossRef]
11. Waugh, R.; Leader, D.J.; Mc Callum, N.; Cadwell, D. Harvesting the potential of induced biological diversity. *Trends Plant Sci.* **2006**, *11*, 71–79. [CrossRef]
12. Ghani, M.; Sharma, S.K. Induction of powdery mildew resistance in gerbera (*Gerbera jamesonii*) through gamma irradiation. *Physiol. Mol. Biol. Plants* **2019**, *25*, 159–166. [CrossRef]
13. Salava, H.; Thula, S.; Mohan, V.; Kumar, R.; Maghuly, F. Application of Genome Editing in Tomato Breeding: Mechanisms, Advances, and Prospects. *Int. J. Mol. Sci.* **2021**, *22*, 682. [CrossRef] [PubMed]
14. Oladosu, Y.; Rafii, M.Y.; Abdullah, N.; Hussin, G.; Ramli, A.; Rahim, H.A.; Miah, G.; Usman, M. Principle and application of plant mutagenesis in crop improvement: A review. *Biotechnol. Biotechnol. Equip.* **2016**, *30*, 1–16. [CrossRef]
15. Spinoso-Castillo, J.L.; Escamilla-Prado, E.; Aguilar-Rincón, C.H.; Morales-Ramos, V.; García-de los Santos, G.; Corona-Torres, T. Physiological response of seeds of three coffee varieties to gamma rays (^{60}Co). *Rev. Chapingo Ser. Hortic.* **2021**, *27*, 101–112. [CrossRef]
16. Hernández-Muñoz, S.; Pedraza-Santos, M.E.; López, P.A.; De La Cruz-Torres, E.; Martínez-Palacios, A.; Fernández-Pavía, S.P.; Chávez-Bárcenas, A.T. Estimulación de la germinación y desarrollo in vitro de *Laelia autumnalis* con rayos gamma. *Rev. Fitotec. Mex.* **2017**, *40*, 271–283. [CrossRef]
17. Abdelnour-Esquivel, A.; Pérez, J.; Rojas, M.; Vargas, W.; Gatica-Arias, A. Use of gamma radiation to induce mutations in rice (*Oryza sativa* L.) and the selection of lines with tolerance to salinity and drought. *Vitr. Cell. Dev. Biol. Plant* **2020**, *56*, 88–97. [CrossRef]
18. Mohamed, E.A.; Osama, E.; Manal, E.; Samah, A.; Salah, G.; Hazem, K.M.; Nabil, E. Impact of gamma irradiation pretreatment on biochemical and molecular responses of potato growing under salt stress. *Chem. Biol. Technol. Agric.* **2021**, *8*, 35. [CrossRef]
19. Özge, Ç.; Alp, A.; Sinan, M.; Çimen, A. Comparison of tolerance related proteomic profiles of two drought tolerant tomato mutants improved by gamma radiation. *J. Biotechnol.* **2021**, *330*, 35–44. [CrossRef]
20. Pardo, A.; Hernández, A.; Méndez, N.; Alvarado, G. Análisis genético, mediante marcadores RAPD, de microbulbos de ajo conservados e irradiados in vitro. *Bioagro* **2015**, *27*, 143–150.
21. Muñoz-Miranda, L.A.; Rodríguez-Sahagún, A.; Acevedo, H.G.J.; Cruz-Martínez, V.O.; Torres-Morán, M.I.; Lépiz-Ildefonso, R.; Aarland, R.C.; Castellanos-Hernández, O.A. Evaluation of somaclonal and ethyl methane sulfonate-induced genetic variation of Mexican oregano (*Lippia graveolens* HBK). *Agronomy* **2019**, *9*, 166. [CrossRef]
22. Ranghoo-Sanmukhiya, V.M. Somaclonal Variation and Methods Used for its Detection. In *Propagation and Genetic Manipulation of Plants*; Siddique, I., Ed.; Springer Nature: Singapore, 2021. [CrossRef]
23. Zietkiewicz, E.; Rafalski, A.; Labuda, D. Genome Fingerprinting by Simple Sequence Repeat (SSR)-Anchores Polimerase Chain Reaction Amplification. *Genomics* **1994**, *20*, 176–183. [CrossRef]

24. Ghorbanpour, M.; Khadivi-Khub, A. Somaclonal variation in callus samples of Plantago major using inter-simple sequence repeat marker. *Caryologia* **2015**, *68*, 19–24. [CrossRef]

25. Martínez-Estrada, E.; Caamal-Velázquez, J.H.; Salinas-Ruiz, J.; Bello-Bello, J.J. Assessment of somaclonal variation during sugarcane micropropagation in temporary immersion bioreactors by intersimple sequence repeat (ISSR) markers. *Vitr. Cell. Dev. Biol. Plant* **2017**, *53*, 553–560. [CrossRef]

26. Kritskaya, T.A.; Kashin, A.S.; Kasatkin, M.Y. Micropropagation and Somaclonal Variation of *Tulipa suaveolens* (Liliaceae) in vitro. *Russ. J. Mar. Biol.* **2019**, *50*, 209–215. [CrossRef]

27. Vitamvas, J.; Viehmannova, I.; Cepkova, P.H.; Mrhalova, H.; Eliasova, K. Assessment of somaclonal variation in indirect morphogenesis-derived plants of *Arracacia xanthorrhiza*. *Pesqui. Agropecu. Bras.* **2019**, *54*, e00301. [CrossRef]

28. Ulvrova, T.; Vitamvas, J.; Cepkova, P.H.; Eliasova, K.; Janovska, D.; Bazant, V.; Viehmannova, I. Micropropagation of an ornamental shrub *Disanthus cercidifolius* Maxim. and assessment of genetic fidelity of regenerants using ISSR and flow cytometry. *Plant Cell Tissue Organ Cult.* **2021**, *144*, 555–566. [CrossRef]

29. Bello-Bello, J.J.; Spinoso-Castillo, J.L.; Arano-Avalos, S.; Martínez-Estrada, E.; Arellano-García, M.E.; Pestryakov, A.; Toledano-Magaña, Y.; García-Ramos, J.C.; Bogdanchikova, N. Cytotoxic, genotoxic, and polymorphism effects on *Vanilla planifolia* Jacks ex Andrews after long-term exposure to Argovit® silver nanoparticles. *Nanomaterials* **2018**, *8*, 754. [CrossRef]

30. Ramírez-Mosqueda, M.A.; Iglesias-Andreu, L.G.; Favián-Vega, E.; Telxeira da Silva, J.A.; Leyva-Ovalle, O.R.; Murguía-González, J. Morphogenetic stability of variegated *Vanilla planifolia* Jacks. plants micropropagated in a temporary immersion system (TIB®). *Rend. Lincei Sci. Fis. Nat.* **2019**, *30*, 603–609. [CrossRef]

31. Pastelín-Solano, M.C.; Salinas, R.J.; González, A.M.T.; Castañeda, C.O.; Galindo, T.M.E.; Bello-Bello, J.J. Evaluation of in vitro shoot multiplication and ISSR marker based assessment of somaclonal variants at different subcultures of vanilla (*Vanilla planifolia* Jacks). *Physiol. Mol. Biol. Plants* **2019**, *25*, 561–567. [CrossRef]

32. Murashige, T.; Skoog, F. A Revised medium for rapid growth and bio assays with tobacco tissue cultures. *Physiol. Plant.* **1962**, *15*, 473–497. [CrossRef]

33. Saitou, N.; Nei, M. The neighbor-joining method: A new method for reconstructing phylogenetic trees. *Mol. Biol. Evol.* **1987**, *4*, 406–425. [CrossRef]

34. Stewart, C.N.; Via, L.E. A rapid CTAB DNA isolation technique useful for RAPD fingerprinting and other PCR applications. *Biotechniques* **1993**, *14*, 748–750. [PubMed]

35. Hernández-Muñoz, S.; Pedraza-Santos, M.E.; López, P.A.; Gómez-Sanabria, J.M.; Morales-García, J.L. Mutagenesis in the improvement of ornamental plants. *Rev. Chapingo Ser. Hortic.* **2019**, *25*, 151–167. [CrossRef]

36. Chahal, G.S.; Gosal, S.S. *Principles and Procedures of Plant Breeding*; Alpha Science International Ltd.: Oxford, UK, 2002; pp. 399–412.

37. Magdy, A.M.; Fahmy, E.M.; Abd EL-Rahman, M.A.A.; Awad, G. Improvement of 6-gingerol production in ginger rhizomes (*Zingiber officinale* Roscoe) plants by mutation breeding using gamma irradiation. *Appl. Radiat. Isot.* **2020**, *162*, 109193. [CrossRef] [PubMed]

38. Hasbullah, N.A.; Taha, R.M.; Saleh, A.; Mahmad, N. Irradiation effect on in vitro organogenesis, callus growth and plantlet development of *Gerbera jamesonii*. *Hortic. Bras.* **2012**, *30*, 252–257. [CrossRef]

39. Calabrese, E.J. Hormesis: Path and progression to significance. *Int. J. Mol. Sci.* **2018**, *19*, 2871. [CrossRef]

40. Jalal, A.; Oliveira, J.J.C.; Ribeiro, J.S.; Fernandes, G.C.; Mariano, G.G.; Trindade, V.D.R.; Reis, A.R. Hormesis in plants: Physiological and biochemical responses. *Ecotoxicol. Environ. Saf.* **2021**, *207*, 111225. [CrossRef]

41. Oliveira, N.M.; de Medeiros, A.D.; de Lima Nogueira, M.; Arthur, V.; de Araújo Mastrangelo, T.; da Silva, C.B. Hormetic effects of low-dose gamma rays in soybean seeds and seedlings: A detection technique using optical sensors. *Comput. Electron. Agric.* **2021**, *187*, 106251. [CrossRef]

42. Liu, H.; Li, H.; Yang, G.; Yuan, G.; Ma, Y.; Zhang, T. Mechanism of early germination inhibition of fresh walnuts (*Juglans regia*) with gamma radiation uncovered by transcriptomic profiling of embryos during storage. *Postharvest Biol. Technol.* **2021**, *172*, 111380. [CrossRef]

43. Iavicoli, I.; Leso, V.; Fontana, L.; Calabrese, E.J. Nanoparticle exposure and hormetic dose–responses: An update. *Int. J. Mol. Sci.* **2018**, *19*, 805. [CrossRef]

44. Beyaz, R.; Kahramanogullari, C.T.; Yildiz, C.; Darcin, E.S.; Yildiz, M. The effect of gamma radiation on seed germination and seedling growth of *Lathyrus chrysanthus* Boiss. under in vitro conditions. *J. Environ. Radioact.* **2016**, *162*, 129–133. [CrossRef]

45. Kapare, V.; Satdive, R.; Fulzele, D.P.; Malpathak, N. Impact of Gamma Irradiation Induced Variation in Cell Growth and Phytoecdysteroid Production in *Sesuvium portulacastrum*. *J. Plant Growth Regul.* **2017**, *36*, 919–930. [CrossRef]

46. Khan, M.M.H.; Rafii, M.Y.; Ramlee, S.I.; Jusoh, M.; Al Mamun, M.; Halidu, J. DNA fingerprinting, fixation-index (Fst), and admixture mapping of selected Bambara groundnut (*Vigna subterranea* [L.] Verdc.) accessions using ISSR markers system. *Sci. Rep.* **2021**, *11*, 14527. [CrossRef] [PubMed]

47. Ramírez-Mosqueda, M.A.; Iglesias-Andreu, L.G. Indirect organogenesis and assessment of somaclonal variation in plantlets of *Vanilla planifolia* Jacks. *Plant Cell Tissue Organ Cult.* **2015**, *123*, 657–664. [CrossRef]

48. Sreedhar, R.V.; Venkatachalam, L.; Bhagyalakshmi, N. Genetic fidelity of long-term micropropagated shoot cultures of vanilla (*Vanilla planifolia* Andrews) as assessed by molecular markers. *Biotechnol. J.* **2007**, *2*, 1007–1013. [CrossRef]

49. Gantait, S.; Mandal, N.; Bhattacharyya, S.; Das, P.K.; Nandy, S. Mass multiplication of *Vanilla planifolia* with pure genetic identity confirmed by ISSR. *Int J. Plant Dev. Biol* **2009**, *3*, 18–23.

50. Manokari, M.; Priyadharshini, S.; Jogam, P.; Dey, A.; Shekhawat, M.S. Meta-topolin and liquid medium mediated enhanced micropropagation via *ex vitro* rooting in *Vanilla planifolia* Jacks. ex Andrews. *Plant Cell Tissue Organ Cult.* **2021**, *146*, 69–82. [CrossRef]

51. Bairu, M.W.; Aremu, A.O.; Van Staden, J. Variación somaclonal en plantas: Causas y métodos de detección. *Plant Growth Regul.* **2011**, *63*, 147–173. [CrossRef]

52. Yang, G.; Luo, W.; Zhang, J.; Yan, X.; Du, Y.; Zhou, L.; Li, W.; Wand, H.; Chen, Z.; Guo, T. Genome-wide comparisons of mutations induced by carbon-ion beam and gamma-rays irradiation in rice via resequencing multiple mutants. *Front. Plant Sci.* **2019**, *10*, 1514. [CrossRef]

53. Holme, I.B.; Gregersen, P.L.; Brinch-Pedersen, H. Induced genetic variation in crop plants by random or targeted mutagenesis: Convergence and differences. *Front. Plant Sci.* **2019**, *10*, 1468. [CrossRef]

54. Jain, S.M. Tissue culture-derived variation in crop improvement. *Euphytica* **2001**, *118*, 153–166. [CrossRef]

55. Nair, R.R.; Ravindran, P.N. Somatic association of chromosomes and other mitotic abnormalities in *Vanilla planifolia* (Andrews). *Caryologia* **1994**, *47*, 65–73. [CrossRef]

56. Bory, S.; Catrice, O.; Brown, S.; Leitch, I.J.; Gigant, R.; Chiroleu, F.; Grisoni, M.; Duval, M.F.; Besse, P. Natural polyploidy in *Vanilla planifolia* (Orchidaceae). *Genome* **2008**, *51*, 816–826. [CrossRef] [PubMed]

57. Predieri, S. Mutation induction and tissue culture in improving fruits. *Plant Cell Tissue Organ Cult.* **2001**, *64*, 185–210. [CrossRef]

58. Villanueva-Viramontes, S.; Hernández-Apolinar, M.; Fernández-Concha, G.C.; Dorantes-Euán, A.; Dzib, G.R.; Martínez-Castillo, J. Wild *Vanilla planifolia* and its relatives in the Mexican Yucatan Peninsula: Systematic analyses with ISSR and ITS. *Bot. Sci.* **2017**, *95*, 169–187. [CrossRef]

 *horticulturae*

Article

Optimum Sterilization Method for In Vitro Cultivation of Dimorphic Seeds of the Succulent Halophyte *Suaeda aralocaspica*

Yu Si [1,2], Yakupjan Haxim [1,3] and Lei Wang [1,2,*]

[1] State Key Laboratory of Desert and Oasis Ecology, Xinjiang Institute of Geography and Ecology, Chinese Academy of Sciences, Urumqi 830011, China; siyu19@mails.ucas.ac.cn (Y.S.); yakoo@ms.xjb.ac.cn (Y.H.)
[2] University of Chinese Academy of Sciences, Beijing 100049, China
[3] Turpan Eremophytes Botanical Garden, Chinese Academy of Sciences, Turpan 838008, China
* Correspondence: egiwang@ms.xjb.ac.cn; Tel.: +86-0991-7823189

Abstract: *Suaeda aralocaspica* is an annual halophyte in the Amaranthaceae in the saline deserts of central Asia. This plant has succulent leaves and grape-like fruits and is a potential horticultural plant. To obtain the efficient sterilization method and optimal culture conditions, two types of seeds produced from a single plant of *S. aralocaspica* were treated with 75% ethanol for different time durations first, and then sodium hypochlorite (NaClO) or mercury chloride ($HgCl_2$), with five different timing treatments were used for second seed surface sterilization. Sterilized seeds were germinated on a Murashige and Skoog (MS) medium at different potential hydrogenation (pH) levels, to examine germination and seedling performance. The results showed that the highest germination percentage of brown seeds was 100% and that of black seeds was 17%. Thus, brown seeds were more suitable for further culture experiments than black seeds. For brown seeds, the sterilization effect of NaClO was better than that of $HgCl_2$, based on the results of seed germination, contamination, and seedling survival. Rinsing with 75% ethanol for 60 s, sterilizing with NaClO for 8 min, and cultivating at pH 8.0 MS for 7 days was the best of all sterilization procedures and cultivation methods tested, which has been successfully applied to *S. aralocaspica* in vitro culture. The optimized protocol described here can be used as the reference for the *Suaeda* genus.

Keywords: sodium hypochlorite; mercury chloride; sterilization time; seed germination

Citation: Si, Y.; Haxim, Y.; Wang, L. Optimum Sterilization Method for In Vitro Cultivation of Dimorphic Seeds of the Succulent Halophyte *Suaeda aralocaspica*. *Horticulturae* **2022**, *8*, 289. https://doi.org/10.3390/horticulturae8040289

Academic Editor: Sergio Ruffo Roberto

Received: 18 February 2022
Accepted: 28 March 2022
Published: 29 March 2022

Publisher's Note: MDPI stays neutral with regard to jurisdictional claims in published maps and institutional affiliations.

1. Introduction

Suaeda aralocaspica is an annual halophyte (Amaranthaceae), with succulent leaves and grape-like fruits, distributed in the saline deserts of central Asia. In China, this plant is mainly distributed in the southern margin of the Junggar Basin in Xinjiang [1–3]. The study of *S. aralocaspica* is mainly focused on the leaf morphology and anatomical structure [4,5], germination characteristics [6–9], and photosynthetic type [10]. *S. aralocaspica* is a single-cell C_4 photosynthetic plant without Kranz anatomy and has two types of chloroplast, called 'Borszczowioid type' [10,11]. This plant can produce two distinct types of seeds on a single plant, which have obvious differences in seed coat color, seed size, dormancy and germination characteristics. The brown seeds have high salt tolerance and are not dormant, while the black seeds have low salt tolerance and non-deep physiological dormancy [12–14]. However, there is no significant difference in growth, mineral nutrient content, and salt tolerance at middle and late growth stages [12]. Besides, short time pre-soaking, with a low concentration of abscisic acid (ABA), promotes the germination and seedling growth of dimorphic seeds of *S. aralocaspica* [15].

There are few molecular studies about *S. aralocaspica*. Compared with four traditionally used reference genes, *GAPDH* and *β-TUB* are stable internal reference genes and more suitable for the subsequent gene research of *S. aralocaspica* under different experimental conditions [16]. The results of KEGG enrichment and gene expression analysis reveal that specific genes and miRNAs are regulated differently between black and brown seeds during germination, which may contribute to the different germination behaviors of dimorphic seeds of *S. aralocaspica* in unpredictable environments [17]. The sequencing and assembly of the *S. aralocaspica* whole genome are finished and the length is 425 Mb. In addition, a complete chloroplast genome is also assembled. *S. aralocaspica* is the first sequenced halophyte and single-cell C_4 plant [18].

Suaeda. aralocaspica can grow normally in typical saline soils (salt content of the topsoil exceeds 10%). Brown seeds can germinate at high salinity (over 1000 mmol/L NaCl). This plant is valuable for the study of salt tolerance, C_4 photosynthesis without Kranz anatomy, and seed heteromorphism. However, this plant is limited to saline deserts in central Asia, and the field sampling is also restricted by the season and the low density of persistent soil seed bank. These factors influence the supplement of *S. aralocaspica* material for experimental research. Thus, an efficient seed sterilization and culture method is essential for the effective supply of this research material.

Obtaining high-quality sterile seeds and seedlings is affected by various factors, such as the type of disinfectant, the time duration of sterilization, the pH of the medium, etc. [19]. Common surface disinfectants include ethanol, sodium hypochlorite, hydrogen peroxide, and mercury chloride [20–22]. As a commonly used medical disinfectant, 70–75% ethanol inactivates some bacteria by infiltrating through their cell membranes to denature various proteins. The bactericidal effect of 75% ethanol, when used together with other disinfectants, is better than that of using only ethanol as a disinfectant [22]. Mung bean seeds are disinfected with 75% ethanol for 30 s and then disinfected with 1.0% NaClO for 10 min [23]. *Salicornia europaea* seeds are treated with different concentrations of sodium hypochlorite (NaClO), mercuric chloride ($HgCl_2$), and hydrogen peroxide (H_2O_2) on a Murashige and Skoog (MS) medium, with different concentrations of the hormone. The optimal sterilization effect of *S. europaea* seeds is after they have been treated with mercuric chloride, with a quality fraction of 0.1% for 10–20 min [24].

Seed sterilization and aseptic seedling cultivation have played a crucial role in subsequent research. Sterile seedlings are the source of explants in the tissue culture system. However, the sterilization methods of *S. aralocaspica* seeds have not been reported. We hypothesized that dimorphic seeds had distinct responses to disinfectants, sterilization time, and pH value of the mediums. Therefore, the present study was conducted to compare different sterilizing protocols, employed for different types of seeds in vitro culture, and to find out the best and most efficient sterilization procedure, based on germination, contamination, and seedling survival, which can be used for the rapid propagation system of *S. aralocaspica*.

2. Materials and Methods

2.1. Seed and Pretreatment

Freshly matured fruits of *S. aralocaspica* were collected from Fukang, Xinjiang in October 2020. All fruits were manually rubbed to remove the fruit coat. Brown seeds and black seeds were hand-sorted before sterilization to ensure uniformity in type. The seeds were rinsed using running water for 30 s to remove the impurities and then dried naturally in the laboratory. Every group of 50 seeds was packed in a 2 mL Eppendorf tube for further use.

2.2. Preparation of Reagents

Precisely measured 36 mL 10% NaClO solution and 64 mL sterilized double distilled H_2O (ddH_2O) were poured into a 150 mL sterile conical flask. Then the mixture was the required solution (100 mL 3.6% NaClO) for this experiment. Accurately weighed 0.1 g $HgCl_2$ reagent powders (Analysis pure) using a calibrated and zeroed electronic balance were poured into a 100 mL sterile beaker. We slowly poured a small amount of sterile ddH_2O and stirred with a glass stick until the powder dissolved completely, and then poured the solution into a 100 mL volumetric flask. Next, took a small amount of the new sterile water rinse beaker and glass stirring bar, and combined the rinse solution into the volumetric flask. Repeated the rinsing step 3–4 times (the total volume of the liquid should be less than 100 mL). Finally, added an appropriate amount (depending on the situation) of sterile water to a constant volume of 100 mL. At this point, the 100 mL solution in the 100 mL volumetric flask was the 0.1% $HgCl_2$ solution required for the experiment. MS medium was poured into a disposable sterile bacterial petri dish of 90 mm (d) × 15 mm (h) after autoclave sterilization and solidified into a flat plate at room temperature. The preparation of all the above reagents was completed in an ultra-clean workbench. It is worth noting that mercuric chloride is toxic, so we should be careful when configuring and using it. Besides, adjusting the pH values of MS medium to 5.0, 6.0, 7.0, 8.0, 9.0 respectively was necessary before autoclave sterilization.

2.3. Sterilization and Germination Procedure

Both types of seeds were first sterilized with 75% ethanol, with a treatment time duration of 30 s, 1 min, 3 min, 5 min, and 8 min. Then seeds were rinsed with sterile distilled water 3 times. Following this, 3.6% NaClO or 0.1% $HgCl_2$ was selected as the secondary sterilizing agent. The soaking time duration of 3.6% NaClO was set to five gradients, including 3, 5, 8, 11, and 15 min and that of 0.1% $HgCl_2$ was also set to five gradients, including 1, 3, 5, 8, and 11 min. Then seeds were cultured on MS mediums with different pH values (Table 1). According to Orthogonal Table L25 (5^6), 3 columns were used in the test. The sterilization experiment for dimorphic seeds of *S. aralocaspica* was designed. To differentiate the treatments of 3.6% NaClO and 0.1% $HgCl_2$ for different types of seeds, 25 treatments of 3.6% NaClO for brown seeds were named N1–N25 (Table S1), and 25 treatments of 3.6% NaClO for black seeds were named n1–n25 (Table S2). Further, 25 treatments of 0.1% $HgCl_2$ brown seeds were named H1–H25 (Table S3), and the 25 treatments of black seeds treated with 0.1% HgCl2 were named h1–h25 (Table S4). Each experimental group in this study was repeated three times and contained 20 seeds. The above operations were completed in an ultraclean workbench.

Table 1. Sterilization programs and parameters of *Suaeda aralocaspica* seeds.

	Factor	Level 1	Level 2	Level 3	Level 4	Level 5
Seed sterilization program (I)	Ethanol	30 s	1 min	3 min	5 min	8 min
	NaClO	3 min	5 min	8 min	11 min	15 min
	pH	5.0	6.0	7.0	8.0	9.0
	Factor	Level 1	Level 2	Level 3	Level 4	Level 5
Seed sterilization program (II)	Ethanol	30 s	1 min	3 min	5 min	8 min
	$HgCl_2$	1 min	3 min	5 min	8 min	11 min
	pH	5.0	6.0	7.0	8.0	9.0

All Petri dishes were incubated in a growth chamber at 25/10 °C under a 14 h light/10 h dark photoperiod for 20 days. A seed was considered to be germinated when the radicle length reached 5 mm. Germinated seeds were recorded every day. The final germination percentage (Equation (1)), contamination percentage (Equation (2)) and seedling survival percentage (Equation (3)) were calculated after 20 days of cultivation.

$$\text{Final germination percentage (\%)} = \text{number of germinated seeds/numbers of tested seeds} \times 100\% \qquad (1)$$

$$\text{Contamination percentage (\%)} = \text{number of seeds contaminated by microorganism/number of tested seeds} \times 100\% \qquad (2)$$

$$\text{Seedling survival percentage (\%)} = \text{number of seedlings without contamination and browning/number of tested seeds} \times 100\% \qquad (3)$$

2.4. Statistical Analysis

All data were expressed as mean $\pm$ s.e. Arcsine transformation was performed before statistical analysis to meet assumptions. Linear mixed models were used to test the significance of main effects (time duration of ethanol, pH, secondary disinfectant type, time duration of secondary disinfectant, and seeds type) on final germination percentage, contamination percentage, and seedling survival percentage. The statistical analysis was performed using SPSS version 16 (SPSS for Windows, Released 2007, Chicago, IL, USA, SPSS Inc.). One-way ANOVA was used to compare treatments. For comparison, least significance difference test (LSD) ($p < 0.05$) was employed. Independent samples T Test was used to analyze differences between brown and black seeds under different secondary disinfectant treatments.

3. Results

3.1. The Effect of Different Treatments on Final Germination Percentage of S. aralocaspica Seeds

Ethanol time and pH had no significant effect on the germination of dimorphic seeds ($p > 0.05$). Secondary disinfectant type ($p < 0.001$), secondary disinfectant time ($p < 0.001$), and seed type ($p < 0.001$) significantly affected the seed germination percentage (Table 2). Compared with the seeds without sterilization treatment (Figure S1), the NaClO treatment did not affect the germination percentage of *S. aralocaspica* seeds. With the increase in the sterilization time duration of 0.1% $HgCl_2$, the germination percentage of brown seeds decreased significantly. The germination percentage of black seeds was low under all treatments and the highest germination percentage was only 16.67%.

Table 2. A mixed model ANOVA on final germination percentage of dimorphic seeds of *Suaeda aralocaspica*.

Source	Numerator df	Denominator df	F	Sig.
Ethanol time	4	285	1.132	0.341
pH value	4	285	0.435	0.783
Secondary disinfectant type	1	285	93.649	0.000
Secondary disinfectant time	4	285	5.753	0.000
Seed type	1	285	1230.597	0.000

When 3.6% NaClO was used, the germination percentages of brown seeds in treatment N11, N15, N16, N17 were 100%, and the germination percentage of other treatments, except N5 and N21, was $\geq$90% (Table 3). The germination percentages of black seeds in treatment n10 and n23 were the highest (16.67%). When 0.1% $HgCl_2$ was used, the germination percentage of brown seeds in treatment H1 and H7 was the highest (93.33%). The germination percentage of brown seeds in the other groups was from 20% to 90%. In contrast, black seeds did not germinate under many treatments. The highest germination percentage of treatment h13 was only 13.33%, and the other groups were less than 10%. The toxicity of 3.6% NaClO on seed germination of *S. aralocaspica* was lower than that of 0.1% $HgCl_2$.

Table 3. Final germination percentages of *Suaeda aralocaspica* seeds cultured with different treatments after 20 days incubation.

Treatment	3.6% NaClO		Treatment	3.6% NaClO		Treatment	0.1% HgCl$_2$		Treatment	0.1% HgCl$_2$	
	Brown Seeds			**Black Seeds**			**Brown Seeds**			**Black Seeds**	
N1	96.67 ± 3.33 aA		n1	10.00 ± 10.00 aB		H1	93.33 ± 3.33 aA		h1	3.33 ± 3.33 abB	
N2	90.00 ± 5.77 aA		n2	3.33 ± 3.33 aB		H2	83.33 ± 8.82 abcA		h2	3.33 ± 3.33 abB	
N3	93.33 ± 3.33 aA		n3	3.33 ± 3.33 aB		H3	86.67 ± 6.67 abA		h3	3.33 ± 3.33 abB	
N4	96.67 ± 3.33 aA		n4	6.67 ± 6.67 aB		H4	53.33 ± 13.33 cdefA		h4	0.00 ± 0.00 bB	
N5	76.67 ± 3.33 bA		n5	10.00 ± 0.00 aB		H5	26.67 ± 3.33 fghA		h5	6.67 ± 6.67 abA	
N6	93.33 ± 6.67 aA		n6	10.00 ± 10.00 aB		H6	73.33 ± 8.82 abcdeA		h6	3.33 ± 3.33 abB	
N7	90.00 ± 5.77 aA		n7	3.33 ± 3.33 aB		H7	93.33 ± 3.33 aA		h7	6.67 ± 6.67 abB	
N8	93.33 ± 6.67 aA		n8	6.67 ± 3.33 aB		H8	90.00 ± 5.77 aA		h8	3.33 ± 3.33 abB	
N9	90.00 ± 0.00 aA		n9	13.33 ± 3.33 aB		H9	50.00 ± 5.77 defghA		h9	6.67 ± 3.33 abB	
N10	93.33 ± 3.33 aA		n10	16.67 ± 12.02 aB		H10	40.00 ± 15.28 fghA		h10	10.00 ± 0.00 abA	
N11	100.00 ± 0.00 aA		n11	6.67 ± 3.33 aB		H11	76.67 ± 14.53 abcdeA		h11	6.67 ± 6.67 abB	
N12	93.33 ± 3.33 aA		n12	10.00 ± 0.00 aB		H12	73.33 ± 12.02 abcdeA		h12	0.00 ± 0.00 bB	
N13	96.67 ± 3.33 aA		n13	3.33 ± 3.33 aB		H13	80.00 ± 10.00 abcdA		h13	13.33 ± 6.67 aB	
N14	90.00 ± 0.00 aA		n14	6.67 ± 6.67 aB		H14	23.33 ± 3.33 fghA		h14	3.33 ± 3.33 abA	
N15	100.00 ± 0.00 aA		n15	0.00 ± 0.00 aB		H15	46.67 ± 6.67 efghA		h15	0.00 ± 0.00 bB	
N16	100.00 ± 0.00 aA		n16	3.33 ± 3.33 aB		H16	73.33 ± 14.53 abcdeA		h16	6.67 ± 3.33 abB	
N17	100.00 ± 0.00 aA		n17	6.67 ± 3.33 aB		H17	80 ± 11.55 abcdA		h17	3.33 ± 3.33 abB	
N18	93.33 ± 3.33 aA		n18	10.00 ± 0.00 aB		H18	46.67 ± 6.67 efghA		h18	6.67 ± 3.33 abB	
N19	93.33 ± 3.33 aA		n19	0.00 ± 0.00 aB		H19	50.00 ± 0.00 defghA		h19	0.00 ± 0.00 bB	
N20	90.00 ± 5.77 aA		n20	3.33 ± 3.33 aB		H20	23.33 ± 6.67 ghA		h20	0.00 ± 0.00 bA	
N21	70.00 ± 0.00 bA		n21	6.67 ± 3.33 aB		H21	83.33 ± 8.82 abcA		h21	0.00 ± 0.00 bB	
N22	93.33 ± 3.33 aA		n22	10.00 ± 0.00 aB		H22	40.00 ± 5.77 fghA		h22	0.00 ± 0.00 bB	
N23	96.67 ± 3.33 aA		n23	16.67 ± 3.33 aB		H23	56.67 ± 14.53 bcdefA		h23	0.00 ± 0.00 bB	
N24	96.67 ± 3.33 aA		n24	6.67 ± 3.33 aB		H24	20.00 ± 0.00 hA		h24	0.00 ± 0.00 bA	
N25	93.33 ± 6.67 aA		n25	10.00 ± 0.00 aB		H25	40 ± 17.32 fghA		h25	3.33 ± 3.33 abA	

Different lowercase letters in each column indicate significant differences among different treatments; different uppercase letters indicate significant difference between dimorphic seeds treated with the same secondary disinfectant.

3.2. The Effect of Different Treatments on Contamination Percentage of S. aralocaspica Seeds

Compared with the seeds that were not disinfected, almost all sterilization treatments had a good sterilizing effect. The effects of ethanol time ($p > 0.05$) and secondary disinfectant time ($p > 0.05$) on the contamination percentage were not significant. The effects of pH ($p < 0.05$), secondary disinfectant type ($p < 0.05$) and seed type ($p < 0.05$) on the contamination percentage were significant ($p < 0.05$) (Table 4).

Table 4. A mixed model ANOVA on contamination percentage of dimorphic seeds of *Suaeda aralocaspica*.

Source	Numerator df	Denominator df	F	Sig.
Ethanol time	4	285	1.182	0.319
pH value	4	285	5.936	0.000
Secondary disinfectant type	1	285	15.211	0.000
Secondary disinfectant time	4	285	0.264	0.901
Seed type	1	285	15.508	0.000

When using 3.6% NaClO as the secondary disinfectant, the brown seed contamination percentages were 56.67% and 53.33% in treatments N21 and N9, respectively, followed by 33.33%, 26.67% and 23.33% in treatments N19, N12 and N5, respectively (Table 5). The black seeds were most polluted in treatments n3 and n24, and the microbial contamination percentage was 6.67% (Table 5). When using 0.1% HgCl$_2$ as the secondary disinfectant, the sterilization effect was more significant, and only three treatments of both seeds were polluted. The highest contamination percentage of brown seeds was 23.33% in treatment

H20, and the contamination percentages in treatments H23 and H24 were 10% and 6.67%, respectively (Table 5). In treatment h1, 16.67% of the black seeds were polluted, which was the most serious, followed by treatment h17 and treatment h21, with the contamination percentage of 3.33%. Other treatment combinations were not polluted by microorganisms (Table 5).

Table 5. Contamination percentage of *Suaeda aralocaspica* seeds cultured with different treatments after 20 days incubation.

Treatment	3.6% NaClO Brown Seeds	Treatment	3.6% NaClO Black Seeds	Treatment	0.1% $HgCl_2$ Brown Seeds	Treatment	0.1% $HgCl_2$ Black Seeds
N1	6.67 ± 6.67 cA	n1	0.00 ± 0.00 aA	H1	0.00 ± 0.00 bA	h1	16.67 ± 16.67 aA
N2	0.00 ± 0.00 cA	n2	3.33 ± 3.33 aA	H2	0.00 ± 0.00 bA	h2	0.00 ± 0.00 bA
N3	6.67 ± 3.33 cA	n3	6.67 ± 6.67 aA	H3	0.00 ± 0.00 bA	h3	0.00 ± 0.00 bA
N4	0.00 ± 0.00 cA	n4	0.00 ± 0.00 aA	H4	0.00 ± 0.00 bA	h4	0.00 ± 0.00 bA
N5	23.33 ± 8.82 bcA	n5	0.00 ± 0.00 aA	H5	0.00 ± 0.00 bA	h5	0.00 ± 0.00 bA
N6	0.00 ± 0.00 cA	n6	0.00 ± 0.00 aA	H6	0.00 ± 0.00 bA	h6	0.00 ± 0.00 bA
N7	10.00 ± 10.00 cA	n7	0.00 ± 0.00 aA	H7	0.00 ± 0.00 bA	h7	0.00 ± 0.00 bA
N8	0.00 ± 0.00 cA	n8	0.00 ± 0.00 aA	H8	0.00 ± 0.00 bA	h8	0.00 ± 0.00 bA
N9	53.33 ± 8.82 abA	n9	3.33 ± 3.33 aB	H9	0.00 ± 0.00 bA	h9	0.00 ± 0.00 bA
N10	0.00 ± 0.00 cA	n10	0.00 ± 0.00 aA	H10	0.00 ± 0.00 bA	h10	0.00 ± 0.00 bA
N11	0.00 ± 0.00 cA	n11	0.00 ± 0.00 aA	H11	0.00 ± 0.00 bA	h11	0.00 ± 0.00 bA
N12	26.67 ± 26.67 abcA	n12	0.00 ± 0.00 aA	H12	0.00 ± 0.00 bA	h12	0.00 ± 0.00 bA
N13	13.33 ± 6.67 cA	n13	0.00 ± 0.00 aA	H13	0.00 ± 0.00 bA	h13	0.00 ± 0.00 bA
N14	0.00 ± 0.00 cA	n14	0.00 ± 0.00 aA	H14	0.00 ± 0.00 bA	h14	0.00 ± 0.00 bA
N15	0.00 ± 0.00 cA	n15	0.00 ± 0.00 aA	H15	0.00 ± 0.00 bA	h15	0.00 ± 0.00 bA
N16	0.00 ± 0.00 cA	n16	0.00 ± 0.00 aA	H16	0.00 ± 0.00 bA	h16	0.00 ± 0.00 bA
N17	20.00 ± 0.00 cA	n17	3.33 ± 3.33 aB	H17	0.00 ± 0.00 bA	h17	3.33 ± 3.33 bA
N18	6.67 ± 3.33 cA	n18	0.00 ± 0.00 aA	H18	0.00 ± 0.00 bA	h18	0.00 ± 0.00 bA
N19	33.33 ± 33.33 abcA	n19	0.00 ± 0.00 aA	H19	0.00 ± 0.00 bbA	h19	0.00 ± 0.00 bA
N20	6.67 ± 6.67 cA	n20	3.33 ± 3.33 aA	H20	23.33 ± 23.33 aA	h20	0.00 ± 0.00 bA
N21	56.67 ± 14.53 aA	n21	3.33 ± 3.33 aA	H21	0.00 ± 0.00 bA	h21	3.33 ± 3.33 bA
N22	0.00 ± 0.00 cA	n22	0.00 ± 0.00 aA	H22	0.00 ± 0.00 bA	h22	0.00 ± 0.00 bA
N23	0.00 ± 0.00 cA	n23	0.00 ± 0.00 aA	H23	10.00 ± 0.00 aA	h23	0.00 ± 0.00 bA
N24	0.00 ± 0.00 cA	n24	6.67 ± 6.67 aA	H24	6.67 ± 3.33 bA	h24	0.00 ± 0.00 bA
N25	3.33 ± 3.33 cA	n25	0.00 ± 0.00 aA	H25	0.00 ± 0.00 bA	h25	0.00 ± 0.00 bA

Different lowercase letters in each column indicate significant differences among different treatments; different uppercase letters indicate significant difference between dimorphic seeds treated with the same secondary disinfectant.

Medium pH ($p > 0.05$) at different levels had no significant effect on seed germination, but pH ($p < 0.05$) significantly affected microbial growth and reproduction. No matter whether NaClO or $HgCl_2$ was selected as the secondary disinfectant, almost no microorganisms could grow on the MS medium under acidic conditions (pH 5.0). A small amount of bacterial or fungal contamination was observed at pH 6.0 to 8.0.

3.3. The Effect of Different Treatments on Seedling Survival Percentage of S. aralocaspica

Only the seedlings without the growth of bacteria and browning can be used as the source of plant material. The available sterile seedlings were obtained after the seeds were sterilized and cultured for 20 days. There were significant differences in the number of available seedlings obtained from different types of disinfectants ($p < 0.001$), ethanol time ($p < 0.001$) and seed type ($p < 0.001$) (Table 6).

Table 6. A mixed model ANOVA on the seedling survival percentage of *Suaeda aralocaspica*.

Source	Numerator df	Denominator df	F	Sig.
Ethanol time	4	285	5.850	0.000
pH value	4	285	1.005	0.405
Secondary disinfectant type	1	285	79.494	0.000
Secondary disinfectant time	4	285	0.447	0.775
Seed type	1	285	45.029	0.000

Seedlings grown from brown seeds had a higher survival percentage than that from black seeds. The seedlings sprouted after sterilization with $HgCl_2$ were short, and most browning deaths cannot be included in the surviving available seedlings. Compared with NaClO, the $HgCl_2$ treatment significantly reduced the seedling survival percentage for brown seeds of *S. aralocaspica*. The survival percentage of seedlings produced by brown seeds in treatment H7 was only 10%. When NaClO was used as the main disinfectant, the highest survival percentage of seedlings from brown seeds in treatments N4 and N7 was 46.67%, and the survival percentage was 43.33% in treatments N3 and N10. Under the treatment of two secondary disinfectants, the survival percentage of seedlings from black seeds was low. The highest survival percentage was only 13.33% under the NaClO treatment n10 and 10% under the $HgCl_2$ treatment h10 (Table 7).

Table 7. Seedling survival percentage of *Suaeda aralocaspica* cultured with different treatments after 20 days incubation.

Treatment	3.6% NaClO Brown Seeds	Treatment	3.6% NaClO Black Seeds	Treatment	0.1% $HgCl_2$ Brown Seeds	Treatment	0.1% $HgCl_2$ Black Seeds
N1	23.33 ± 8.82 abcdA	n1	10.00 ± 10.00 aA	H1	0.00 ± 0.00 cA	h1	0.00 ± 0.00 bA
N2	16.67 ± 12.02 cdA	n2	0.00 ± 0.00 aA	H2	3.33 ± 3.33 bcA	h2	0.00 ± 0.00 bA
N3	43.33 ± 3.33 abA	n3	3.33 ± 3.33 aB	H3	3.33 ± 3.33 bcA	h3	0.00 ± 0.00 bA
N4	46.67 ± 3.33 aA	n4	6.67 ± 6.67 aA	H4	0.00 ± 0.00 cA	h4	0.00 ± 0.00 bA
N5	23.33 ± 8.82 abcdA	n5	3.33 ± 3.33 aA	H5	0.00 ± 0.00 cA	h5	6.67 ± 6.67 abA
N6	26.67 ± 6.67 abcdA	n6	6.67 ± 6.67 aA	H6	6.67 ± 3.33 abA	h6	0.00 ± 0.00 bA
N7	46.67 ± 3.33 aA	n7	0.00 ± 0.00 aB	H7	10.00 ± 0.00 aA	h7	3.33 ± 3.33 abA
N8	20.00 ± 0.00 bcdA	n8	3.33 ± 3.33 aB	H8	0.00 ± 0.00 cA	h8	3.33 ± 3.33 abA
N9	23.33 ± 12.09 abcdA	n9	6.67 ± 3.33 aA	H9	0.00 ± 0.00 cA	h9	3.33 ± 3.33 abA
N10	43.33 ± 3.33 abA	n10	13.33 ± 13.33 aA	H10	0.00 ± 0.00 cA	h10	10.00 ± 0.00 aA
N11	6.67 ± 3.33 dA	n11	3.33 ± 3.33 aA	H11	3.33 ± 3.33 bcA	h11	0.00 ± 0.00 bA
N12	13.34 ± 8.82 cdA	n12	6.67 ± 3.33 aA	H12	3.33 ± 3.33 bcA	h12	0.00 ± 0.00 bA
N13	10.00 ± 10.00 cdA	n13	3.33 ± 3.33 aA	H13	0.00 ± 0.00 cA	h13	3.33 ± 3.33 abA
N14	20.00 ± 10.00 bcdA	n14	6.67 ± 6.67 aA	H14	0.00 ± 0.00 cA	h14	3.33 ± 3.33 abA
N15	26.67 ± 14.53 abcdA	n15	0.00 ± 0.00 aA	H15	0.00 ± 0.00 cA	h15	0.00 ± 0.00 bA
N16	13.33 ± 3.33 cdA	n16	3.33 ± 3.33 aA	H16	0.00 ± 0.00 cA	h16	3.33 ± 3.33 abA
N17	6.67 ± 3.33 dA	n17	0.00 ± 0.00 aA	H17	0.00 ± 0.00 cA	h17	0.00 ± 0.00 aA
N18	10.00 ± 10.00 dA	n18	0.00 ± 0.00 aA	H18	0.00 ± 0.00 cA	h18	6.67 ± 3.33 abA
N19	20 ± 11.55 bcdA	n19	0.00 ± 0.00 aA	H19	3.33 ± 3.33 bcA	h19	0.00 ± 0.00 bA
N20	20.00 ± 10.00 bcdA	n20	0.00 ± 0.00 aA	H20	3.33 ± 3.33 bcA	h20	0.00 ± 0.00 bA
N21	0.00 ± 0.00 dA	n21	3.33 ± 3.33 aA	H21	0.00 ± 0.00 cA	h21	0.00 ± 0.00 bA
N22	10.00 ± 5.77 cdA	n22	3.33 ± 3.33 aA	H22	0.00 ± 0.00 cA	h22	0.00 ± 0.00 bA
N23	13.33 ± 3.33 cdA	n23	3.33 ± 3.33 aA	H23	0.00 ± 0.00 cA	h23	0.00 ± 0.00 bA
N24	6.67 ± 3.33 dA	n24	3.33 ± 3.33 aA	H24	3.33 ± 3.33 bcA	h24	0.00 ± 0.00 bA
N25	13.33 ± 8.82 cdA	n25	3.33 ± 3.33 aA	H25	0.00 ± 0.00 cA	h25	0.00 ± 0.00 bA

Different lowercase letters in each column indicate significant differences among different treatments; different uppercase letters indicate significant difference between dimorphic seeds treated with the same secondary disinfectant.

4. Discussion

Although the ecology, physiology, and molecular biology of *Suaeda* species have been studied extensively, there is no culture system in vitro for further study of the molecular mechanism. The effective acquisition of high-quality sterile explant material is the key to the subsequent tissue culture [19]. Our study takes the first step of this process by comparing the sterilization effects of different disinfectants and their effects on seed germination percentage.

Compared with the brown seeds, the black seeds were not easily contaminated by microorganisms, which might be due to the protective effect of the black and dense seed coat on the surface of the black seeds. 75% ethanol needs to be used with other disinfectants, for using it solely has an incomplete and unsatisfactory sterilization effect [21,22], and 0.1% $HgCl_2$ has a good sterilizing effect because Hg^{2+} can combine with negatively charged proteins to desaturated bacterial proteins and inactivate enzymes. NaClO solution is much milder than mercury chloride and is often used to sterilize tissue culture explants [20–28]. When the same disinfectant is used, the contamination rate will decrease, and the death rate will increase with the extension in sterilization time. Under natural conditions or abiotic stresses, the germination percentages of brown seeds of *S. aralocaspica* were much higher than that of black seeds [7,9,15], which was consistent with the germination results after our sterilization treatment. The seed type had a significant effect on the three evaluation indexes of germination percentage, bacterial growth percentage, and survival percentage. When treating explants, different sanitizer and sterilization time was used for sterilization, and the effect was obviously different. The results showed that with the prolongation of ethanol infiltration time, the browning number of seedlings from brown seeds of *S. aralocaspica* increased and the survival percentage decreased.

In our study, mercury chloride and sodium hypochlorite were used to disinfect with 75% ethanol. Sodium hypochlorite has strong oxidation, and long sterilization time means it is easy to cause plant browning. When 3.6% NaClO was used as the main disinfectant, N8 had the best comprehensive effect on brown seeds, which were soaked in 75% ethanol for 60 s, and then sterilized with 3.6% NaClO for 8 min, and finally inoculated into a pH 8.0 MS medium. Black seeds grew well under treatment 6 (n6), which was disinfected with 75% ethanol for 1 min and then treated with 3.6% NaClO for 3 min. Although mercury chloride can be effectively sterilized, it also has strong toxicity, causing irreversible browning damage to plants [19,20,22]. When 0.1% $HgCl_2$ was used as the main disinfectant, H7 had the best comprehensive effect on brown seeds (75% ethanol for 1 min + 0.1% $HgCl_2$ for 5 min + pH 8.0 MS). h6 had the best comprehensive effect on black seeds, which was 75% ethanol for 1 min + 0.1% $HgCl_2$ for 1 min + pH 6.0 MS.

It was found that a large number of browning seedlings appeared on the 8th day of culture, and the whole germination and growth process was completed on the 7th day. Therefore, the culture time can be shortened to 7 days to reduce energy and costs. Some studies found that adding anti-browning agents, such as vitamin C, activated carbon to the medium, or improving the activity of polyphenol oxidase and the antioxidant system enzyme Mars could effectively inhibit seedling browning [29–31]. In this experiment, we did not take special measures to prevent seedling browning, which could be optimized in further research.

5. Conclusions

In summary, this study shows that brown seeds with a high germination percentage should be chosen as a source of sterile explants. The best sterilization method entailed 75% ethanol 60 s + 3.6% NaClO 8 min, placed in an MS medium with pH 8.0 for 7 days. At present, this method has been successfully applied to the seed sterilization of *S. aralocaspica* in vitro, and the pollution-free percentage can reach 100%, when ignoring the pollution caused by improper operation. The sterilization and cultivation method was successful for *S. aralocaspica* and may be also applied to other *Suaeda* species. In the future, other combinations of sterilization methods should be tested for *S. aralocaspica* seeds, and the exact mechanism of the sterilization effects needs to be fully understood.

Supplementary Materials: The following supporting information can be downloaded at: https://www.mdpi.com/article/10.3390/horticulturae8040289/s1, Figure S1: The final germination percentage and contamination percentage of dimorphic seeds that washed with sterile water and incubated at different pH conditions, Table S1: 25 different treatments with NaClO as the secondary disinfector to sterilize brown seeds of *Suaeda aralocaspica*, Table S2: 25 different treatments with NaClO as the secondary disinfector to sterilize black seeds of *Suaeda aralocaspica*, Table S3: 25 different treatments with 0.1% $HgCl_2$ as the secondary disinfector to sterilize brown seeds of *Suaeda aralocaspica*, Table S4: 25 different treatments with 0.1% $HgCl_2$ as the secondary disinfector to sterilize black seeds of *Suaeda aralocaspica*.

Author Contributions: Conceptualization, L.W. and Y.S.; methodology, Y.S. and Y.H.; software, Y.S.; validation, L.W. and Y.S.; formal analysis, Y.S.; investigation, Y.S.; resources, L.W.; data curation, Y.S.; writing—original draft preparation, Y.S.; writing—review and editing, L.W., Y.S. and Y.H.; visualization, Y.S.; supervision, L.W.; project administration, L.W.; funding acquisition, L.W. All authors have read and agreed to the published version of the manuscript.

Funding: This research was funded by the National Natural Science Foundation of China (grant number 32171514) and the Tianshan elite program of Xinjiang Uygur Autonomous Region (grant number Y970000333).

Institutional Review Board Statement: Not applicable.

Informed Consent Statement: Not applicable.

Data Availability Statement: The data presented in this study are available on request from the corresponding author.

Conflicts of Interest: The authors declare no conflict of interest.

References

1. Flora of China Editorial Committee. *Flora of China*; Science Press: Beijing, China, 2003.
2. Xi, J.B.; Zhang, F.S.; Tian, C.Y. *Halophytes in Xinjiang*; Science Press: Beijing, China, 2006.
3. Schütze, P.; Freitag, H.; Weising, K. An Integrated Molecular and Morphological Study of the Subfamily *Suaedoideae Ulbr.* (Chenopodiaceae). *Plant Syst. Evol.* **2003**, *239*, 257–286. [CrossRef]
4. Koteyeva, N.K.; Voznesenskaya, E.V.; Berry, J.O.; Cousins, A.B.; Edwards, G.E. The Unique Structural and Biochemical Development of Single Cell C_4 Photosynthesis along Longitudinal Leaf Gradients in *Bienertia Sinuspersici* and *Suaeda aralocaspica* (Chenopodiaceae). *J. Exp. Bot.* **2016**, *67*, erw082. [CrossRef] [PubMed]
5. Lara, M.V.; Chuong, S.D.; Akhani, H.; Andreo, C.S.; Edwards, G.E. Species Having C_4 Single-Cell-Type Photosynthesis in the Chenopodiaceae Family Evolved a Photosynthetic Phosphoenolpyruvate Carboxylase Like That of Kranz-Type C_4 Species. *Plant Physiol.* **2006**, *142*, 673–684. [CrossRef]
6. Wang, L.; Huang, Z.; Baskin, C.C.; Baskin, J.M.; Dong, M. Germination of Dimorphic Seeds of the Desert Annual Halophyte *Suaeda aralocaspica* (Chenopodiaceae), a C_4 Plant without Kranz Anatomy. *Ann. Bot.* **2008**, *102*, 757–769. [CrossRef]
7. Wang, H.L.; Tian, C.Y.; Wang, L. Germination of Dimorphic Seeds of *Suaeda aralocaspica* in Response to Light and Salinity Conditions during and after Cold Stratification. *PeerJ* **2017**, *5*, e3671. [CrossRef]
8. Park, J.; Okita, T.W.; Edwards, G.E. Salt Tolerant Mechanisms in Single-Cell C_4 Species *Bienertia Sinuspersici* and *Suaeda aralocaspica* (Chenopodiaceae). *Plant Sci.* **2009**, *176*, 616–626. [CrossRef]
9. Wang, L.; Baskin, J.M.; Baskin, C.C.; Cornelissen, J.H.C.; Dong, M.; Huang, Z. Seed Dimorphism, Nutrients and Salinity Differentially Affect Seed Traits of the Desert Halophyte *Suaeda aralocaspica* via Multiple Maternal Effects. *BMC Plant Biol.* **2012**, *12*, 170. [CrossRef]
10. Voznesenskaya, E.; Franceschi, V.R.; Kiirats, O.; Freitag, H.; Edwards, G.E. Kranz Anatomy Is Not Essential for Terrestrial C. *Nature* **2001**, *414*, 543–546. [CrossRef]
11. Freitag, H.; Stichler, W. A Remarkable New Leaf Type With Unusual Photosynthetic Tissue in a Central Asiatic Genus of Chenopodiaceae. *Plant Biol.* **2000**, *2*, 154–160. [CrossRef]
12. Wang, L.; Jiang, L.; Tian, C.Y. Effects of NaCl on the Growth and Mineral Nutrien Content of Plants Grown from Dimorphic Seeds of *Suaeda aralocaspica*. *Arid Zo. Res.* **2018**, *35*, 510–514.
13. Song, Y.G.; Li, L.; Zhang, X.M.; Pan, X.L.; Zeng, X.H. Differences of Seed Coat Structure and Ions Content between Dimorphic Seeds of *Borszczowia aralocaspica*. *Bull. Bot. Res.* **2012**, *32*, 290–295.

14. Song, Y.G.; Li, L.; Zeng, X.H.; Deng, M.; Zhang, X.M. Responses of the Germination on Dimorphic Seeds of *Suaeda aralocaspica* to Salt Stress. *Acta Prataculturae Sin.* **2014**, *23*, 192–198.
15. Cao, J.; Li, X.; Wang, C.; Wang, L.; Lan, X.; Lan, H. Effects of Exogenous Abscisic Acid on Heteromorphic Seed Germination of *Suaeda Aralocaspica*, a Typical Halophyte of Xinjiang Desert Region. *Acta Ecol. Sin.* **2015**, *35*, 6666–6677.
16. Cao, J.; Wang, L.; Lan, H.Y. Validation of Reference Genes for Quantitative RT-PCR Normalization in *Suaeda Aralocaspica*, an Annual Halophyte with Heteromorphism and C4 Pathway without Kranz Anatomy. *PeerJ* **2016**, *4*, e1697. [CrossRef]
17. Wang, L.; Wang, H.-L.; Yin, L.; Tian, C.-Y. Transcriptome Assembly in *Suaeda aralocaspica* to Reveal the Distinct Temporal Gene/MiRNA Alterations between the Dimorphic Seeds during Germination. *BMC Genomics* **2017**, *18*, 806. [CrossRef]
18. Wang, L.; Ma, G.; Wang, H.; Cheng, C.; Mu, S.; Quan, W.; Jiang, L.; Zhao, Z.; Zhang, Y.; Zhang, K.; et al. A Draft Genome Assembly of Halophyte *Suaeda aralocaspica*, a Plant That Performs C4 Photosynthesis within Individual Cells. *Gigascience* **2019**, *8*, giz116. [CrossRef]
19. Barampuram, S.; Allen, G.; Krasnyanski, S. Effect of Various Sterilization Procedures on the in Vitro Germination of Cotton Seeds. *Plant Cell Tissue Organ Cult.* **2014**, *118*, 179–185. [CrossRef]
20. Lu, L.M.; An, Y. Effects of Different Disinfectant on Sterilization Effect and Germination of Tobacco Seeds. *Seed* **2012**, *31*, 93–95. [CrossRef]
21. Ma, M.; Zhao, L.; Tang, S.; Chen, X.; Qin, R. The Effects of Different Disinfection Methods on Seed Germination and Study on the Environmental Bacteria in Safflower (*Carthamus tinctorius* L.). *Crops* **2018**, *34*, 162–167. [CrossRef]
22. Yuan, Y. Selection and Disinfection of Tissue Culture Explants of *Toona sinensis* Roem. *Anhui Agric.* **2020**, *26*, 19–20, 57.
23. Wang, S.; Fan, B.J.; Wang, Y. Effects of Different Disinfection Methods on Callus Culture of Mung Bean (*Vigna radiata*). *J. Anhui Agric. Sci.* **2021**, *49*, 29–31, 50.
24. Zhou, F.; Tang, N.; Chen, Q.Z.; Hua, C.; Mao, S.G. Effect of Different Disinfector on Seeds Germination and Seedlings Growth of *Salicornia bigelovii* Torr. *Seed* **2009**, *28*, 31–33.
25. Zhu, Y.L.; An, D.D.; Zeng, X.C. Study on Seed Surface Sterilization Technology and Aseptic Germination Conditions of *Ammodendron argenteum*(Pall.) Kuntze. *J. Xinjiang Norm. Univ. Sci. Ed.* **2014**, *33*, 17–20. [CrossRef]
26. Talei, D.; Valdiani, A.; Abdullah, M. A Rapid and Effective Method for Dormancy Breakage and Germination of King of Bitters (*Andrographis paniculata* Nees.) Seeds. *Maydica* **2012**, *57*, 98–105.
27. Zhang, B.H.; Wang, Y.; Yao, C.B. Plant Regeneration via Somatic Embryogenesis in Cotton. *Plant Cell Tissue Organ Cult.* **2012**, *60*, 89–94. [CrossRef]
28. Daud, N.H.; Jayaraman, S.; Mohamed, R. Methods Paper: An Improved Surface Sterilization Technique for Introducing Leaf, Nodal and Seed Explants of *Aquilaria malaccensis* from Field Sources into Tissue Culture. *Asia-Pac. J. Mol. Biol. Biotechnol.* **2012**, *20*, 55–58.
29. Li, Y.; Feng, J.-T.; Wang, Y.-H.; Cui, H.; Zhang, X. Effects of Browning Inhibitors on Callus Growth and Secondary Metabolites Production in *Tripterygium wilfordii* Hook F. *Plant Sci. J.* **2010**, *28*, 224–228. [CrossRef]
30. Ji, Y.Z.; Lei, Y.; Li, X.L.; Wu, L.H.; Zhao, Z. Establishment of the Tissue Culture and Regeneration System of *Daphne tangutica*. *J. Northwest For. Univ.* **2018**, *33*, 94–99.
31. Fu, Z.Z.; Chen, J.; Xu, P.P.; He, S.L.; Wang, Z. Relationship Between Tissue Browning and the Contents of Phenolicsand Related Enzymes Activity in *Peaonia suffruticosa*. *J. Northwest For. Univ.* **2011**, *26*, 66–69.

Article

In Vitro Growth Responses of Ornamental Bananas (*Musa* sp.) as Affected by Light Sources

Wagner A. Vendrame [1,*], Cassandre Feuille [1], David Beleski [1] and Paulo Mauricio Centenaro Bueno [2]

[1] Environmental Horticulture Department, Institute of Food and Agricultural Sciences, University of Florida, Gainesville, FL 32611, USA; c.feuillec@gmail.com (C.F.); dbeleski@ufl.edu (D.B.)

[2] Instituto Federal do Paraná, Palmas 8555, PR, Brazil; paulo.bueno@ifpr.edu.br

* Correspondence: vendrame@ufl.edu

Abstract: Light-emitting diodes (LEDs) have become very popular for the production of horticultural crops. LEDs represent an alternative lighting source to regular fluorescent (FL) bulbs, increasing the quality of plants and minimizing production costs. LEDs also provide selective light intensity and quality, suitable for commercial micropropagation. The objective of this study was to evaluate the growth and development of in vitro ornamental bananas under different light sources. Two ornamental banana varieties were selected for this study: Musa 'Little Prince' and Musa 'Truly Tiny'. Light quality and intensity of three different light sources were evaluated: LED-1 (116 µmol m^{-2} s^{-1}), LED-2 (90 µmol m^{-2} s^{-1}), and FL (100 µmol m^{-2} s^{-1}). Length and biomass of plantlets were greater under LED-1 compared to FL but not significantly different from LED-2. The fresh and dry weight of shoots and roots, number of leaves, and number and length of roots were not significantly different between treatments. Chlorophyll content was greater under LEDs. Leaf number and stomata number and size were greater under FL. Our results indicate that shoot length and biomass could be improved by optimizing light quality and intensity. Different responses to light sources between the two banana varieties also indicated a genotype effect.

Keywords: banana; micropropagation; light intensity; light quality; plant growth and development; leaf anatomy

Citation: Vendrame, W.A.; Feuille, C.; Beleski, D.; Bueno, P.M.C. In Vitro Growth Responses of Ornamental Bananas (*Musa* sp.) as Affected by Light Sources. *Horticulturae* **2022**, *8*, 92. https://doi.org/10.3390/horticulturae8020092

Academic Editor: Sergio Ruffo Roberto

Received: 21 December 2021
Accepted: 14 January 2022
Published: 20 January 2022

Publisher's Note: MDPI stays neutral with regard to jurisdictional claims in published maps and institutional affiliations.

1. Introduction

Light is one of the primary factors that affect in vitro plant morphogenesis [1,2]. Artificial light sources, including fluorescent lamps, high-pressure sodium lamps, metal halide lamps, and incandescent lamps, among others, have been widely used for plant tissue culture research and commercial micropropagation of several crops [3]. Cool-white fluorescent lamps remain the most used type of light source for micropropagation [4,5]. While popular, cool-white fluorescent light sources have a wide spectrum distribution (350 to 750 nm) and therefore are of low quality for promoting plant growth. In addition, energy consumption is increased, representing the second-highest cost in micropropagation after labor [6,7]. Fluorescent lights also emit heat, which can cause damage and photo-stress to plants [3]. Therefore, more efficient and cost-effective light systems that promote in vitro plant growth and development are necessary. Light-emitting diodes (LEDs) have become very popular in agriculture, particularly for the production of horticultural crops, and have been widely used in microgravity studies aiming at space life support systems [8,9]. More recently, LEDs have been incorporated into in vitro plant systems [10]. LEDs represent an alternative lighting source to regular fluorescent bulbs, increasing the quality of in vitro plantlets and minimizing the per plant production costs. LEDs provide selective light intensity and quality, are suitable for commercial micropropagation, and also allow the control of photosynthetically active radiation (PAR), providing optimal conditions for plant growth and development, including improved morphology and metabolism [11].

Additional advantages of LEDs include the longer life and smaller size of light-emitting diodes compared to fluorescent bulbs, with very little heat generation [3]. In addition, LEDs may influence the growth and production of secondary metabolites in plants, such as those reported for *Pfaffia glomerata* [12]. The spectral distribution of the light affects plant growth mainly via photosynthesis, and it affects plant morphology. Plant morphology is also affected by photon flux density; therefore, there may be an interaction between these effects. Understanding these effects and these interactions is required to control plant growth and morphology using LEDs [13].

Because of the high sterility and polyploidy of cultivated or domesticated banana varieties and hybrids [14], asexual propagation methods are the primary means for banana production. However, many diseases, such as black sigatoka, fusarium wilt, banana bunchy top virus, burrowing nematodes, and banana weevil borer [15], are associated with traditional propagation techniques [16–18]. Therefore, micropropagation has become a common practice for the production of bananas, thus assisting with the production of clean, disease-free plant material [16,19,20]. The tissue culture of bananas also allows large-scale mass in vitro clonal propagation of elite banana varieties, free of diseases. The use of LEDs in micropropagation could provide higher-quality banana plants with the benefit of reduced energy costs. LEDs have been studied previously for banana in vitro cultures [21–23] with a focus either on specific wavelengths, LED ratios, or lower intensities (45 to 75 μmol m^{-2} s^{-1} PPFD).

Therefore, the main objective of this study was to evaluate the growth and development of in vitro ornamental bananas as affected by different light sources, including simple LED and traditional fluorescent lighting. In addition to the growth and development parameters of two in vitro ornamental banana varieties, relative chlorophyll content, leaf stomata, and anatomy characteristics were also evaluated.

2. Materials and Methods

2.1. Plant Material and Culture Establishment

Two varieties of ornamental bananas, Musa 'Little Prince' and Musa 'Truly Tiny', were obtained from AgriStarts, Inc. (Apopka, FL, USA). They are dwarf varieties of dwarf Cavendish, with a compact habit and thick dark green leaves. Musa 'Little Prince' has thick trunks and is generally used as an ornamental indoors or as a landscape plant outdoors. Musa 'Truly Tiny' is extremely dwarf and produces the smallest edible fruits in the world. Both are desirable as ornamental plants and commonly produced via tissue culture, and have similar growth rates. In vitro plantlets were established in Murashige and Skoog (MS) culture medium [24] supplemented with benzylaminopurine (BAP; 8.8 μM) and sucrose (30 g L^{-1}). The pH was adjusted to 5.8, and the medium was solidified with agar (agar–agar, Sigma Aldrich Inc., St. Louis, MO, USA) at 8 g L^{-1}. A volume of 50 mL of medium was dispensed into baby food jars, and the medium was autoclaved at 121 °C and 20 psi for 20 min. In vitro banana plantlets measuring about 5 cm in height had their leaves and roots trimmed to 2 cm shoot-tips, which were used as explants. Baby food jars with polypropylene lids were sealed with parafilm, and cultures were placed in a growth chamber under controlled environmental conditions of 27 ± 2 °C and a 16 h photoperiod at 100 μmol m^{-2} s^{-1} PPFD. In vitro shoots were subcultured to fresh MS medium at four-week intervals.

2.2. Light Sources

Three different light sources were evaluated: two provided by LED lighting, LED-1 at 116 μmol m^{-2} s^{-1} (Philips GreenPower DR/B 3:1 150 43W) and LED-2 at 90 μmol m^{-2} s^{-1} (Philips GreenPower DR/W 3:1 150 33W); and one provided by fluorescent lighting (Philips 9A fluorescent bulbs 40W) at 100 μmol m^{-2} s^{-1} (FL). Photoperiod was 16/8 h (light/dark). The spectral energy distribution for the lighting used in this study is shown in Figure 1: LED-1 showed peak emissions at 440 nm and 650 nm (Figure 1A), LED-2 showed peak emissions at 440 nm and 670 nm (Figure 1B); and fluorescent lighting (FL) showed a broader spectrum with peaks in the green (550 nm), blue (440 nm), and some additional peaks in between (490 nm, 590 nm, 610 nm, and 710 nm) (Figure 1C). The intensity and composition of all light sources were measured using an LI-180 Li-Cor spectrometer (Li-Cor, Lincoln, NE, USA).

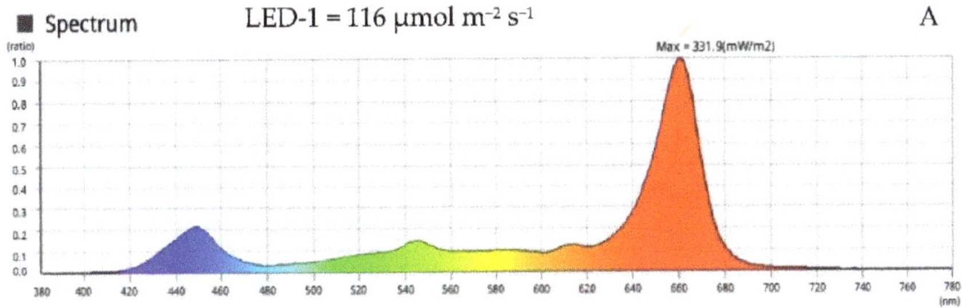

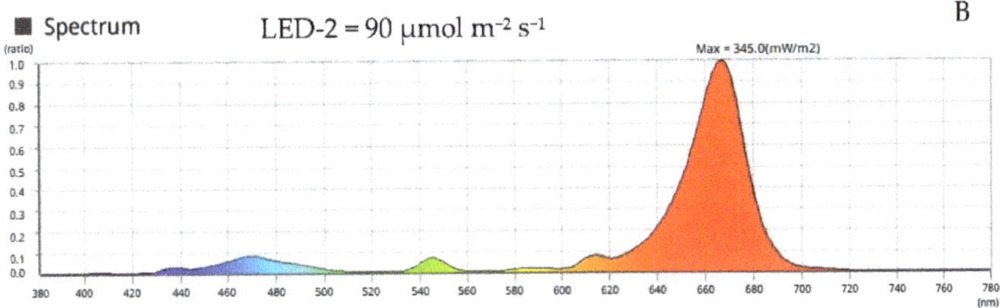

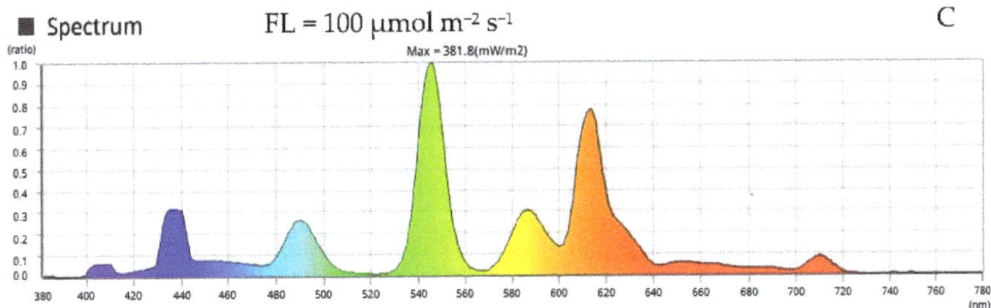

Figure 1. Intensity and composition of lighting as photosynthetically active radiation (PAR) for the different light treatments in this study: (**A**) LED-1 = 116 µmol m^{-2} s^{-1}; (**B**) LED-2 = 90 µmol m^{-2} s^{-1}; and (**C**) fluorescent light (FL) = 100 µmol m^{-2} s^{-1}. While both LED lights peak in the red range (660–670 nm) with some blue peaks (440–470 nm), fluorescent light has a broader distribution with peaks in the green (550 nm), blue (440 nm), and some additional peaks in between (490 nm, 590 nm, 610 nm, and 710 nm). Maximum irradiance (mW m^{-2}) is shown for each spectrum. Measurements were obtained with an LI-180 Li-Cor spectrometer.

2.3. In Vitro Growth and Development

Explants from all treatments were evaluated four weeks after in vitro establishment for shoot length, root length and number, plantlet fresh and dry weight, shoot fresh and dry weight, root fresh and dry weight, root length, and number determined. Dry weight was determined by oven-drying plantlets at 70 °C until they reached constant weight. Five random plants were selected per treatment.

2.4. Relative Chlorophyll Content

Relative chlorophyll content was evaluated as SPAD value by placing the third expanded leaf of each plantlet, counted from top downwards, in a portable SPAD-502 chlorophyll meter (SPAD-502, Minolta Co., Ltd., Tokyo, Japan). Five random plantlets were selected per treatment.

2.5. Stomata Analysis

The middle third portion of the third and fourth fully expanded leaves were cut into approximately 1 cm × 1 cm sections. Impressions of the leaves were obtained by placing leaf sections on top of a thin layer of super glue (Elmer's Products, Inc., Westerville, OH, USA), spread over microscope glass slides, and removing the leaf after drying. Impressions were obtained for both the adaxial and abaxial surfaces of the leaves. Stomata observations were performed under an optical Leica DMLB microscope (Leica microsystems, Buffalo, NY, USA) at 200× magnification. Images were recorded using a SPOT 4.7 digital camera coupled to the microscope and analyzed using the SPOT basic software (SPOT Imaging, Diagnostic Instruments, Inc., Sterling Heights, MI, USA). The number of stomata was counted under the microscope for both the abaxial and adaxial surfaces of leaves. Three different areas of 5 mm in diameter were selected as replicates for observations for each light treatment. Results were expressed as a means of counts of the three areas per mm^2. The length and width of three randomly selected stomata were measured for both the adaxial and abaxial surfaces of each leaf in three different sections of each selected leaf. Five random plantlets were selected per treatment.

2.6. Anatomical Observations

Anatomical studies were conducted using the middle-third section of the second completely developed leaf. Leaf cross-sections were obtained by freehand sectioning using a steel blade, fixed in 70% FAA (formaldehyde-acetic acid-ethyl alcohol 70%) for 48 h, and then preserved in ethanol 70% (v/v). Following discoloration in sodium hypochlorite (1–1.25% active chlorine), triple-rinsing in distilled water, and staining in toluidine blue (0.05% w/v), leaf sections were subsequently fixed on semi-permanent slides with glycerinated water [25]. The slides were examined and photographed under the same light Leica DMLB microscope and the SPOT 4.7 idea digital camera and software, as described earlier. The images were evaluated by assessing five fields per repetition for each variable analyzed. The thickness of the abaxial surface epidermis (ASE), adaxial epidermis (ABE), abaxial hypodermis (AH), adaxial hypodermis (AbH), palisade parenchyma (PP), and spongy parenchyma (SP) were determined. Five random plants were selected per treatment.

2.7. Experimental Design and Statistical Analysis

The experimental design was completely randomized and consisted of six treatments (three light sources × two banana varieties), with five replications per treatment. Each replication consisted of 3 baby food flasks containing 1 in vitro plantlet per flask for a total of 90 experimental units. Data were collected and submitted to analysis of variance (ANOVA) using the R statistical analysis program, with means compared by LSD (Least Significant Difference) test at the 5% level of significance.

3. Results

3.1. In Vitro Growth and Development

The light source had no significant effect on most in vitro growth and development parameters evaluated for both banana varieties, including stem diameter, shoot and root fresh weight, leaf number, shoot and root length, and shoot and root dry weight (Table 1). However, 'Little Prince' plantlets produced under LED-1 and LED-2 showed larger overall plantlet fresh weight and shoot length compared to plantlets under fluorescent lights, while 'Truly Tiny' plantlets showed higher fresh weight only under LED-1, but no differences in shoot length.

Table 1. Evaluation of light quality and light level effects on in vitro growth and development parameters for two ornamental banana varieties; Musa 'Little Prince' and Musa 'Truly Tiny'. LED-1 = 116 µmol m^{-2} s^{-1}; LED-2 = 90 µmol m^{-2} s^{-1}; fluorescent light (FL) = 100 µmol m^{-2} s^{-1}.

Banana Variety	Light Source	Stem Diameter * (mm)	Plant Fresh Weight (g)	Shoot Fresh Weight (g)	Root Fresh Weight (g)	Number of Leaves	Shoot Length (cm)	Root Length (cm)	Number of Roots	Shoot Dry Weight (g)	Root Dry Weight (g)
Little Prince	LED-1	5.32 ± 0.50 a	2.84 ± 0.14 a	1.63 ± 0.25 a	0.19 ± 0.08 a	3.6 ± 0.89 a	7.45 ± 0.67 a	3.02 ± 0.30 a	6.4 ± 2.07 a	0.24 ± 0.03 a	0.01 ± 0.01 a
	LED-2	5.24 ± 0.44 a	2.62 ± 0.42 ab	1.94 ± 0.54 a	0.34 ± 0.09 a	4.0 ± 1.00 a	7.33 ± 0.76 a	3.70 ± 0.82 a	8.2 ± 2.17 a	0.28 ± 0.07 a	0.03 ± 0.01 a
	FL	5.08 ± 0.71 a	2.21 ± 0.39 c	1.44 ± 0.05 a	0.16 ± 0.07 a	4.2 ± 0.84 a	5.87 ± 0.82 b	3.21 ± 0.25 a	7.8 ± 3.19 a	0.24 ± 0.02 a	0.02 ± 0.01 a
Truly Tiny	LED-1	4.28 ± 0.55 a	2.31 ± 0.16 bc	1.27 ± 0.19 a	0.25 ± 0.10 a	3.6 ± 0.89 a	6.67 ± 0.30 ab	4.21 ± 0.34 a	5.6 ± 1.52 a	0.18 ± 0.03 a	0.02 ± 0.01 a
	LED-2	3.94 ± 0.37 a	1.79 ± 0.21 d	1.01 ± 0.27 a	0.27 ± 0.07 a	3.6 ± 1.14 a	6.46 ± 0.67 b	4.38 ± 0.92 a	5.2 ± 1.79 a	0.12 ± 0.03 a	0.02 ± 0.01 a
	FL	4.36 ± 0.92 a	1.91 ± 0.38 cd	1.39 ± 0.62 a	0.34 ± 0.16 a	4.0 ± 1.00 a	6.02 ± 0.47 b	4.60 ± 0.36 a	6.0 ± 2.12 a	0.21 ± 0.10 a	0.02 ± 0.01 a

* Means within columns followed by the same letter are not significantly different at $p \leq 0.05$ according to the LSD (Least Significance Difference) test. The tests were conducted separately for each parameter evaluated.

3.2. Relative Chlorophyll Content

Significant differences were shown in relative chlorophyll content among the different light sources and banana varieties (Figure 2). In general, the relative chlorophyll content of plants under LED lighting showed greater SPAD values compared to plantlets under fluorescent lamps. SPAD values were 36.84 for 'Little Prince' and 41.54 for 'Truly Tiny' under LED-1, 31.62 for 'Truly Tiny' and 40.92 for 'Little Prince' under LED-2, and 24.12 for 'Little Prince' and 24.56 for 'Truly Tiny' under fluorescent lamps (Figure 2). However, relative chlorophyll content was not significantly different between plantlets of both varieties under LED-1 and LED-2, but higher than FL, with the exception of 'Truly Tiny' under LED-2, which had similar values to both varieties under FL. The leaves of plantlets under LED-1 and LED-2 showed a dark green color, while the leaves of plantlets under FL had a yellowish-green color.

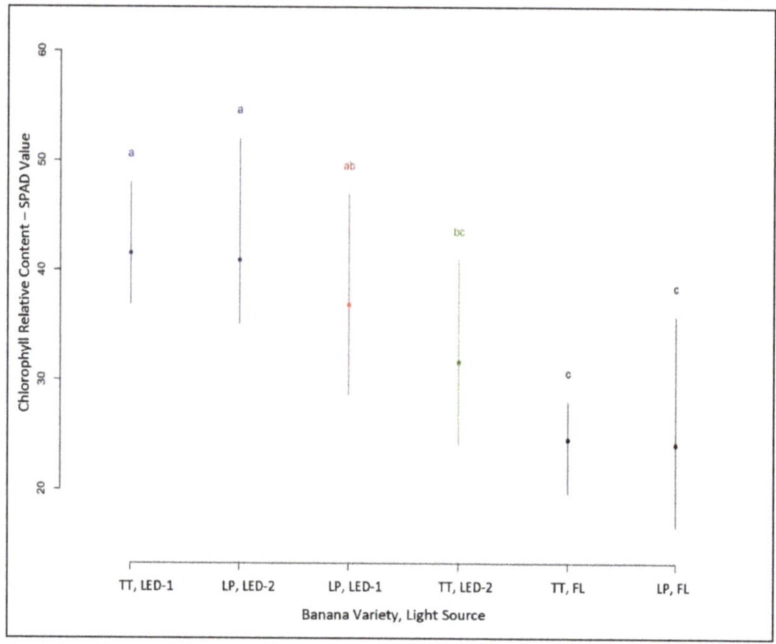

Figure 2. Relative chlorophyll content of two banana varieties, Musa Little Prince (LP) and Musa Truly Tiny (TT) after 4 weeks of in vitro culture under different light sources, LED-1 (116 μmol m^{-2} s^{-1}) LED-2 (90 μmol m^{-2} s^{-1}) and FL (100 μmol m^{-2} s^{-1}). Values with the same letter are not significantly different at α = 0.05 level based on LSD mean separation. Bars represent standard error.

3.3. Stomata Analysis

In this study, stomata were present in both adaxial and abaxial leaf surfaces of both banana varieties under the different light sources, thus characterizing them as amphistomatic (Figure 3). Light source influenced stomata number of both banana varieties (Table 2). Some significant differences were observed between lighting treatments, following the same pattern observed for relative chlorophyll content. Overall, there were no significant differences in the effect of light sources on the adaxial number of stomata (AdSN) for both banana varieties, except for LED-2 and FL, which were slightly higher than LED-1 (Table 2). For the adaxial surface, plantlets had an average leaf stomata number per mm^2 of 22.8 and 24.4 under LED-1, 28.4 and 25.8 for LED-2, and 30.0 and 32.0 under FL for Musa 'Little Prince' and Musa 'Truly Tiny', respectively (Table 2). A similar trend was observed for the abaxial number of stomata per mm2 (AbSN) for both varieties with 75.4 and 92.0 under

LED-1, 67.6 and 94.2 under LED-2 and 83.6 and 101.6 under FL for Musa 'Little Prince' and Musa 'Truly Tiny', respectively (Table 2). The size of the stomata evaluated by length and width showed some variation for both adaxial and abaxial surfaces under the different light sources and for both varieties. The length of the adaxial surface stomata (AdSL) was higher for 'Little Prince' under FL but not significantly different for 'Truly Tiny' under LED-2 and FL (Table 2). The length of the abaxial surface stomata (AbSL) was higher for both varieties under LED-2 and FL (Table 2). The width of the adaxial surface stomata (AdSW) was higher for both varieties under FL, but for the abaxial surface stomata (AbSW), width showed no significant differences (Table 2).

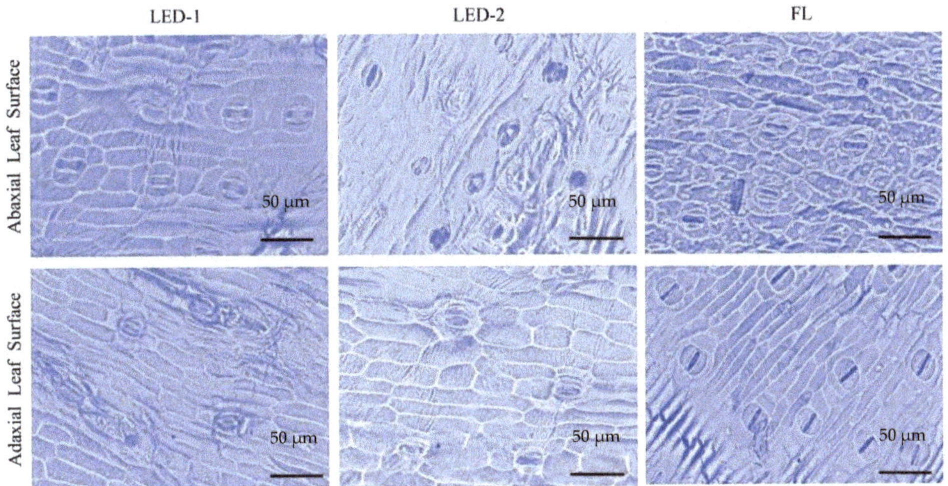

Figure 3. Stomatal prints of abaxial and adaxial leaf surfaces of in vitro banana leaves (Musa 'Little Prince' and 'Truly Tiny') under different light sources. LED-1 = 116 µmol m^{-2} s^{-1}; LED-2 = 90 µmol m^{-2} s^{-1}, fluorescent lighting (FL) = 100 µmol m^{-2} s^{-1}. Bars = 50 µm.

Table 2. Effects of light sources on stomata number (per mm^2) and size (µm) in leaves of two ornamental banana varieties grown *in vitro*; Musa 'Little Prince' and 'Truly Tiny'. LED-1 = 116 µmol m^{-2} s^{-1}; LED-2 = 90 µmol m^{-2} s^{-1}; fluorescent light (FL) = 100 µmol m^{-2} s^{-1}.

Banana Variety	Light Source	AdSN * (per mm^2)	AbSN (per mm^2)	AdSL (µm)	AbSL (µm)	AdSW (µm)	AbSW (µm)
Little Prince	LED-1	22.8 ± 7.40 b	75.4 ± 14.08 cd	25.18 ± 2.57 b	21.41 ± 3.19 c	10.13 ± 0.94 c	10.35 ± 1.25 a
	LED-2	28.4 ± 7.67 ab	67.6 ± 12.03 d	25.63 ± 1.11 b	27.08 ± 2.47 ab	11.13 ± 0.91 bc	11.35 ± 1.08 a
	FL	30.0 ± 6.20 ab	83.6 ± 11.35 bcd	28.51 ± 3.08 a	27.16 ± 1.15 ab	12.87 ± 0.97 a	11.47 ± 1.58 a
Truly Tiny	LED-1	24.4 ± 2.51 b	92.0 ± 5.74 abc	25.00 ± 1.89 b	24.81 ± 1.65 b	10.16 ± 0.61 b	10.11 ± 1.72 a
	LED-2	25.8 ± 4.66 ab	94.2 ± 4.32 ab	27.72 ± 2.19 ab	27.68 ± 2.46 a	10.00 ± 1.79 c	11.05 ± 0.68 a
	FL	32.0 ± 3.54 a	101.6 ± 21.55 a	28.45 ± 0.98 a	25.64 ± 1.44 ab	12.02 ± 1.20 ab	9.97 ± 1.18 a

* Means within columns followed by the same letter are not significantly different at $p \leq 0.05$ according to the LSD (Least Significance Difference) test. The tests were conducted independently within each parameter (adaxial and abaxial stomata number, adaxial and abaxial stomata size). AdSN = Adaxial Stomata Number; AbSN = Abaxial Stomata Number; AdSL = Adaxial Stomata Length; AbSL = Abaxial Stomata Length; AdSW = Adaxial Stomata Width; AbSW = Abaxial Stomata Width.

3.4. Anatomical Observations

Some parameters of leaf anatomy of in vitro plantlets were significantly affected by the light source, including abaxial epidermis (AbE), and abaxial (AbH) and adaxial (AdH) hypodermis for both varieties (Table 3). However, no significant differences were observed for adaxial epidermis (AdE) and for palisade (PP) and spongy (SP) parenchyma for both varieties (Table 3). The abaxial epidermis (AbE) was significantly thinner than the adaxial epidermis (AdE) (Table 3). As for the organization of mesophyll, the banana is dorsiventral or bifacial, with the palisade parenchyma oriented towards the adaxial epidermis and immediately below the adaxial hypodermis and the spongy parenchyma facing the abaxial epidermis (Figure 4). The palisade parenchyma was comprised of closely packed cylindrical cells, exhibiting one to two layers of cells, while the spongy parenchyma was comprised of two to three layers of spongy mesophyll cells, packed with inconspicuous air spaces (Figure 4). The cells were not well defined and had more or less a round shape under FL (Figure 4). However, no statistical differences were observed between most parameters evaluated.

Table 3. Effects of light source on leaf anatomy of two ornamental banana varieties grown in vitro; Musa 'Little Prince' and 'Truly Tiny'. LED-1 = 116 μmol m^{-2} s^{-1}; LED-2 = 90 μmol m^{-2} s^{-1}; fluorescent light (FL) = 100 μmol m^{-2} s^{-1}.

Banana Variety	Light Source	AbE (μm)	AdE (μm)	AbH (μm)	AdH (μm)	PP (μm)	SP (μm)
Little Prince	LED-1	15.53 ± 4.89 b	17.63 ± 5.89 a	59.02 ± 14.25 ab	59.17 ± 5.69 c	35.87 ± 9.23 a	38.91 ± 9.38 ab
	LED-2	17.70 ± 2.79 b	16.40 ± 5.92 a	52.38 ± 39.51 b	69.31 ± 13.52 abc	39.63 ± 10.75 a	41.12 ± 10.29 b
	FL	20.66 ± 6.31 ab	20.55 ± 5.28 a	59.82 ± 4.46 ab	64.33 ± 7.75 bc	42.71 ± 11.53 a	38.40 ± 11.25 ab
Truly Tiny	LED-1	19.68 ± 7.05 ab	15.07 ± 2.39 a	61.61 ± 7.71 ab	82.90 ± 14.51 a	43.71 ± 11.57 a	42.10 ± 8.66 ab
	LED-2	17.73 ± 5.80 b	15.08 ± 4.37 a	69.01 ± 18.48 ab	78.79 ± 20.14 ab	35.39 ± 7.94 a	33.88 ± 9.39 ab
	FL	25.10 ± 5.54 a	18.15 ± 3.82 a	62.76 ± 14.47 a	69.96 ± 15.37 abc	47.95 ± 13.36 a	52.33 ± 17.81 a

* Means within columns followed by the same letter are not significantly different at $p \leq 0.05$ according to the LSD (Least Significance Difference) test. The tests were conduct independently within each parameter (adaxial surface epidermis, abaxial surface epidermis, adaxial hypodermis, abaxial hypodermis, palisade parenchyma, spongy parenchyma). AbE = Abaxial Epidermis; AdE = Adaxial Epidermis; AbH = Abaxial Hypodermis, AdH = Adaxial Hypodermis; PP = Palisade Parenchyma; SP = Spongy Parenchyma.

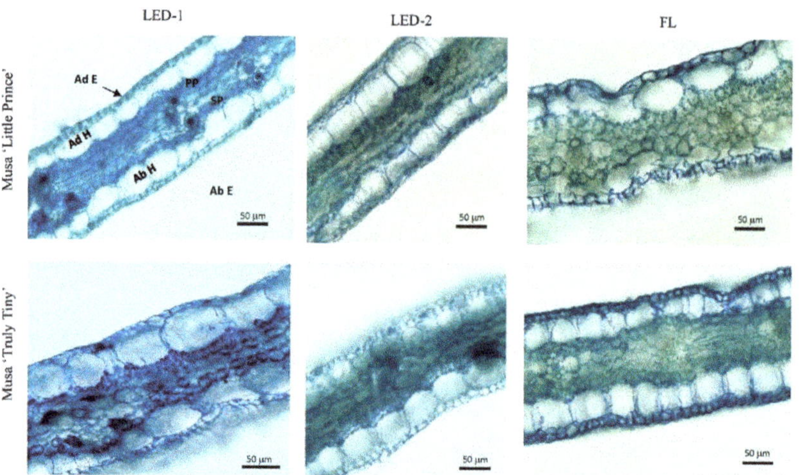

Figure 4. Cross-section of in vitro banana leaves (Musa 'Little Prince' and 'Truly Tiny') under different light sources. LED-1 = 116 μmol m^{-2} s^{-1}; LED-2 = 90 μmol m^{-2} s^{-1}, fluorescent lighting (FL) = 100 μmol m^{-2} s^{-1}. AdE = adaxial epidermis; AbE = abaxial epidermis; AdH = adaxial hypodermis; AbH = abaxial hypodermis; PP = palisade parenchyma; SP = spongy parenchyma. Bar = 50 μm.

4. Discussion

The maintenance of the quality of in vitro plantlets requires an environment that provides optimum conditions for growth and development [26]. Light is among some of the main factors influencing the growth and development of in vitro plantlets, including light quality, intensity, and the photoperiod [27]. In this study, our goal was to evaluate the effect of different light sources on in vitro growth and development of two ornamental banana varieties. However, for most parameters evaluated, no significant differences were observed. The stem diameter of in vitro banana plantlets in our study was not significantly affected by the light source. This contrasts with similar studies showing an increase in stem diameter of in vitro banana plantlets under LED lighting (45, 60, and 75 μmol m^{-2} s^{-1} PPF) compared to fluorescent lights (45 μmol m^{-2} s^{-1} PPF) [22], although light intensities evaluated were lower as compared to the light intensities evaluated in our study [22]. This could indicate that lower light levels are more desirable to induce a larger stem diameter in in vitro banana plantlets. Another aspect was the spectral distribution evaluated [22], where LED lighting had a 9:1 = red:blue ratio and FL had an 8:2 = red:blue ratio. Our study had similar red:blue ratios for both LEDs compared to FL, which had almost an inverse red:blue ratio and a peak in the green (Figure 1), indicating that the light spectral distribution might have influenced the results observed. Similar results were reported in rapeseed in vitro plantlets under LED lighting [28]. In contrast, the stem diameter of Cymbidium in vitro plantlets was not affected by different light sources [29], thus showing similarities to our study. In our study, however, stem diameter had a correlation with banana varieties, whereas 'Little Prince' showed a larger average diameter than 'Truly Tiny', suggesting a genotype effect. Such differences in responses confirm that the influences and mechanisms related to light quality, intensity, and photoperiod in plants are rather specific to plant species or cultivars. A similar genotype effect has been shown for two annatto (*Bixa orellana*) cultivars grown under different irradiance and light quality [2].

Banana in vitro plantlet fresh weight was slightly greater under LEDs compared to FL, but differences were not statistically significant between LED-1 and LED-2 for both varieties. This trend was observed when banana in vitro plantlets were cultured under LED lighting [22] as well as in rapeseed [28], where no significant differences in fresh weight of plantlets under different LED lighting were observed.

Shoot fresh and dry weight of banana in vitro plantlets were not significantly different under different light sources. Similarly, no significant differences were observed for root fresh and dry weight for both banana varieties. However, similar studies with in vitro banana [30] and Doritaenopsis [31] plantlets under LED reported higher shoot and root fresh weight.

Because the spectral energy distribution of red and blue lights aligns with chlorophyll absorption, it is generally accepted that LED lighting improves plant growth and development by enhancing the net photosynthetic rate [32–34]. However, in our study, the effect of LED lighting on shoot length was significantly different from FL only for one banana variety, 'Little Prince', reinforcing the concept of a genotype effect, as previously indicated.

There were no statistically significant differences among plantlets for root length under the different light sources evaluated in our study. These results contrast with those on banana [23] and rapeseed [28], where root length was significantly different between different light sources. However, these studies showed no significant differences among plantlet stem lengths under different light treatments, thus showing similarities with our study.

In our study, the relative chlorophyll content of banana in vitro plantlets grown under LED lighting was higher compared to fluorescent lighting. Similar results for in vitro banana plantlets were reported [35], showing that higher chlorophyll values were obtained when banana plantlets were cultured under LED lights. Similar behavior has been reported for other species, including chrysanthemum [11], Doritaenopsis [32], and rapeseed [28].

The general morphology of plantlets proved quite homogeneous under all lighting treatments evaluated. No differences were observed for leaf number among the different light treatments, and our results are similar to those reported in bananas [35] and Cymbidium. However, in similar studies with banana and Anacardium othonianum plantlets, a higher number of leaves was observed under fluorescent lamps [15,30,36]. However, strawberry in vitro plantlets showed a higher number of leaves in plantlets cultured under

LED compared to fluorescent lighting [37]. The great diversity in responses in vitro among different species as affected by light intensity and quality illustrates the complexity of the topic and the need for continued studies to narrow down the best parameters for proper growth and development of plants in vitro.

Light sources influenced leaf stomata in our study. The number of stomata was generally higher on the abaxial than the adaxial leaf surface. Stomata of plantlets grown under the LED lighting were ellipsoid and closed, compared to open stomata and guard cells with round shapes in plantlets under FL (Figure 3). In general, banana plantlets grown under FL had larger numbers of stomata in both adaxial and abaxial leaf surfaces and similar higher values for length and width of stomata in adaxial leaf surfaces (Figure 3). However, for the abaxial leaf surfaces, while the length was slightly higher for FL, no differences were observed for width. Comparable results were reported in chrysanthemums [11]. However, another study with bananas [35] indicated the increased formation of stomata on both leaf surfaces (adaxial and abaxial) when banana plantlets were grown under LED compared to FL lighting.

In the present study, the epidermis of the adaxial leaf surface was thicker than the abaxial epidermis. These results corroborate a similar study of bananas in vitro [38], where the adaxial surface was thicker than the abaxial surface in leaves of in vitro-derived bananas under acclimatization. However, most of the anatomical leaf features measured in the current study were similar among the different light sources regardless of the banana variety. Similar results have been reported in pepper, showing a similar thickness in anatomical leaf features [39]. High light intensity also affected the anatomic structure in sugarcane leaves, including both adaxial and abaxial sides, and in the mesophyll and bulliform cell number [40]. However, no differences were observed in anatomic structures for ornamental bananas in our study, except for the thickness of adaxial compared to abaxial leaf surfaces.

5. Conclusions

Our results revealed that shoot mass and length could be promoted by controlling light quality and intensity. However, the effect of light quality and intensity related to plant growth and development (stem diameter, shoot and root fresh weight, shoot and root dry weight, root length, root number, and leaf number) were not evident despite contrasting reports from other studies cited here. This leaves additional work to be performed to better address the effects of light sources, intensity, and quality on the in vitro growth and development of different banana varieties. Although not directly evaluated in this study, the number of in vitro shoots produced per explant were higher for some of the cultures grown under the LED lightings, with multiple shoots produced, while no shoots were formed in cultures grown under FL. Subsequent studies will address such parameters. Different responses as demonstrated between the two banana varieties (Musa 'Little Prince' and Musa 'Truly Tiny') indicated a genotype effect as affected by different light environments. LED lighting affected the relative chlorophyll content as well as stomata size in banana in vitro plantlets. Based on the responses for both banana varieties to the different light sources evaluated in this study, we suggest that LEDs at 90 μmol m^{-2} s^{-1} could be a suitable selection for the micropropagation of ornamental bananas.

Author Contributions: Conceptualization, W.A.V.; methodology, W.A.V., C.F., D.B. and P.M.C.B.; validation, W.A.V.; formal analysis, W.A.V.; investigation, C.F. and P.M.C.B.; resources, W.A.V. and D.B.; data curation, C.F.; writing—original draft preparation, C.F.; writing—review and editing, W.A.V., D.B. and P.M.C.B.; supervision, W.A.V. and D.B.; project administration, W.A.V. and D.B.; funding acquisition, W.A.V. All authors have read and agreed to the published version of the manuscript.

Funding: This research project was funded under the USAID-AREA project at the University of Florida, Project Number G000260- 60290000-201-3300-00118870.

Institutional Review Board Statement: Not applicable.

Informed Consent Statement: Not applicable.

Data Availability Statement: Not applicable.

Acknowledgments: This study was supported by the USAID-AREA project under the Department of Environmental Horticulture, Institute of Food and Agricultural Sciences, the University of Florida.

Conflicts of Interest: The authors declare no conflict of interest.

References

1. Rajapakse, N.C.; Shahak, Y. Light-quality manipulation by horticulture industry. *Annual Plant Rev. Light Plant Dev.* **2008**, *30*, 290.
2. Faria, D.V.; Correia, L.N.F.; Souza, M.V.C.; Ríos-Ríos, A.M.; Vital, C.E. Irradiance and light quality affect two annatto (*Bixa orellana* L.) cultivars with contrasting bixin production. *J. Photochem. Photobiol.* **2019**, *197*, 111549. [CrossRef]
3. Gupta, S.D.; Jatothu, B. Fundamentals and applications of light-emitting diodes (LEDs) in in vitro plant growth and morphogenesis. *Plant Biotechnol. Rep.* **2013**, *7*, 211–220. [CrossRef]
4. Miyashita, Y.; Kitaya, Y.; Kozai, T.; Kimura, T. Effects of red and far-red light on the growth and morphology of potato plantlets in vitro: Using light emitting diode as a light source for micropropagation. *Environ. Eff. Control. Plant Tissue Cult.* **1994**, *393*, 189–194. [CrossRef]
5. Kodym, A.; Zapata-Arias, F.J. Natural light as an alternative light source for the in vitro culture of banana (*Musa acuminata* cv.'Grande Naine'). *Plant Cell Tis. Org. Cult.* **1998**, *55*, 141–145. [CrossRef]
6. Dooley, J.H. *Influence of Lighting Spectra on Plant Tissue Culture*; American Society of Association Executives: Washington, DC, USA, 1991; No. 7530.
7. Standaert-de Metsenaere, R.E.A. Economic considerations. *Micropropag. Technol. Appl.* **1991**, 123–140.
8. Mitchell, C. Plant lighting in controlled environments for space and earth applications. *Acta Hortic.* **2012**, *956*, 23–36. [CrossRef]
9. Olle, M.; Viršilė, A. The effects of light-emitting diode lighting on greenhouse plant growth and quality. *Agric. Food Sci.* **2013**, *22*, 223–234. [CrossRef]
10. Nhut, D.T.; Nam, N.B. Light-emitting diodes (LEDs): An artificial lighting source for biological studies. In Proceedings of the Third International Conference on the Development of Biomedical Engineering in Vietnam, Ho Chi Minh City, Vietnam, 11–14 January 2010; IFMBE Proceedings. Van Toi, V., Khoa, T.Q.D., Eds.; Springer: Berlin/Heidelberg, Germany, 2010; Volume 27, pp. 134–139.
11. Kim, S.-J.; Hahn, E.-J.; Heo, J.-W.; Paek, K.-Y. Effects of LEDs on net photosynthetic rate, growth and leaf stomata of chrysanthemum plantlets in vitro. *Sci. Hortic.* **2003**, *101*, 143–151. [CrossRef]
12. Silva, T.D.; Batista, D.; Fortini, E.A.; de Castro, K.M.; Felipe, S.H.S.; Fernandes, A.M.; Sousa, R.M.D.J.; Chagas, K.; da Silva, J.V.S.; Correia, L.N.D.F.; et al. Blue and red light affects morphogenesis and 20-hydroxyecdisone content of in vitro Pfaffia glomerata accessions. *J. Photochem. Photobiol. B Biol.* **2019**, *203*, 111761. [CrossRef]
13. Jishi, T. LED lighting technique to control plant growth and morphology. In *Smart Plant Factory*; Kozai, T., Ed.; Springer: Singapore, 2018; pp. 211–222.
14. Ortiz, R. Conventional banana and plantain breeding. In Proceedings of the VII International Symposium on Banana: ISHS-ProMusa Symposium on Bananas and Plantains: Towards Sustainable Global Production, Salvador, Brazil, 14 October 2011; International Society for Horticultural Science: Leuven, Belgium, 2011; Volume 986, pp. 177–194.
15. Nguyen, Q.T.; Kozai, T. Growth of In vitro banana (*Musa* spp.) shoots under photomixotrophic and photoautotrophic conditions. *Vitr. Cell. Dev. Biol.-Anim.* **2001**, *37*, 824–829. [CrossRef]
16. Wong, W.C. In vitro propagation of banana (*Musa* spp.): Initiation, proliferation and development of shoot-tip cultures on defined media. *Plant Cell Tissue Organ Cult.* **1986**, *6*, 159–166. [CrossRef]
17. Bairu, M.W.; Stirk, W.A.; Doležal, K.; van Staden, J. The role of topolins in micropropagation and somaclonal variation of banana cultivars 'Williams' and 'Grand Naine' (*Musa* spp. AAA). *Plant Cell Tis. Org. Cult.* **2008**, *95*, 373–379. [CrossRef]
18. Makara, A.M.; Rubaihayo, P.R.; Magambo, M.J.S. Carry-over effect of Thidiazuron on banana in vitro proliferation at different culture cycles and light incubation conditions. *African J. Biotechnol.* **2010**, *9*, 3079–3085.
19. Ahirwar, M.K.; Sananda, M.; Singh, M.K.; Chaitali, S.; Singh, R.P. A high frequency plantlets regeneration protocol for banana (*Musa paradisiaca* L.) micropropagation. *Asian J. Hort.* **2012**, *7*, 397–401.
20. Yusnita, Y.; Danial, E.; Hapsoro, D. In vitro shoot regeneration of indonesian bananas (*Musa* spp.) cv. ambon kuning and raja bulu, plantlet acclimatizationand field performance. *AGRIVITA J. Agric. Sci.* **2015**, *37*, 51–58. [CrossRef]
21. Nhut, D.; Hong, L.; Watanabe, H.; Goi, M.; Tanaka, M. Growth of banana plantlets cultured in vitro under red and blue light-emitting diode (led) irradiation source. *Acta Hortic.* **2002**, *575*, 117–124. [CrossRef]
22. Nhut, D.T.; Takamura, T.; Watanabe, H.; Tanaka, M. Efficiency of a novel culture system by using light-emitting diode (LED) on in vitro and subsequent growth of micropropagated banana plantlets. *Acta Hort.* **2003**, *161*, 121–127. [CrossRef]
23. Trivedi, A.; Sengar, R.S. Effect of various light emittimg diodes on growth and photosynthetic pigments of banana (*Musa acuminata*) CV. Grande naine in vitro plantlets. *Int. J. Chem. Stud.* **2017**, *5*, 1819–1821.
24. Murashige, T.; Skoog, F. A revised medium for rapid growth and bioassays with tobacco tissue cultures. *Physiol. Plant.* **1962**, *15*, 473–497. [CrossRef]
25. Kraus, J.E.; Arduin, M. *Manual Básico de Métodos em Morfologia Vegetal*; EDUR: Seropédica, Brazil, 1997; p. 198.
26. Chen, C. Fluorescent Lighting Distribution for Plant Micropropagation. *Biosyst. Eng.* **2005**, *90*, 295–306. [CrossRef]

27. Aitken-Christie, J.; Kozai, T.; Smith, M.A.L. (Eds.) *Automation and Environmental Control in Plant Tissue Culture*; Springer Science & Business Media: Berlin/Heidelberg, Germany, 2013; p. 586.

28. Li, H.; Tang, C.; Xu, Z. The effects of different light qualities on rapeseed (*Brassica napus* L.) plantlet growth and morphogenesis in vitro. *Sci. Hortic.* **2013**, *150*, 117–124. [CrossRef]

29. Tanaka, M.; Takamura, T.; Watanabe, H.; Endo, M.; Yanagi, T.; Okamoto, K. In vitro growth of Cymbidium plantlets cultured under super bright red and blue light-emitting diodes (LEDs). *J. Hort. Sci. Biotechnol.* **1998**, *73*, 39–44. [CrossRef]

30. Nhut, D.T.; Don, N.T.; Tanaka, M. Light-emitting diodes as an effective lighting source for in vitro banana culture. In *Protocols for Micropropagation of Woody Trees and Fruits*; Springer: Dordrecht, The Netherlands, 2007; pp. 527–541.

31. Shin, K.S.; Murthy, H.N.; Heo, J.W.; Hahn, E.J.; Paek, K.Y. The effect of light quality on the growth and development of in vitro cultured Doritaenopsis plants. *Acta Physiol. Plant.* **2008**, *30*, 339–343. [CrossRef]

32. Goins, G.D.; Yorio, N.C.; Sanwo, M.M.; Brown, C.S. Photomorphogenesis, photosynthesis, and seed yield of wheat plants grown under red light-emitting diodes (LEDs) with and without supplemental blue lighting. *J. Exp. Bot.* **1997**, *48*, 1407–1413. [CrossRef] [PubMed]

33. Hoenecke, M.; Bula, R.; Tibbitts, T. Importance of 'Blue' Photon Levels for Lettuce Seedlings Grown under Red-light-emitting Diodes. *HortScience* **1992**, *27*, 427–430. [CrossRef]

34. Tripathy, B.C.; Brown, C.S. Root-Shoot Interaction in the Greening of Wheat Seedlings Grown under Red Light. *Plant Physiol.* **1995**, *107*, 407–411. [CrossRef]

35. Nascimento Vieira, L.; de Freitas Fraga, H.P.; dos Anjos, K.G.; Puttkammer, C.C.; Scherer, R.F.; da Silva, D.A.; Guerra, M.P. Light-emitting diodes (LED) increase the stomata formation and chlorophyll content in *Musa acuminata* (AAA) 'Nanicão Corupá' in vitro plantlets. *Theor. Experim. Plant Physiol.* **2015**, *27*, 91–98. [CrossRef]

36. Rosa, M.; Neto, A.R.; Marques, V.D.O.; Silva, F.G.; De Assis, E.S.; Costa, A.C.; Dantas, L.A.; Pereira, P.S. Variations in photon flux density alter the morphophysiological and chemical characteristics of Anacardium othonianum Rizz. in vitro. *Plant Cell Tissue Organ Cult.* **2019**, *140*, 523–537. [CrossRef]

37. Nhut, D.T.; Takamura, T.; Watanabe, H.; Okamoto, K.; Tanaka, M. Responses of strawberry plantlets cultured in vitro under superbright red and blue light-emitting diodes (LEDs). *Plant Cell Tissue Organ Cult.* **2003**, *73*, 43–52. [CrossRef]

38. Costa, F.D.S.; de Castro, E.M.; Pasqual, M.; Pereira, J.E.S.; Oliveira, C.D. Alterações anatômicas de bananeiras micropropagadas em resposta a aclimatização ex vitro. *Ciência Rural.* **2009**, *39*, 386–392. [CrossRef]

39. Schuerger, A.C.; Brown, C.S.; Stryjewski, E.C. Anatomical Features of Pepper Plants (*Capsicum annuum* L.) Grown under Red Light-emitting Diodes Supplemented with Blue or Far-red Light. *Ann. Bot.* **1997**, *79*, 273–282. [CrossRef] [PubMed]

40. Neto, A.R.; Chagas, E.A.; Costa, B.N.S.; Chagas, P.C.; Vendrame, W.A. Photomixotrophic growth response of sugarcane plantlets with increased light intensity and design of culture vessel. *Vitr. Cell Dev. Biol. Plant* **2020**, *56*, 504–514. [CrossRef]

horticulturae 

Review

De Novo Shoot Development of Tropical Plants: New Insights for *Syngonium podophyllum* Schott.

Camelia Sava Sand and Maria-Mihaela Antofie *

Faculty of Agricultural Sciences, Food Engineering and Environment Protection, Lucian Blaga University of Sibiu, 7-9 Dr. Ioan Ratiu, 550012 Sibiu, Romania
* Correspondence: mihaela.antofie@ulbsibiu.ro

Abstract: *Syngonium podophyllum* Schott. cv. 'White Butterfly' is recognized as a valuable ornamental plant, and today it is also an important plant species of medicinal interest due to its high contents of phenolic compounds. The purpose of this article is to review the main scientific publications from our laboratory with regard to new scientific achievements dealing with *Syngonium* species or topics of interest, such as callus formation and further de novo shoot regeneration. The principles and stages necessary to start an industrial-level micropropagation protocol are discussed based on our experience. Different media compositions induced different morphogenetic responses inside the callus—particularly those related to the development of xylematic elements in the organogenetic areas, such as those for rooting, protocorms, and de novo shoot formation. The re-evaluation of old histological images revealed for the first time that xylematic elements are constantly closely positioned to all organogenetic centers, and that their development is closely dependent on the composition of the culture medium. Separate protocorms can be identified only when xylematic tracheary elements are well developed and closely connected to them. The formation of protocorms is strongly dependent on the mineral composition of the culture medium and the balance of plant growth regulators.

Keywords: *Syngonium*; callus; shoots; minerals; cysteine; xylem; protocorms

Citation: Sand, C.S.; Antofie, M.-M. De Novo Shoot Development of Tropical Plants: New Insights for *Syngonium podophyllum* Schott. *Horticulturae* **2022**, *8*, 1105. https://doi.org/10.3390/horticulturae8121105

Academic Editor: Alicja Tymoszuk

Received: 31 August 2022
Accepted: 23 November 2022
Published: 25 November 2022

Publisher's Note: MDPI stays neutral with regard to jurisdictional claims in published maps and institutional affiliations.

1. Introduction

The organogenesis of shoots is an essential developmental step in the more complex process of plants' morphogenesis for in vitro micropropagation [1]. More than a century ago, Gottlieb Haberlandt stated that plant cells may undergo totipotency—a complex physiological process that, under appropriate conditions, implies somatic plant cells expressing their full genetic information and having the ability to develop into a whole new plant [2]. In 1935, George Avery defined the auxin concentration gradient in plants as being expressed much higher from the tip to the base of the leaf; he also mentioned the 1928 discoveries by Frits Warmolt Went regarding the quantitative determination of auxins using oat seedlings [3]. This discovery was considered to be pivotal in his working group [4] and was followed by a series of other discoveries related to plant growth regulators, which were later underlined by Folke Skoog in 1951 [5].

We should also highlight the contribution of Philip White using tobacco callus tissue culture and, furthermore, his scientific success in obtaining fully developed de novo shoots [6]. The year 1938 can be considered to have brought new scientific evidence proving that an apparently undifferentiated tissue, under controlled conditions, may undergo some differentiation processes that will eventually produce tobacco buds. The principle of totipotency was confirmed by Philip White in 1934 [7]. Moreover, it took only a few steps to prove over time that more and more plant species had become subjects for study due to this ability to produce callus and furthermore to produce the entire plant body from callus tissues.

In the meantime, scientists have raised questions such as the following: What types of donor plant tissues are most suitable to be used for in vitro cultures? How are shoots formed from callus tissues? What processes take place inside the callus? Why do some areas undergo de novo shoot development while others do not? [8–10]. Later, in 1970, based on hundreds of scientific results as well as their own experience, Frederick Steward and collaborators stated that all somatic plant cells can potentially express totipotency under appropriate conditions [11].

Attila Feher recently published a relevant scientific article related to the review of the use of terms in this scientific field [12]. While callus was defined as a growing mass of non-organized cells in 1957 [13], Attila Feher presented new insights related to the need to redefine the structure and organization of callus by taking into account molecular makers already described in *Arabidopsis thaliana* for the processes of dedifferentiation, differentiation, and transdifferentiation. Feher agrees that callus should no longer be considered to be an unorganized tissue but, rather, a result of transdifferentiation processes, further supporting the 2011 observations of Kaoru Sugimoto, Sean Gordon, and Elliot Meyerowitz [14]. Based on their scientific results, auxin-induced callus formation appears to express lateral roots markers instead. Moreover, there seem to be relevant differences between different types of calli, which originate from different plant tissues according to Attila Feher [12]. In this regard, it is also essential to mention the range in the levels of plant growth regulators used to induce callogenesis, as well as to further scientifically substantiate the understanding of organogenetic processes.

Kaoru Sugimoto and collaborators also stated that callus is rather heterogeneous in its composition, being formed of different cell types, and only a few callus cells may ultimately be able to undergo organogenesis and somatic embryogenesis [14]. In this regard, Attila Feher considers callus to be the result of cell reprograming due to environmental conditions, and a single cell can become totipotent if that cell forms a completely new plant via somatic embryogenesis [12]. Based on these scientific findings, the idea arises that callus is not a totipotent or pluripotent tissue [14] but, rather, an intermediary living structure developed under stress factors, which is not controlled by the plant itself, acting as a new independent body that can contain different types of cells, only some of which can become totipotent or pluripotent, being involved in regenerating the whole plant body or only certain organs [12].

The analysis of the main genetic factors supporting de novo shoot formation has been of great interest among scientists [15]. Today, it is well established that auxin-induced de novo shoot formation may originate from perivascular cells, similar to the xylem-pole pericycle cells that initiate the development of lateral roots in *A. thaliana* [16]. These authors also define the concept of a "small founder group of cells" involved in shoot organogenesis. Furthermore, they mention that in order to fully understand the molecular mechanisms, it is necessary to apply synergic cytological, molecular, and genomic approaches [16].

Later, other authors concluded that de novo organogenesis is a complex process that takes place via three logical stages: (1) the activation of regenerating cells, (2) the acquisition of competency, and (3) initial de novo establishment of the apical meristem that can stimulate shoot formation [17].

Today, based on accumulated scientific information on the subject, the genetic pathway for shoot regeneration without plant growth regulators—by altering the expression levels of a group of several genes controlling the process—has already been patented [18].

In direct connection with the process of callus formation at the sites of plants' wounds, abundant relevant scientific evidence has been accumulated. Thus, it is well established that wounded tissues or organs are able to form callus to seal the wounds, which may indicate either adventitious organogenesis or somatic embryogenesis, based on distinct developmental pathways that require specific decisions on the cell fate [19]. These authors suggest that organogenesis at the callus level depends on the balance between three factors: (1) the components of the lipid transfer pathway, (2) reactive oxygen species (ROSs) homeostasis, and (3) cell expansion.

In the history of plant biotechnology, scientists have also carried out intensive histological studies in order to understand the processes taking place inside the callus. One of the first interesting studies carried out using a camera lucida was published in 1923 by Cora Beals [20]. The scientific accuracy of this article remains outstanding today in terms of the interpretation of its results, with the author explaining her hypothesis that shoots must originate from the division of cambium cells. After almost a century, this hypothesis was corroborated by new evidence in molecular biology published by Sugimoto et al. in 2011 [14]. The original study by Cora Beals was later cited by Tsvi Sachs in 1981 [21] to scientifically ground the development process of transport vessels and, furthermore, to hypothesize the need for polar contact with leaves, roots, and a source of auxin for vascular differentiation. This hypothesis is still valid today, after more than 50 years, based on scientific evidence published by Jinwu Deng and collaborators [19].

However, a clear relationship between the callus areas where xylematic elements are developed and de novo shoot development is not yet well established for histological studies. This finding led us to assume that the constant accumulation of scientific evidence could create new opportunities to re-evaluate various old experimental data, to reassess previously obtained scientific results, and to further support new theories [12,14,19]. Therefore, such studies presenting histological analyses of plant callus could now be revised to re-evaluate their scientific results and, furthermore, to provide new insights with respect to the current scientific knowledge. For example, de novo shoot formation from callus should also be assessed as a process based on the pluripotency that is expressed in the callus cells and that is highly dependent on microenvironmental factors (i.e., genetic factors as well as chemical and physical factors from outside the inoculum) [14].

This should be relevant if such studies are to be targeted at the species level. Therefore, we consider that a further analysis of callus's histological sections could provide new insights to better understand de novo shoot organogenesis and, furthermore, to support the connectivity between this process and the development of xylematic elements.

A species of high interest for plant biotechnology at the global level is *Syngonium podophyllum* Schott., which was first micropropagated by Lynn Miller and Toshio Murashige in 1976 [22]. The term for the genus *Syngonium* was first used in 1829 by H.W. Schott; later, in 1851, the same author described *S. podophyllum* as a species of the Araceae family originating in Central America. The species was described as a liana with sagittate- or hastate-shaped leaves located in tropical rainforests. In 1981, Thomas Croat described the geographical distribution of this species as ranging between Mexico, Brazil, Guiana, and Bolivia [23]. For more than 50 years, the species has also spread in Africa, Australia, and Asia, according to the distribution map published by the Global Biodiversity Information Facility [24].

The increase in commercial demand worldwide creates the opportunity for this species to be studied and introduced for industrial micropropagation [25] and, furthermore, to be considered one of the most valuable indoor ornamental plants in the world [26]. Moreover, today, it is known that syngonium is also a source of valuable phenolic compounds with different potential uses in medicine [27–29]. Generally, a major consequence of intense trade is to facilitate the spread of species to new environments. Therefore, today, this species is considered to have a moderate-to-high potential risk of invading new territories and threatening the conservation of biodiversity in newly occupied ecosystems [30].

In Romania, our laboratory has been interested in developing an industrial protocol for *Syngonium* micropropagation since 1996. However, as a former communist country, the lack of access to scientific literature created the opportunity to study this species as completely new for biotechnology. The first goal of this article was to present a series of principles to follow in order to develop an industrial-scale micropropagation protocol. The second goal was to review the whole technological flow for micropropagation stage by stage, based on the previous experience of our laboratory, with the main focus on the formation of callus and de novo shoots. Thus, we reviewed the main results of experiments conducted in our laboratory on *Syngonium* between 1996 and 2004 and published after 1998, in accordance with the current scientific knowledge, to provide new insights specifically related to de novo shoot formation from callus. Our third goal was to present the results of a histological study involving morphogenetic calli. Particular emphasis was placed on the study of callus areas occupied by xylematic elements and organogenesis (i.e., the formation of shoots, roots, and protocorms).

2. Materials and Methods

This article is a review of the main scientific articles published regarding the study of the different stages required for in vitro micropropagation of *Syngonium podophyllum* Schott. cv. 'White Butterfly' by our laboratory, and with reference to the current literature, in our attempt to improve our understanding of de novo shoot formation from callus.

This article is structured in three parts. The first part is dedicated to discussing the main principles followed for setting an industrial micropropagation protocol, along with underlying opportunities and obstacles based on our experience.

In the second part, we further discuss the materials and methods applied for developing an industrial micropropagation protocol, as described in three published articles [31–33] as well as in a doctoral thesis [34]. All three scientific articles were published in the English language—two of them in Romanian journals and one in a journal from the United States. The thesis was published in the Romanian language. All stages—including the starting plant material, sterilization, meristem culture, initiation, micropropagation, and acclimation—are discussed with respect to the existing scientific knowledge on the species. Wounding stress, callus formation, and histological studies of calli are discussed as relevant in order to further contribute to our understanding of the process of de novo shoot formation in syngonium.

In our search strategy implemented in Google Scholar, we included all scientific articles studying the species *Syngonium podophyllum* whose full text was freely available in the English language [35]. In this strategy, the year 1976 was set as the starting point, when the first published article was recorded [22]. There are also some valuable closed-access articles available online, which are mentioned in this paper as citations in the required contexts. However, in choosing the most relevant scientific articles for discussion in this article, the focus was on the following keywords: callus development, callus histology, and de novo shoot formation.

The scientific names of the plant species were all validated against recognized plant taxonomic databases [36–38].

3. Results and Discussion

3.1. Principles for Starting In Vitro Micropropagation of Syngonium podophyllum Schott. Cv. 'White Butterfly'

Syngonium podophyllum Schott. cv. 'White butterfly' was studied as a representative of the Araceae family in Romania for 8 years, between 1996 and 2004. The objective was to develop a micropropagation protocol for industrial purposes, initiated in 1996 as an important goal to increase the indoor plant trade offerings of the Glasshouse Complex Codlea in Brasov County, Romania. This complex functioned for 20 years between 1988 and 2008, after which it was forced to close due to a nationwide economic crisis. It should be noted that the profit of the state-owned company was outstanding, and the decline was mainly the result of political issues [39,40].

In implementing an industrial micropropagation protocol for *Syngonium*, we followed three major principles: (1) to maintain long-term genetic stability for all micropropagation stages, (2) to provide the most cost-effective technology, and (3) to constantly upgrade the technology according to the latest scientific achievements.

To adhere to the first principle, based on our laboratory experience, we used (a) meristem culture to avoid contamination by viruses and other microorganisms (this protocol had already been implemented for carnations, among other species), (b) the simplest and cheapest possible culture media and avoidance of long-term propagation (i.e., by using low hormone quantities and avoiding mutagenesis), and (c) a very rigorous control of technological factors—such as light intensity and photoperiod, day/night temperature, humidity, air sterilization, and ventilation—similar to existing technology for carnation micropropagation, and for the entire workflow from laboratory to greenhouse.

The second principle aimed to obtain the simplest and most cost-effective industrial protocol possible based on the available reagents in the laboratory and greenhouse, as well as the skills developed by laboratory personnel. A highly rigorous control system was implemented for all technological factors (e.g., light intensity and photoperiod, day/night temperature, humidity, air sterilization, ventilation, electricity supply, heating/cooling during the winter/summer seasons, and disease/pest control in the greenhouse). We also applied rigorous control of all personnel activities and created bypass-type plans for potential technological remedies (e.g., virus detection in case of infection in the laboratory, disease/pest control in the greenhouse, contingencies in case of delays to the import of reagents or other materials).

The third principle, referring to the constant upgrading of technology according to the latest scientific achievements, was implemented with the limitation of restricted access to the newest scientific information due to prohibitive costs. We should also mention the total lack of access to the scientific journals published between 1996 and 2000, during which time the scientific literature library of the Romanian Academy was not open-access. In this regard, three relevant peer-reviewed scientific articles were published in the English language on subjects related to the callogenesis and micropropagation of the species [31–33], which scientifically substantiated a Ph.D. thesis that was publicly defended after six years of research in 2002 [34].

These principles discussed above are consistent with those applied and discussed by other authors [41,42].

In order to develop a completely new micropropagation protocol for newly introduced ornamental species, it was considered relevant at the time to devise some experimental tests for ensuring the most appropriate balance between auxin and cytokinin, as well as the best mineral and vitamin compositions for use in the culture media. In this case, it was preferred to start using the plant growth regulators already available in our laboratory, such as benzyl-aminopurine (BAP), indole-3-acetic acid (IAA), naphthyl acetic-acid (NAA), and 2,4-dichlorophenoxyacetic acid (2,4-D). All of the other reagents (i.e., minerals, vitamins, sucrose, and agar) were provided by Merck, Sigma, and Difco, and they were already in use in our laboratory in the industrial protocol for carnation micropropagation (*Dianthus* sp.), as well as other species, cultivars, and hybrids (i.e., *Gerbera x hybrida* Hort., *Chrysanthemum* × *morifolium* (Ramat.) Hemsl., *Nephrolepis exaltata var. hirsutula* (*G. Forst.*) *Baker*, sp., *Cymbidium floribundum* Lindl., *Saintpaulia ionantha* H.Wendl., *Sequoia gigantea* Endl., *Solanum tuberosum* L.). Based on the experience of our laboratory, these reagents were considered to be the most reliable in terms of costs as well as for midterm preservation. The simplified micropropagation protocol was published in *Aroideana*—the Journal of the International Aroid Society—in 2004 [33].

3.2. Micropropagation Protocol Review

3.2.1. In Vitro Culture Initiation

Four-year-old, healthy, certified mother plants of *Syngonium* imported from the Netherlands were used as donor plants. These plants were maintained in culture pots under controlled greenhouse conditions for more than 2 years to ensure that there was no phytosanitary contamination. It should be noted that using healthy mother plants is essential for starting plant meristem cultures according to Toshio Murashige and Folke Skoog [43]. Explants as stem fragments comprising the nodal segment (i.e., less than 5–6 cm in length) were well washed under running tap water for 2 h. Nodal explants were further immersed in an aqueous solution of Tween 20 (2–3 drops/L) under continuous shaking. After 10 min, all nodal explants were surface-sterilized in the laminar hood flow for 30 s using a sterile solution of 0.1% $HgCl_2$, followed by three consecutive rinses in sterile water for 10 min, each under continuous shaking. Each nodal explant was placed on sterilized filter paper before taking the meristem under stereomicroscope [33,34].

Sterilized meristems were cultivated on a solidified, modified Murashige and Skoog (MS62) [43] culture medium and sub-cultivated after 10 days under the same conditions to prevent browning of meristems, implementing good laboratory practices to ensure a virus-free process similar to that used for carnations [44]. At this stage, the MS62 inoculation medium was supplemented with BAP (30 mg/L), IAA (1 mg/L), and sucrose (30 mg/L). Merck agar (7 g/L) was added before adjustment of the pH to 5.8, and after sterilization it was reduced to 5.4. The success rate for initiating meristem cultures was 100%, with all of the inoculated meristems being viable [33,34]. This was also due to the disinfecting method used, as well as to the skills of the staff members, who had almost 16 years of practice performing this activity. We should also note that some authors have successfully used antibiotics for in vitro micropropagation of *Syngonium* into liquid culture media at an industrial scale [45].

Meristematic domes with the first or second leaf primordia at a maximum size of 0.2 mm were taken under stereomicroscope in sterile conditions during springtime. *Syngonium* in vitro cultures have also been used by other professional groups [22,46–63] (see Table 1). In 1976, Lynn Miller and Toshio Murashige published the first protocol for micropropagation of *Syngonium*, starting with 0.2–0.4 mm axillary meristems taken under stereomicroscope [22].

Table 1. Balance of plant growth regulators and relevant observations related to in vitro micropropagation of *S. podophyllum* from our review of scientific articles whose full text was freely available on Google Scholar.

References	Plant Growth Regulators for Initiation (mg/L)	Plant Growth Regulators for Multiplication (mg/L)	Observations
Miller and Murashige 1976 [22]	3 mg/L IAA + 2 mg/L 2iP	2 mg/L IAA + 30 mg/L 2iP	No callus described; solidified and liquid culture media; complete technology described
Scaramuzzi et al., 1992 [46]	1 mg/L IAA + 5 mg/L Kin	1 mg/L IAA + 5 mg/L BAP	Callus formation induced by high levels of cytokinins; cytological and histological studies; complete technology described
Salame and Zieslin 1994 [47]	3 mg/L IAA + 2 mg/L 2iP	2 mg/L IAA + 30 mg/L 2iP	Peroxidase analysis for in vitro plant wound-stress study
Watad et al., 1997 [48]	3 mg/L IAA + 2 mg/L 2iP	2 mg/L Kin	Using interfacial membrane rafts for liquid culture media
Rajeevan et al., 2002 [49]	0.5–2 mg/L BAP	2 mg/L BAP + 0.5–2 mg/L Kin	Callus was also induced at high cytokinin levels
Chan et al., 2003 [50]	Not specified	2 mg/L BAP or 2 mg/L IBA + 2 mg/L BAP	No callus described
Schwertner and Zaffari 2003 [51]	1 mg/L BAP + 1 mg/L IAA	1–4 mg/L BAP	The best multiplication rate for 4 mg/L BAP; callus was obtained when 4 mg of BAP was added
Hassanein 2004 [52]	1 mg/L BAP	1 mg/L BAP	Shoot multiplication
Chen and Henny 2006 [53]	Citing Miller and Murashige 1976 [22]	Citing Miller and Murashige 1976 [22]	Review of scientific literature on the micropropagation of *Syngonium*
Zhang et al., 2006 [54]	80 mg/L adenine	0.2 mg/L NAA + 2 mg/L BAP	No callus formation; the study is relevant for somatic embryogenesis

Table 1. *Cont.*

References	Plant Growth Regulators for Initiation (mg/L)	Plant Growth Regulators for Multiplication (mg/L)	Observations
Wang et al., 2007 [55]	80 mg/L adenine	0.2 mg/L NAA + 2 mg/L BAP	No callus formation; the study is relevant for somatic embryogenesis
Cui et al., 2008 [56]	0.1 mg/L NAA + 0.2 mg/L TDZ	1 mg/L NAA + 2 mg/L CPPU or 2 mg/L TDZ	Callus description; protocorms and histological study; the study is relevant for industry
Rajesh et al., 2011 [57]	Not specified	20 mg/L BAP	The average shoot formation was similar to our results: 9.5 shoots/explant
Teixeira Da Silva et al., 2014 [58]	Not specified	Not specified	Ploidy study of in vitro plantlets
Kalimuthu and Prabakaran 2014 [59]	0.5–3 mg/L BAP + 200 mg/L NaH$_2$PO$_4$/0.2 mg/L NAA/0.2 mg/L TDZ	0.5–3 mg/L BAP + 200 mg/L NaH$_2$PO$_4$/0.2 mg/L NAA/0.2 mg/L TDZ	The best results were obtained for the combination 1 mg/L BAP + 200 mg/L NaH$_2$PO$_4$
Teixeira Da Silva 2015 [60]	Citing Wang et al., 2007 [55]	Citing Cui et al., 2008 [56]	Ploidy study of in vitro plantlets
Moumita et al., 2016 [61]	MS62	2 mg/L BAP + 0.5 mg/L NAA	The best formula for shoot multiplication
Kane 2018 [62]	Citing Miller and Murashige 1976 [22]	Citing Miller and Murashige 1976 [22]	The chapter provides an activity for students' education
Sharifi et al., 2022 [63]	MS62; not in English	1 mg/L BAP + 3 mg/L Kin	The authors focused on testing shooting success

Abbreviations: IAA: indole-3-acetic acid; 2iP: 6-(γ,γ-dimethylallylamino)purine; Kin: kinetin; BAP: 6-benzylaminopurine; IBA: indole-3-butyric acid; NAA: 1-naphthaleneacetic acid; CPPU: (2-chloro-4-pyridyl)-N$'$-phenylurea; TDZ: thidiazuron, or N-phenyl-N$'$-1,2,3-thiadiazol-5-ylurea.

Further stages in the development of meristems are discussed in our previous papers [33,34]. After 12 weeks of cultivation in the same culture medium, it was generally possible to consider that all inocula were sufficiently developed (i.e., a diameter ranging between 0.5 and 1 cm) to the first multiplication stage by splitting clusters of buds into two to four pieces. The first multiplication stage was carried out in MS62 culture medium supplemented with BAP (1 mg/L), Kin (3 mg/L), and IAA (0.1 mg/L). After a total of 20 weeks from the initiation of the process (i.e., the day of meristem inoculation), it was possible to isolate clearly differentiated shoots (i.e., heights ranging between 10 and 15 mm) to begin the multiplication experiments, and only these shoots were recorded to assess the multiplication rate. All shoots and buds less than 10 mm in height were considered to be insufficiently developed to enter the economic workflow at the industrial scale. Consequently, these calli were transferred to new culture media in order to ensure further development of buds and elongation of shoots.

The balance of plant growth regulators used in our laboratory for meristem cultures was in favor of cytokinin (i.e., 30 mg of BAP and 1 mg of IAA) [33,34]. Other authors used different balances of plant growth regulators, as well as different regulators. For example, Lynn Miller and Toshio Murashige used a hormone balance in favor of auxin (i.e., 2 mg of 2iP and 3 mg of IAA) [22]. Other teams used a balance in favor of cytokinin [49,51,52,56–59]. Later, other authors used new synthetic regulatory substances, again more inclined towards cytokinin, such as 1 mg of NAA and 2 mg of (2-chloro-4-pyridyl)-N$'$-phenylurea (CPPU) or 2 mg of N-phenyl-N$'$-1,2,3-thiadiazol-5-ylurea (TDZ) [56,59] (see Table 1).

As shown Table 1, very different types and balances of plant growth regulators have been used in different laboratories to initiate meristem cultures, indicating that this species is very reactive towards different culture conditions and very easy to culture in vitro. We should also note the contributions to developing a micropropagation protocol by relevant researchers who are cited by others but whose articles were not available. For example, Scaramuzzi and coworkers [46] cited an unavailable paper published by Makino and Makino in 1978 (e.g., the article was entitled "Propagation of Syngonium podophyllum cultivars through tissue culture", and published in In Vitro, volume 14, page 357). Some other articles were also unavailable or written in languages that do not use the Latin alphabet.

The decision to incline the balance of the plant growth regulators in favor of cytokinin was based on our observations related to the well-expressed apical dominance of the main shoots of the pots' donor plants. Apical dominance has long been known to be supported by the high internal levels of auxin secreted by young organs [64]. However, the interruption of this auxin gradient flow can be achieved during the cutting-off of the meristem, and this process is confirmed by the further addition of auxin to the meristem's culture medium, as reported by Lynn Miller and Toshio Murashige [22]. This idea was not tested in our laboratory, and it should be taken into consideration for further studies.

3.2.2. Micropropagation and Culture Media Testing

In the 28th week of in vitro cultivation, we began the testing of auxins and vitamins using an MS62 basal medium composition [33,34]. The highest recorded multiplication rate was ~9.17 shoots/explant in the case of an auxin–cytokinin ratio of 0.5 mg/L NAA to 1 mg/L BAP. It should be noted that only shoots over 7–9 mm in height were recorded.

In the 36th week of in vitro cultivation, we conducted the cytokinin tests, and based on our results the rate of multiplication was slightly higher than 9.85 shoots/explant [33,34]. This multiplication rate was slightly higher than that obtained on the same solidified MS62 culture medium by Lynn Miller and Toshio Murashige (i.e., 7.9 shoots/explant, with their minimum height not defined) [22]. However, Miller and Murashige obtained a higher multiplication rate when liquid culture medium was used (i.e., 26 –> 60 shoots/explant). Comparable multiplication rates were also reported by other authors [46,49,51,53,57,59,61,63].

In our laboratory, a multiplication rate of 9 shoots/explant was considered to be excellent for solid culture medium use when recording only shoots of over 7–8 mm in height, and we also found a low capacity for infection compared to liquid culture media. At the end of this stage, the height of the shoots used for industrial-scale micropropagation was set to be longer than 7–8 mm. Other authors obtained the same multiplication rate using different compositions of plant growth regulators and different cytokinin–auxin ratios [57]. The clusters comprising shoots shorter than 7 mm in length, as well as buds, were transferred for further multiplication and elongation of the shoots on the same culture medium, with a decreased cytokinin–auxin ratio [33]. However, increasing the shoots' multiplication rate to more than 60 shoots/explant should be more cost-effective for starting the technological flow and for producing a large number of shoots in a very short time. Some other teams were also successful in obtaining somatic embryos, completing all morphogenetic programs for the species, but without callus generation [54,55].

A multiplication rate of nine plantlets/explant up to the greenhouse stage generated total costs of 1 USD/100 pot plants, and in 1998 this was considered to be profitable. Under present conditions, such a value is no longer worthwhile; furthermore, the technology needs to be updated for the new energy requirements. Starting with the third stage of the technological flow, different experimental tests were initiated. Relevant stages related to Syngonium micropropagation are presented in Figure 1.

Figure 1. Different stages in the micropropagation of *Syngonium podophyllum* cv. 'White Butterfly'. From left to right, de novo shoots over 7 mm in height were separately transferred to solidified MS62 culture medium for elongation of shoots up to the moment of transfer for acclimation. The culture medium's composition is described in 2004 [33].

3.3. Callus Induction and Development

3.3.1. Wound Stress and Callus Initiation

In 2000, based on contract no. 6127/2000 signed with the Romanian Ministry of Research, it was possible to develop an experimental test to understand the relationship between wound stress and callus initiation for shoot formation. The expected results were to be used to improve the *Syngonium* micropropagation protocol [33,34]. In that experiment, we used fragments of in vitro petioles and leaves, and salicylic acid (SA) and the synthetic hormone 2,4-D were tested as signaling molecules. SA was recognized at the time for its role in preventing wound signaling in plants [65,66], while 2,4-D had long been recognized for its role in inducing callus formation based on the experiment conducted by John Torrey and Kenneth Thimann in 1949, who sprayed this herbicide on the surface of stumps of the tropical sicklebush tree (*Dichrostachys nutans* Benth.) [67]. They observed that a ring of callus tissue developed at the cambium level after a couple of weeks. We should also note that the use of the synthetic plant growth regulator 2,4-D for in vitro plant experiments was successfully applied after 1945 [68–70].

Today, it is already recognized and scientifically proven that wound stress is responsible for the production of callus tissue in order to seal the wounded site and to facilitate adventitious organogenesis or somatic embryogenesis [19].

3.3.2. Experimental Design

The tested culture media were composed of two parts: the first part was a solidified MS62 culture medium supplemented with BAP (1 mg/L), which was covered with the second part, comprising 2 mL of liquid culture medium with the same composition and supplemented with 2,4-D (0.1 mg/L) and/or SA (0.1 mg/L). The entire experiment was set up in Petri dishes of 6 cm in diameter. The inoculum was fixed to the surface of the solidified culture medium, and a thin liquid culture medium covering the inoculum ensured rapid and uniform contact with the used signaling molecules.

3.3.3. Total Peroxidase Activity

The total peroxidase activity (TPA) was assessed at 72 h according to the method of Dana Iordăchescu and Ioan Florea Dumitru published in 1988 [71]. The size of the leaf fragments was 0.5 cm × 0.5 cm, while the petiole fragments were 1 cm in length. As a general control, unwounded shoots were used. In this case, our experimental hypothesis was based on the knowledge that wound stress is already expressed during in vitro cultivation of plants, according to various authors [72,73]. Moreover, it has long been observed that a burst of phenolic compounds is generally released at the wound sites of plants, and that this process is immediately followed by the activation of membrane-bound peroxidases that are required to initiate healing processes [65,74].

- Effects of 2,4-D:

After 72 h, the petiole fragments treated with 2,4-D exhibited a 1.84-fold increase in TPA compared to control shoots (i.e., intact shoots). In this case, the auxin and the supplemented cytokinin clearly contributed to an increase in TPA to initiate cell proliferation, as the first and most obvious event was recorded four weeks later. In the case of control petioles, we found a 1.6-fold decrease in the TPA activity compared to explants in the presence of auxin, suggesting that auxin activated peroxidase activity in the petiole fragments. In fact, upon the increase in TPA activity at the wound site, cell proliferation began as an essential process for the survival of the plant inoculum. The calli or fragments of plants can further act to continue the plants' morphogenesis [10]. In the case of leaf fragments, 2,4-D induced a 2.45-fold increase in TPA activity compared to intact leaves and a 1.25-fold increase compared to the control leaf fragments. Due to the larger leaf surface for wounding, the release of phenolic compounds [27–29] might have been responsible for the activation of enzymes with peroxidase activity, which should greatly enhance the wound stress response of leaf fragments in the presence of 2,4-D.

- Effects of salicylic acid:

For petiole fragments, SA induced a 4.26-fold increase in TPA activity compared to petioles taken from intact shoots and a 1.46-fold increase compared to the control petiole fragments. This value was more than double the effect of 2,4-D at the wound site and is consistent with previous scientific results [60]. In the case of leaf explants, SA induced a ~1.2-fold increase in TPA activity compared to the intact leaves but, conversely, a 1.6-fold decrease compared to control leaf fragments. Here, SA induced TPA activity at half the rate of 2,4-D. Moreover, the TPA value in fragmented leaves was almost equal to that in intact leaves from intact shoots. It can be concluded that the exogenously applied SA may minimize the expression of TPA in syngonium at the leaf level, but it can induce overexpression in the fragments of petioles.

- Effects of 2,4-D and salicylic acid:

When both compounds (2,4-D and SA) were exogenously applied, it appeared that the TPA activity was similar to that obtained in the fragmented petioles or leaves subjected only to 2,4-D. Thus, the effects of SA appeared to be hidden by the presence of 2,4-D.

3.3.4. Callus Initiation from Fragmented Petioles and Leaves

The entire experiment was observed continuously for 8 weeks. In the first three weeks, leaf and petiole fragments subjected to 2,4-D became etiolated compared to controls and those treated with SA, which remained green. All petiole fragments subjected to SA developed hypertrophy at the cut part of the petiole following the circular cambium, which also spread along the entire length of the fragment. However, some of these samples only developed callus at the wound site—white and friable, in small quantities—and later in the cambial tissue as well. Among the two types of explants, only petiole fragments were able to produce a proliferative callus for syngonium at the basipetal part of the petiole, in a similar manner to that described for other species [62]. In contrast, based on the conditions of this experiment, it was not possible to induce callus formation at the leaf fragment level.

Moreover, other authors were successful in initiating and studying callus formation based on a specific hormone balance [46,49,51–53,56,63]. However, it should be noted that a burst of phenolic compounds may also affect callus formation [75].

In another experiment conducted by Jin Cui and collaborators on the same cultivar but under different culture conditions (i.e., different plant growth regulators and different balances), the proliferation of callus was also possible at the leaf level [56].

Re-evaluating these results, it can be concluded that wound stress is also responsible for a clear sequence of events, i.e., increased peroxidase activity, followed by the expression of hypertrophy at the wound sites and, subsequently, by callus development. Hormone balance and mineral composition are the major inductors of callus formation, and different organ fragments may respond differently. The lack of callus formation in fragmented leaves under our experimental conditions may have been due to the culture medium composition, which was not supportive of the expression of hypertrophy, although it was proven to express the highest peroxidase activity. This subject should be of interest when investigating callus initiation for different purposes [46,49,51–53,56,63].

3.4. Review of Histological Studies on Syngonium Calli Obtained with Different Media Compositions

In 1998, the very first scientific article dedicated to the study of the effects of medium composition on in vitro morphogenesis was published in Romania in the English language [31].

The first hypothesis of the experiment was based on the statement that morphogenesis in plants is a very complex process that is strictly controlled in time and space by its genetic and external factors [76]. In this regard, we considered that a histological study was needed to reveal morphogenetic events such as de novo morphogenesis of shoots and roots—as proposed for other species by various authors—to better describe and understand the in vitro morphogenesis of syngonium [77,78]. To this end, young shoots of 8–10 mm in height were tested by cultivating them in two types of cultivation media—MS62 [39] and Nitsch 1969 (N69) [79]—as basal mineral compositions [31,34].

The second hypothesis considered that, with *Syngonium* being a tropical plant, it would be interesting to study the effects of a completely different mineral composition in parallel to that of the MS62 culture medium. We considered the fact that essential elements would be at different concentrations in the original tropical rainforest environment. Therefore, for successful in vitro multiplication at a low cost, it was also considered relevant to test various culture media whose mineral composition could be modified. At the time, we also considered that the electrolytic strength of the MS62 culture medium should be different compared to that of N69. Furthermore, it may play an important role in organogenetic processes such as de novo shoot formation and rooting (i.e., different osmotic pressure). In the case of N69's mineral composition, calcium was at half the usual concentration, while zinc and manganese were increased (from 8.6 to 10 mg/L and from 16.9 to 18.94 mg/L, respectively). In addition to these changes, cobalt and iodine were absent, boron was supplemented to an increased concentration (from 6.2 to 10 mg/L), and the concentrations of phosphate and nitrate were reduced (from 85 to 68 mg/L and from 825 to 720 mg/L, respectively). Later publications provided highly accurate descriptions of relevant differences in mineral changes for tropical forests [80,81].

3.4.1. Experimental Design and Histological Method

The basal mineral culture media (i.e., MS62 and N69) were supplemented only with MS62 vitamins and the same hormonal balance: 2,4D (0.1 mg/L), BAP (1 mg/L), and Kin (3 mg/L). The molar ratio was inclined towards cytokinin which, according to our previous experience, was needed to test the morphogenetic process. Based on the aforementioned working hypotheses for implementing this experiment, the morphogenetic callus obtained after 8 weeks of cultivation was processed for histological study.

For histological analysis, all callus samples were collected at the same time and immersed in Navashin's fixation solution at room temperature, as recommended by other authors [82], and then in liquid paraffin. After the samples solidified, they were sectioned at 8–10 μm and colored on slides with a solution of hematoxylin–eosin in order to enhance the contrast between different tissue structures [83].

The published conclusions of this experiment revealed that different medium compositions induced distinct responses in syngonium callus organogenesis: MS62 culture media induced the formation of a softer, yellowish callus expressing complete organogenesis (i.e., shoots and roots), as opposed to N69, which induced the formation of a harder, greener callus expressing only shoot formation [31]. In 1992, Scaramuzzi and collaborators in 1992 published the first histological analysis of the morphogenetic callus, the findings of which were similar to our results [46]. Later, a similar yellowish callus was also obtained and described by Jin Cui and collaborators [56].

3.4.2. Histological Analysis of Callus: Xylematic Elements and Protocorms

The histological analysis of calli obtained in MS62 and N69 culture media revealed some interesting peculiarities that should be further discussed. In 1998, we underlined the significant differences that appeared for organogenetic responses in relation to "xylematic-like structures" (XLSs) detected by our team in the structures of the calli [31]. As we now have full access to additional scientific articles, we felt it important to investigate the scientific meaning of the "xylematic-like structures" once more. We found that, in 1995, such structures had already been described as procambial cells for *A. thaliana* by Simona Baima et al. [84]. The authors proved at the time that the auxin IAA involved in vascular development modulates the expression of *Athb-8*—a gene responsible for the regulation of vascular development in *A. thaliana*. In that period, the term "vascular bundle" was well established, having also been coined early in 1920 for histological studies of callus by Robert John Harvey-Gibson and Elsie Horsman in Liverpool (UK) [85].

The scientific quest to understand how vascular bundles or strands are formed yielded more results after 1980, and yet some 20 years ago the mechanism was still largely unknown [86]. In 1987, Harry Klee and collaborators stated that auxins are able to induce differentiation of xylem tracheary elements (XTEs) in suspension cultures of certain species [87]. This is also consistent with our observations of syngonium, as we also used the term xylematic-like elements (XLEs) for tracheary elements observed on histological slides, as a precautionary step. However, the fact that auxins trigger vascular bundle formation was scientifically established in 2000 [88]. The potential role of XLEs had not been discussed previously, but in trying to explain the hard consistency of the callus obtained in N69 culture medium (mineral composition), we consider that this revision is needed.

A closer analysis of the slides for the softer calli obtained in the MS62 culture medium (see Figure 2) clearly reveals that they are of smaller dimensions and constantly appear in the vicinity of organogenetic polarized structures and embedded in the larger parenchymatic cells. The XLEs may be considered to be protoxylem elements—a topic that had not been discussed previously—but it is obvious that they appear constantly in the vicinity of de novo shoots and roots, as well as that of the nodular-like structures. This constant positioning of the XLEs may be due to their potential role in supporting the development of new organogenetic structures. Their high density around the polarized organogenetic structures may support this hypothesis, as they are not observed in callus zones where organogenesis is not occurring. Moreover, this is consistent with the study by Thomas Berleth and collaborators published in 2000 [88], as well as considering the role that auxin may play in the establishment of plant cells' polarity and oriented differentiation which, in turn, are needed for aligning vascular differentiation, among other functions. Following this logic, XLEs develop in the parenchymatic tissue of the callus to support the further development of organogenetic centers in order to accomplish their final morphogenetic objective: whole-plant formation. This hypothesis is supported by recent results that are based on the analysis of molecular markers, supporting the idea that a callus is a group of pluripotent cells where—under appropriate conditions—the promotion of auxin self-production and the enhancement of cytokinin sensitivity are also required for organogenesis [89]. In our case, culture conditions ensured the availability of auxins as well as that of cytokinin. The left-hand image in Figure 2 presents a cross-section of the upper part of the shoot tips, as well as other organogenetic areas, in circular shapes that may occur at the basal parts of different organogenetic areas of the callus. Similar images were published by Scaramuzzi [46], further substantiating our results.

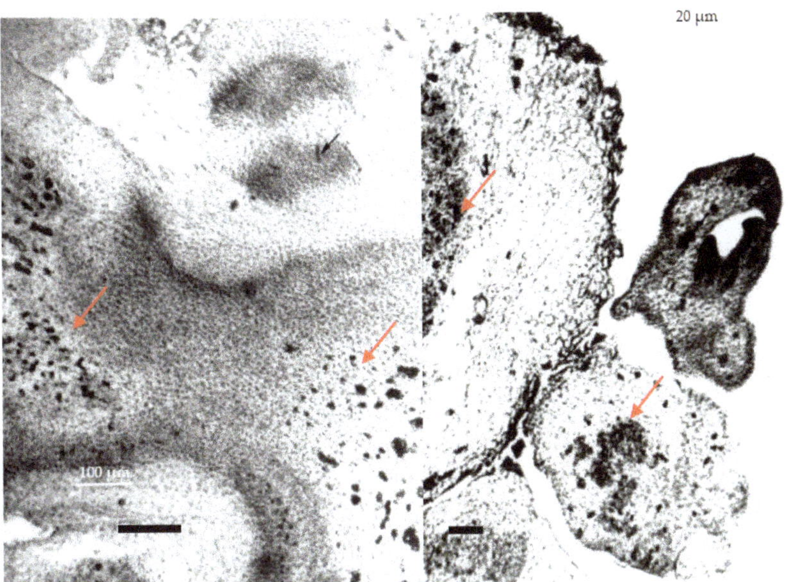

Figure 2. Histological analysis of slide images taken for *S. podophyllum* callus obtained in the MS62 culture medium. This was a full organogenetic callus producing roots and shoots. Nodular-like structures where organogenesis is taking place can be observed (in the tips of the right-hand shoot formation and in a transversal section of the upper part of the left-hand shoot formation; black arrows). Xylematic-like elements or bundles can be observed in the vicinity of all organogenetic centers (see red arrows; magnification ×270 (left) and ×135 (right); bars = 100 µm).

By changing the basic mineral formula, e.g., for the N69 mineral composition, the harder greener callus revealed the formation of well-developed XTEs during the histological study (see Figures 3 and 4). By analyzing different histological images taken at the time, we further observed that these XTEs were constantly arranged in areas positioned at the bases of meristematic domes, as well as those of developing leaves or shoots. Small XLSs could also be observed close to areas comprising polarized nodular-like shapes recognized for their relevance in the initiation of organogenesis.

A re-evaluation of old slides revealed the presence of protocorms (see Figure 3) that were very well differentiated in a transversal section, which were not mentioned previously [31]. Different-sized images of such protocorms were frequently seen in the slides. It should be noted that protocorms were also described by other authors in different culture conditions [56,60].

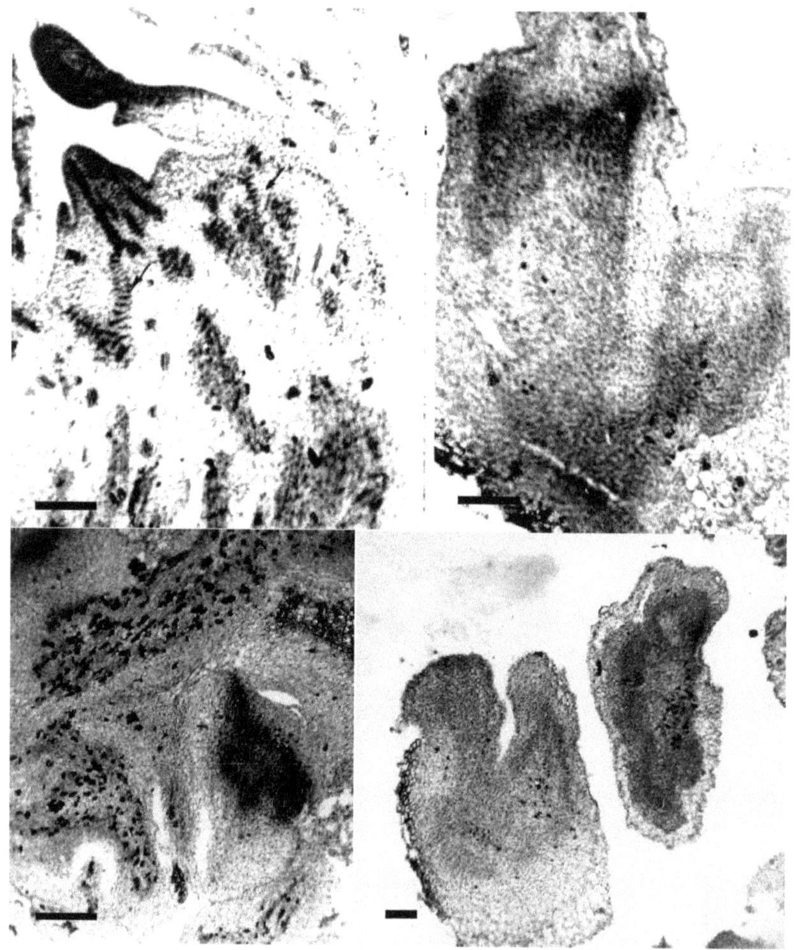

Figure 3. Histological analysis of slide images taken for *S. podophyllum* callus obtained in the N69 culture medium. Shoots and protocorms forming callus were observed. Nodular-like structures could also be observed (lower-left image), where organogenesis was taking place. The black arrows indicate very well-developed XTEs in the vicinity of a meristematic dome and de novo shoot formation. The meristematic dome is very well expressed, with well-developed, asymmetric, primordial leaves. In the upper- and lower-right images, protocorms can be observed in transversal section (magnification ×135 (upper and lower left) and ×270 (upper right), bars = 100 μm).

In the left-hand images (Figure 3), a transversal section of a meristematic dome is very clear, and the asymmetry of its contour reveals the initial stages of leaf formation. Again, XTEs are well developed nearby and entering the meristematic dome. Beneath the meristematic area, there are certain callus zones where XTEs reside among parenchymatic cells. No visible root formation was observed for the N69 mineral composition in 1998. Based on the reassessment of these images, it appears that the same auxin–cytokine ratio and the same vitamin composition but different mineral composition can change the pluripotency and totipotency of callus cells. While the MS62 mineral composition supported separate and complete organogenesis (i.e., shoots and roots), rooting was no longer observed when changing to the N69 mineral composition. Instead, protocorms were very well developed and expressed at a high density, appearing in almost half of the investigated images.

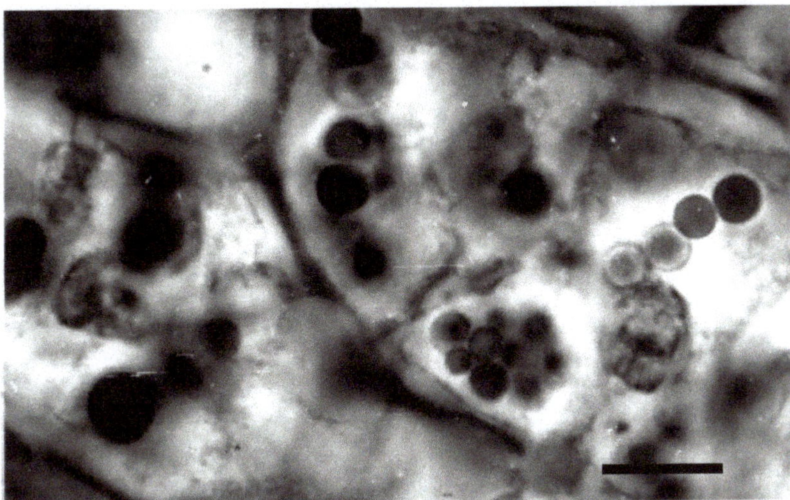

Figure 4. Histological analysis of slide images taken for callus cells of *S. podophyllum*. The small cells in the vicinity of the organogenetic areas present many starch inclusions (magnification ×45,000, bar = 10 µm).

Another easily noticed difference when we analyzed the XLEs is that they were less developed in MS62 compared to the N69 mineral composition. Therefore, it might be possible that the lack of or decrease in the concentration of certain mineral elements is responsible for this type of organogenesis, as was already stated some 47 years ago for *Antirrhinum majus* [90]. In another study, the increased concentration of boron appeared to influence shoot regeneration rather than contributing to increasing the number of shoots/explants, taking into account the results of recent studies on date palm [91]. The high expression of XTEs can also be supported by recent findings [92] and, among others, by the recognized toxicity of boron to plants [93]. The Lewis theory regarding boron's regulatory role in lignin synthesis [94]—highly supported by other scientists [95]—seems to be consistent with our findings regarding the cultivation of *Syngonium* callus in N69 culture medium. However, there are also too many other variables related to the changes in the compositions of other minerals, as mentioned previously. The lack of or decrease in the concentration of cobalt in N69 compared to MS62 was recently associated with decreased shooting in *Cucumis sativus* [96]. An interesting experiment was published by Renata Garcia and collaborators in 2011, where they used a mineral composition that was different from the basal MS62 and similar to N69 for callus cultivation of *Passiflora suberosa* [97]. They also observed the formation of a more compact callus and a significant decrease in shoot formation per explant. However, rooting was not impaired for this species.

3.4.3. Histological Analysis of Callus: Parenchymatic Cells Full of Starch Inclusions

A closer view of the small cells residing in the parenchymatic callus tissue and located very close to organogenetic centers and XLEs revealed that they contained many starch inclusions (Figure 4). This effect may be related to the high metabolic activity that is required in these zones to support the development of new organogenetic centers or XLEs. This observation further supports the findings of previous authors working on different species [98,99]. Recently, it was proven that there is a close relationship between lignin metabolic processes and the metabolism of starch and sucrose as the main factors associated with callus regeneration [100], which could be further investigated for different processes of cell differentiation.

3.4.4. Biochemical Analysis

By analyzing the spectra of electrophoretic peroxidases (POXs) published in 1998, we found clear differences between these two types of callus, further supporting our previous statement that N69 culture medium does not support root organogenesis [31,34]. Based on that analysis, 13 POXs' electrophoretic bands were described for a complete morphogenetic callus obtained in the MS62 culture medium (i.e., roots and shoots). The 6th electrophoretic band belonged to green tissues and the 12th belonged to roots. The 4th, 7th, 11th, and 13th appeared to mark the morphogenetic callus without visible specialized organs. The 3rd electrophoretic band appeared in all samples, and the 12th was the single band that was missing in the callus cultivated in N69 medium [31,34].

The continued development of morphogenetic processes revealed that N69 basal mineral medium was able to produce 42.56 propagules as buds and shoots per explant after 8 weeks of cultivation, while MS62 basal mineral medium induced 26.06 buds and shoots per explant. This result is consistent with the idea that the mineral composition is relevant both for increasing de novo shoot development per explant and for improving shoots' elongation.

3.5. Effects of the Vitamins MS62 and N69 on the De Novo Proliferation of Shoots

Another experiment was conducted to further study the effects of the vitamins MS62 and N69 on shoot multiplication and shoot height, as well as the effects of the composition of the plant growth regulators (but using only an MS62 basal mineral composition). Data on this topic were also published in 2004 [33,34]. In this experiment, we used NAA at different concentrations (0, 0.1, 0.5, and 1 mg/L) and a constant concentration of BAP (1 mg/L). The analysis of the results revealed that the best balance of plant growth regulators for the MS62 culture medium was 0.1 mg/L NAA and 1 mg/L BAP, which produced an average of 9 shoots/explant, considering only shoots > 7 mm in height. When increasing the auxin to 1 mg/L, the shoot multiplication rate decreased significantly compared to the control (i.e., no auxin) and the culture medium supplemented with 0.1 mg/L NAA. In this case, apical dominance of the main shoot was also observed (i.e., an average of 25.5 mm). All the culture media induced the formation of both shoots and roots, as well as a healthy appearance of the shoots. Again, the results of this experiment suggest that the best formula for de novo shoot formation was that provided by the MS62 basal culture medium [34].

3.6. Effects of Cysteine on Callus Formation in N69 Medium

A follow-up experiment was carried out to test the effects of cysteine. Figure 5 shows that a greener callus developed on the N69 medium in the presence of cysteine [34]. Moreover, visible root development and shoot elongation were observed in the presence of cysteine. The purpose of this experiment was to understand whether shoot regeneration in this species may be influenced by the potential release of phenolic compounds—as established in other species according to several reports [101,102]. It should also be noted that cysteine has long been recognized for its action against the release of phenolic compounds, and it has been recommended to be used for in vitro micropropagation of plant species [103,104]. Moreover, the fact that *Syngonium* produces phenols was first observed by Lynn Miller and Toshio Murashige [22]. These observations were later confirmed by several teams of scientists, proving that this ornamental species also has a high therapeutic value—mainly due to phenolic compounds with high antioxidant activity [27–29].

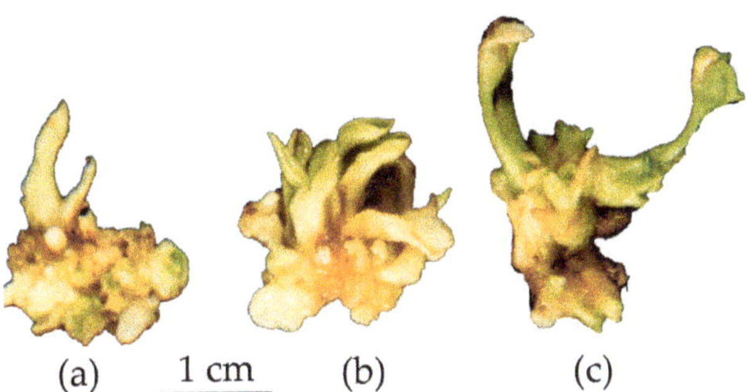

Figure 5. Different types of *Syngonium podophyllum* calli: (**a**) Control callus obtained in MS62 medium without cysteine. (**b**) Control callus obtained in N62 culture medium without cysteine. (**c**) Callus obtained in N69 culture medium supplemented with cysteine. It can be seen that the callus was much greener when it was cultivated in N69 culture medium, and that the shoots were more elongated in the presence of cysteine.

Later, in 2001, Jose Carlos Lorenzo et al. proved that the addition of cysteine reduced shoot formation and the excretion of phenols in the in vitro cultivation of sugarcane [105]. These authors also acknowledged several examples that showed contrasting results in developmental processes for other species and in vitro systems. Positive effects for micropropagation were also observed for grapevine (*Vitis vinifera*) [106], air potato (*Dioscorea bulbifera*) [107], and *Prosopis* species [108]. The elongation of shoots was also observed under cysteine treatment in *Prosopis* species [108]. In the case of the common bean (*Phaseolus vulgaris*), shoot elongation was also observed [109].

De novo shoot formation is a complex process that is highly desirable for all in vitro plant systems, especially to reduce costs for industry. *Syngonium* was investigated in vitro by various research teams before and after our publications [22,46–63]. Most of the teams obtained de novo shoots by using the same basal mineral and vitamin composition of MS62, but using different plant growth regulators as well as different balances between cytokinin and auxin.

The histological study of the callus may provide new insights regarding the stages needed for de novo development of the shoots [31,46,48,49,51–53,57,61]. Moreover, by studying the wound stress of the species, we may provide further evidence for the relevance of the hormone balance used for initiating in vitro callus development [32]. The chain of events consisting of increasing the peroxidase activity, hypertrophy of the wound sites, and callus formation starting from the cambium in petioles is already consistent with findings for other studied plant species [19,54].

Whether the callus can be well established in an in vitro culture to produce de novo shoots rests largely on the use of appropriate mineral, vitamin, and plant growth regulator balances. In this regard, the histological analysis of both types of calli that were cultivated—in MS62 mineral composition culture media and in N69 mineral composition culture media—revealed significant differences that were supported by biochemical analysis of peroxidases.

The aforementioned results raise the following question: is it possible for *Syngonium* calli cultivated in N69 media to also assist in the increased accumulation of phenolic compounds responsible for the development of XTEs? Here, we must underline the clear difference between XLEs developed for the calli cultivated in MS62 and those obtained in N69, as observed in all investigated slides. It should be noted that it has already been proven that cysteine plays a direct role in decreasing the accumulation of phenolic compounds and supporting organogenesis in the hybrid *Miscanthus* × *giganteus* [110].

It has been proven that the addition of an extra nitrogen source can have a positive influence on the development of xylematic bundles and on pigment accumulation in paperwhite (*Narcissus tazetta*) [111]. The direct relationship between lignin production and the development and maturation of tracheary elements upon apoptosis has also been well documented [112]. Additionally, it is possible that the tracheary elements observed in the tiny calli could be formed as a result of phenol accumulation under more stressful in vitro culture conditions. Moreover, it was observed that an increased thickening of the xylematic vessels of candy leaf (*Stevia rebaudiana*) took place under cold stress [113]. Previous studies also proved that the production of xylematic vessels is stimulated under different stress factors, e.g., physical and chemical [114].

In 2018, Shokoofeh Hajihashemi and Omolbanin Jahantigh put forward the hypothesis that an increase in the development of xylematic vessels may further support water transport, among other processes [111]. They also cited an earlier observation relating to the diameter and frequency of xylem vessels, which are critical determinants of water conductance [115]. In this regard, the clear and consistent positioning of the xylematic tracheary elements close to de novo shoot development areas and other meristematic structures provides rapid access to nutrients and water, as previously stated by other authors [114]. Regarding the positive effects of cysteine in supporting shoot elongation, it should be noted that it was proven to stimulate shoot elongation for *Petunia x hybrida* by acting on the gibberellic acid pathway [116,117]. Conversely, by inhibiting the synthesis of cysteine, the opposite effects were observed in cockspur (*Echinochloa crus-galli*) [118].

All of these findings are relevant to further support the improvement of the technology for industrial-scale *Syngonium* production in a more cost-efficient manner, and theoretical study of de novo shoot formation can contribute to this technological improvement. Cysteine may play an important role in decreasing the effects of phenolic compounds discharged under these stressful conditions when cultivating *Syngonium* in N69 culture medium, as well as improving de novo shoot elongation [119]. Thus, the addition of cysteine in the pre-acclimation phase can more quickly reinforce the formation of more adaptable plantlets for acclimation [120,121].

4. Conclusions

Syngonium is a very reactive species that is easy to cultivate in vitro. Different balances of plant growth regulators can be used to successfully initiate meristem cultures and the first stages of micropropagation of this species. Moreover, the different vitamin and mineral compositions tested could not completely impede de novo shoot development. All manner of morphogenetic programs have been studied since 1976, including shoot formation, rooting, callogenesis, protocorm formation, and embryogenesis. The main basal mineral medium composition used was MS62. Based on our histological analysis of calli originating from MS62 and N69, some obvious differences were observed in their structure. Thus, we observed the constant positioning of xylematic elements in the callus zones adjacent to meristematic and nodule-like areas, as well as organogenetic centers. Additionally, protocorms of *Syngonium* were identified with N69, and their formation was associated with very well-developed tracheary elements, constituting a novelty of this study. Conversely, the organogenetic centers observed in the callus originating from MS62 revealed less-differentiated xylematic elements, but these were also constantly in the presence of and closely connected with meristems, roots, and shoots. Based on these observations, the culture medium influences the development of specific areas in the callus structure. Each of these areas may contain pluripotent and totipotent cells, and their specific arrangement inside the callus can further support organogenetic processes, including de novo shoot formation. The addition of cysteine to the culture medium in the pre-acclimation phase can further support the success of de novo shoot development for acclimation. Moreover, lessons learned from all experiments performed on *Syngonium*—including the principles applied to implement industrial-scale micropropagation—may further support the production of phenolic compounds relevant for in vitro systems at the industrial scale.

Author Contributions: Conceptualization, M.-M.A. and C.S.S.; methodology, M.-M.A.; validation, M.-M.A. and C.S.S.; formal analysis, M.-M.A.; investigation, M.-M.A.; resources, M.-M.A.; data curation, M.-M.A.; writing—original draft preparation, M.-M.A.; writing—review and editing, M.-M.A.; visualization, C.S.S.; supervision, C.S.S. All authors have read and agreed to the published version of the manuscript.

Funding: This research received no external funding.

Data Availability Statement: Not applicable.

Acknowledgments: This paper is dedicated to the memory of Aurelia Brezeanu, the mentor in the formation of Maria-Mihaela Antofie, who provided much counsel and inspiration to us both, for her pioneering work, valuable contributions, and untiring efforts in developing the science of plant cell, tissue, and organ cultures in Romania from 1970 to 2022.

Conflicts of Interest: The authors declare no conflict of interest.

References

1. Duclercq, J.; Sangwan-Norreel, B.; Catterou, M.; Sangwan, R.S. De novo shoot organogenesis: From art to science. *Trends Plant Sci.* **2011**, *16*, 597–606. [CrossRef]
2. Haberlandt, G. Uber die Statolithefunktion der Starkekoner. *Ber. Dtsch. Bot. Ges.* **1902**, *20*, 189–195.
3. Avery, G.S. Differential distribution of a phytohormone in the developing leaf of Nicotiana, and its relation to polarized growth. *Bull. Torrey Bot. Club* **1935**, *62*, 313–330. [CrossRef]
4. Koepfli, J.B.; Thimann, K.V.; Went, F.W. Phytohormones: Structure and physiological activity. I. *J. Biol. Chem.* **1938**, *122*, 763–780. [CrossRef]
5. Skoog, F. *Plant Growth Substances*; Skoog, F., Ed.; University of Wisconsin: Madison, WI, USA; William Byrd Press, Inc.: Richmond, VA, USA, 1951; pp. 123–140.
6. White, P.R. Cultivation of excised roots of dicotyledonous plants. *Am. J. Bot.* **1938**, *25*, 348–356. [CrossRef]
7. White, P.R. Potentially unlimited growth of excised tomato root tips in a liquid medium. *Plant Physiol.* **1934**, *9*, 585–600. [CrossRef] [PubMed]
8. Hussey, G. Totipotency in tissue explants and callus of some members of the Liliaceae, Iridaceae, and Amaryllidaceae. *J. Exp. Bot.* **1975**, *26*, 253–262. [CrossRef]
9. Krakowsky, M.D.; Lee, M.; Garay, L.; Woodman-Clikeman, W.; Long, M.J.; Sharopova, N.; Frame, B.; Wang, K. Quantitative trait loci for callus initiation and totipotency in maize (*Zea mays* L.). *Theor. Appl. Genet.* **2006**, *113*, 821–830. [CrossRef]
10. Vasil, I.K.; Vasil, V. Totipotency and embryogenesis in plant cell and tissue cultures. *Vitr.* **1972**, *8*, 117–125. [CrossRef]
11. Steward, F.C.; Ammirato, P.V.; Mapes, M.O. Growth and development of totipotent cells: Some problems, procedures, and perspectives. *Ann. Bot.* **1970**, *34*, 761–787. [CrossRef]
12. Fehér, A. Callus, dedifferentiation, totipotency, somatic embryogenesis: What these terms mean in the era of molecular plant biology? *Front. Plant Sci.* **2019**, *10*, 536. [CrossRef] [PubMed]
13. Skoog, F.; Miller, C.O. Chemical regulation of growth and organ formation in plant tissues cultured in vitro. *Symp. Soc. Exp. Biol.* **1957**, *11*, 118–130. [PubMed]
14. Sugimoto, K.; Gordon, S.P.; Meyerowitz, E.M. Regeneration in plants and animals: Dedifferentiation, transdifferentiation, or just differentiation? *Trends Cell Biol.* **2011**, *21*, 212–218. [CrossRef]
15. Tian, X.; Zhang, C.; Xu, J. Control of cell fate reprogramming towards de novo shoot organogenesis. *Plant Cell Physiol.* **2018**, *59*, 713–719. [CrossRef]
16. Rosspopoff, O.; Chelysheva, L.; Saffar, J.; Lecorgne, L.; Gey, D.; Caillieux, E.; Colot, V.; Roudier, F.; Hilson, P.; Berthomé, R.; et al. Direct conversion of root primordium into shoot meristem relies on timing of stem cell niche development. *Development* **2017**, *144*, 1187–1200. [CrossRef] [PubMed]
17. Sang, Y.L.; Cheng, Z.J.; Zhang, X.S. Plant stem cells and de novo organogenesis. *New Phytol.* **2018**, *218*, 1334–1339. [CrossRef] [PubMed]
18. Scheres, B.; Heidstra, R.; Pijnenburg, M.C.C.; De Both, M.T.J. U.S. Genes for Hormone-Free Plant. Regeneration. Patent Application No. 17/081,832, 2021.
19. Deng, J.; Sun, W.; Zhang, B.; Sun, S.; Xia, L.; Miao, Y.; He, L.; Lindsey, K.; Yang, X.; Zhang, X. GhTCE1-GhTCEE1 dimers regulate transcriptional reprogramming during wound-induced callus formation in cotton. *Plant Cell* **2022**, *34*, koac252. [CrossRef]
20. Beals, C.M. An histological study of regenerative phenomena in plants. *Ann. Mo. Bot. Gard.* **1923**, *10*, 369–384. [CrossRef]
21. Sachs, T. The control of the patterned differentiation of vascular tissues. In *Advances in Botanical Research*; Woolhouse, H.W., Ed.; Academic Press: London, UK; John Innes Institute: Norwich, UK, 1981; Volume 9, pp. 151–262. [CrossRef]
22. Miller, L.R.; Murashige, T. Tissue culture propagation of tropical foliage plants. *Vitr. -Plant* **1976**, *12*, 797–813. [CrossRef]
23. Croat, T.B. A revision of Syngonium (Araceae). *Ann. Mo. Bot. Gard.* **1981**, *68*, 565–651. [CrossRef]
24. GBIF Secretariat. *Syngonium podophyllum* Schott GBIF Backbone Taxonomy. In *Checklist Dataset*; GBIF Secretariat: Copenhagen, Denmark, 2021. Available online: https://doi.org/10.15468/39omeiaccessedviaGBIF.org (accessed on 6 October 2022).

25. Chen, J.; McConnell, D.B.; Henny, R.J. The world foliage plant industry. *Chron. Hortic.* **2005**, *45*, 9–15.
26. Bhatt, D. Commercial Micropropagation of Some Economically Important Crops. In *Agricultural Biotechnology: Latest Research and Trends*; Kumar Srivastava, D., Kumar Thakur, A., Kumar, P., Eds.; Springer: Singapore, 2021; pp. 1–36. [CrossRef]
27. Kumar, S.; Dwivedi, A.; Kumar, R.; Pandey, A.K. Preliminary evaluation of biological activities and phytochemical analysis of *Syngonium podophyllum* leaf. *Natl. Acad. Sci. Lett.* **2015**, *38*, 143–146. [CrossRef]
28. Hossain, M.S.; Uddin, M.S.; Kabir, M.T.; Begum, M.M.; Koushal, P.; Herrera-Calderon, O. In vitro screening for phytochemicals and antioxidant activities of *Syngonium podophyllum* l.: An incredible therapeutic plant. *Biomed. Pharmacol. J.* **2017**, *10*, 1267–1277. [CrossRef]
29. Uddin, M.S.; Hossain, M.S.; Kabir, M.T.; Rahman, I.; Tewari, D.; Jamiruddin, M.R.; Al Mamun, A. Phytochemical screening and antioxidant profile of *Syngonium podophyllum* schott stems: A fecund phytopharmakon. *J. Pharm. Nutr. Sci.* **2018**, *8*, 120–128. [CrossRef]
30. Brunel, S. Pathway analysis: Aquatic plants imported in 10 EPPO countries. *EPPO Bull.* **2009**, *39*, 201–213. [CrossRef]
31. Antofie, M.M.; Brezeanu, A.; Maximilian, C.; Angheluta, R. Histological and biochemical analysis of morphogenetic callus from *Syngonium podophyllum* Schott. *Acta Horti Bot. Bucur.* **1998**, *27*, 123–130.
32. Antofie, M.M.; Carasan, M.E.; Brezeanu, A. Total cellulare peroxidase activity associated with wounding stress on *Syngonium podophyllum* (SCHOTT) in vitro tissue culture. *Proc. Inst. Biol. Suppl. Rev. Roum. Biol.* **1999**, *II*, 339–348.
33. Antofie, M.M.; Brezeanu, A. In vitro developmental peculiarities of Syngonium. *Aroideana* **2004**, *27*, 174–181.
34. Antofie, M.M. Study of Factors Involved in Morphogenetic Potential Expression of Some In Vitro Cultured Ornamental Plant Species. Ph.D. Thesis, Bucharest University, Bucharest, Romania, 2002; p. 239.
35. Google Scholar. Advanced Scholar Search Tips. 2022. Available online: http://scholar.google.com/scholar/refinesearch.html (accessed on 8 October 2022).
36. The International Plant Names Index. 2016. Available online: http://www.ipni.org (accessed on 23 August 2022).
37. Kew, Royal Botanic Gardens. Medicinal Plant Names Services. 2022. Available online: http://mpns.kew.org/mpns-portal/ (accessed on 30 October 2022).
38. The Plant List, 2013, Version 1.1. Available online: http://www.theplantlist.org/ (accessed on 10 November 2022).
39. Șandru, D. Criza din 1929–1933 și criza actuală. *Sfera Politicii* **2009**, *133*, 53–60.
40. Sava, C.S.; Antofie, M.M. The first clustering centre on plant biotechnology in Romania. *Studia Univ. Babes-Bolyai Biol.* **2016**, *61*, 125–131.
41. De Fossard, R.A. Principles of plant tissue culture. In *Tissue Culture as A Plant Production System for Horticultural Crops*; Zimmerman, R.H., Griesbach, R.J., Hammerschlag, F.A., Lawson, R.H., Eds.; Springer: Dordrecht, The Netherlands, 1986; pp. 1–13.
42. Kozai, T.; Smith, M.A.L. Environmental control in plant tissue culture—General introduction and overview. In *Automation and Environmental Control in Plant Tissue Culture*; Aitken-Christie, J., Kozai, T., Smith, M.A.L., Eds.; Springer: Dordrecht, The Netherlands, 1995; pp. 301–318.
43. Murashige, T.; Skoog, F.A. Revised Medium for Rapid Growth and Bio Assays with Tobacco Tissue Cultures. *Physiol. Plant.* **1962**, *15*, 473–497. [CrossRef]
44. Pop, I.V.; Sandu, C.; Constantinovici, D. Results regarding the development and experimentation of a system for production of virus-free propagating material in carnation. *Acta Hortic.* **1994**, *377*, 335–340. [CrossRef]
45. Levin, R.; Stav, R.; Alper, Y.; Watad, A.A. In vitro multiplication in liquid culture of Syngonium contaminated with *Bacillus* spp. and Rathayibacter tritici. *Plant Cell Tissue Organ Cult.* **1996**, *45*, 277–280. [CrossRef]
46. Scaramuzzi, F.; Apollonio, G.; Giodice, L.; D'Emerico, S. Propagation in vitro de deux especes de la famille des Araceae, *Syngonium podophyllum* Schott et *Scindapsus aureus* Engl., par culture et repiquages repetes de divers explants vegetatifs. *Comptes Rendus Séances Soc. Biol. Ses Fil.* **1992**, *186*, 125–138.
47. Salame, N.; Zieslin, N. Peroxidase activity in leaves of *Syngonium podophyllum* following transition from in vitro to ex vitro conditions. *Biol. Plant* **1994**, *36*, 619–622. [CrossRef]
48. Watad, A.A.; Raghothama, K.G.; Kochba, M.; Nissim, A.; Gaba, V. Micropropagation of *Spathiphyllum* and *Syngonium* is facilitated by use of interfacial membrane rafts. *HortScience* **1997**, *1*, 307–308. [CrossRef]
49. Rajeevan, P.K.; Sheeja, K.; Rangan, V.V.; Murali, T.P. Direct organogenesis in *Syngonium podophyllum*. Floriculture research trend in India. In Proceedings of the National Symposium on Indian Floriculture in the New Millennium, Indian Society of Ornamental Horticulture, Lal-Bagh, Bangalore, India, 25–27 February 2002; pp. 353–354.
50. Chan, L.K.; Tan, C.M.; Chew, G.S. Micropropagation of the Araceae ornamental plants. *Acta Hortic.* **2003**, *616*, 383–390. [CrossRef]
51. Schwertner, A.B.S.; Zaffari, G.R. Micropropagação de singônio. *Rev. Bras. Hortic. Ornam.* **2003**, *9*, 135–142.
52. Hassanein, A.M. A study on factors affecting propagation of shade plant-*Syngonium podophyllum*. *J. Appl. Hortic.* **2004**, *6*, 30–34. [CrossRef]
53. Chen, J.; Henny, R.J. Somaclonal variation: An important source for cultivar development of floriculture crops. *Floric. Ornam. Plant Biotech* **2006**, *2*, 244–253.
54. Zhang, Q.; Chen, J.; Henny, R.J. Regeneration of *Syngonium podophyllum* 'Variegatum' through direct somatic embryogenesis. *Plant Cell Tissue Organ Cult.* **2006**, *84*, 181–188. [CrossRef]
55. Wang, X.; Li, Y.; Nie, Q.; Li, J.; Chen, J.; Henny, R.J. In vitro culture of *Epipremnum aureum*, *Syngonium podophyllum*, and *Lonicera macranthodes*, three important medicinal plants. *Acta Hortic.* **2007**, *756*, 155–162. [CrossRef]

56. Cui, J.; Liu, J.; Deng, M.; Chen, J.; Henny, R.J. Plant regeneration through protocorm-like bodies induced from nodal explants of *Syngonium podophyllum* 'White Butterfly'. *HortScience* **2008**, *43*, 2129–2133. [CrossRef]

57. Rajesh, A.M.; Yathindra, H.A.; Reddy, P.V.; Sathyanarayana, B.N.; Harshavardhan, M.; Kantharaj, Y. In vitro regeneration and ex vitro studies in syngonium (*Syngonium podophyllum*). *J. Ecobiol.* **2011**, *20*, 103–106.

58. Teixeira Da Silva, J.A.; Giang, D.T.; Dobránszki, J.; Zeng, S.; Tanaka, M. Ploidy analysis of *Cymbidium, Phalaenopsis, Dendrobium* and *Paphiopedillum* (Orchidaceae), and *Spathiphyllum* and *Syngonium* (Araceae). *Biologia* **2014**, *69*, 750–755. [CrossRef]

59. Kalimuthu, K.; Prabakaran, R. In vitro Micropropagation of *Syngonium podophyllum*. *Int. J. Pure App. Biosci.* **2014**, *2*, 88–92.

60. Da Silva, J.A.T. Response of *Syngonium podophyllum* L. "White Butterfly" shoot cultures to alternative media additives and gelling agents, and flow cytometric analysis of regenerants. *Nusant. Biosci.* **2015**, *7*, 26–32. [CrossRef]

61. Moumita, M.; Pinaki, A.; Sukanta, B. Micropropagation of commercially feasible foliage ornamental plants. *Indian Agric.* **2016**, *60*, 39–44.

62. Kane, M.E. Micropropagation of Syngonium by shoot culture. In *Plant Tissue Culture Concepts and Laboratory Exercises*, 2nd ed.; Trigiano, R.N., Gray, D.J., Eds.; Routledge: Boca Raton, FL, USA, 2018; pp. 87–95. [CrossRef]

63. Sharifi, A.; Moradiyan, M.; Khadem, A.; Kharrazi, M. Optimization Plant Growth Regulation Type and Culture Medium Salts in the Micropropagation of Syngonium (*Syngonium podophyllum* L.). *J. Hortic. Sci.* **2022**, *35*, 647–660.

64. Cline, M.G. Apical dominance. *Bot. Rev.* **1991**, *57*, 318–358. [CrossRef]

65. Pena-Cortés, H.; Albrecht, T.; Prat, S.; Weiler, E.W.; Willmitzer, L. Aspirin prevents wound-induced gene expression in tomato leaves by blocking jasmonic acid biosynthesis. *Planta* **1993**, *191*, 123–128. [CrossRef]

66. Klessig, D.F.; Malamy, J. The salicylic acid signal in plants. *Plant Mol. Biol.* **1994**, *26*, 1439–1458. [CrossRef] [PubMed]

67. Torrey, J.G.; Thimann, K.V. Application of herbicides to cut stumps of a woody tropical weed. *Bot. Gaz.* **1949**, *111*, 184–192. [CrossRef]

68. Hildebrandt, A.C.; Riker, A.J.; Duggar, B.M. Growth in vitro of excised tobacco and sunflower tissue with different temperatures, hydrogen-ion concentrations and amounts of sugar. *Am. J. Bot.* **1945**, *32*, 357–361. [CrossRef]

69. Stewart, W.S.; Ebeling, W. Preliminary Results with the use of 2, 4-Dichlorophenoxy-Acetic Acid as a Spray-Oil Amendment. *Bot. Gaz.* **1946**, *108*, 286–294. [CrossRef]

70. Livingston, G.A. In vitro tests of abscission agents. *Plant Physiol.* **1950**, *25*, 711–721. [CrossRef]

71. Iordăchescu, D.; Dumitru, I.F. Edit. *Univ. Bucureşti Bucureşti Rom.* **1988**, *106–107*, 133–137.

72. Kevers, C.; Coumans, M.; Coumans-Gillès, M.F.; Caspar, T.H. Physiological and biochemical events leading to vitrification of plants cultured in vitro. *Physiol. Plant* **1984**, *61*, 69–74. [CrossRef]

73. Gaspar, T.; Franck, T.; Bisbis, B.; Kevers, C.; Jouve, L.; Hausman, J.F.; Dommes, J. Concepts in plant stress physiology. Application to plant tissue cultures. *Plant Growth Regul.* **2002**, *37*, 263–285. [CrossRef]

74. Anike, F.N.; Konan, K.; Olivier, K.; Dodo, H. Efficient shoot organogenesis in petioles of yam (*Dioscorea* spp.). *Plant Cell Tissue Organ Cult.* **2012**, *111*, 303–313. [CrossRef]

75. Michalak, A. Phenolic compounds and their antioxidant activity in plants growing under heavy metal stress. *Pol. J. Environ. Stud.* **2006**, *15*, 523–530.

76. Tran Thanh Van, K.M. Control of morphogenesis in in vitro cultures. *Annu. Rev. Plant Physiol.* **1981**, *32*, 291–311. [CrossRef]

77. Pieron, S.; Boxus, P.; Dekegel, D. Histological study of nodule morphogenesis from *Cichorium intybus* L. leaves cultivated in vitro. *Vitr. Cell Dev. Biol.-Plant* **1998**, *34*, 87–93. [CrossRef]

78. Sujatha, M.; Mukta, N. Morphogenesis and plant regeneration from tissue cultures of *Jatropha curcas*. *Plant Cell Tissue Organ Cult.* **1996**, *44*, 135–141. [CrossRef]

79. Nitsch, J.P.; Nitsch, C. Haploid plants from pollen grains. *Science* **1969**, *163*, 85–87. [CrossRef] [PubMed]

80. Galloway, J.N.; Townsend, A.R.; Erisman, J.W.; Bekunda, M.; Cai, Z.; Freney, J.R.; Martinelli, L.A.; Seitzinger, S.P.; Sutton, M.A. Transformation of the nitrogen cycle: Recent trends, questions, and potential solutions. *Science* **2008**, *320*, 889–892. [CrossRef] [PubMed]

81. Hedin, L.O.; Brookshire, E.J.; Menge, D.N.; Barron, A.R. The nitrogen paradox in tropical forest ecosystems. *Annu. Rev. Ecol. Evol. Syst.* **2009**, *40*, 613–635. [CrossRef]

82. Darlington, C.D.; La Cour, L.F. *The Handling of Chromosomes*, 3rd ed.; George Allen & Unwin Ltd.: London, UK, 1960; p. 141.

83. Johansen, D.A. *Plant Microtechnique*; McGraw-Hill Book Company, Inc.: London, UK, 1940; 530p.

84. Baima, S.; Nobili, F.; Sessa, G.; Lucchetti, S.; Ruberti, I.; Morelli, G. The expression of the Athb-8 homeobox gene is restricted to provascular cells in *Arabidopsis thaliana*. *Development* **1995**, *121*, 4171–4182. [CrossRef] [PubMed]

85. Harvey-Gibson, R.J.; Horsman, E. XIX.—Contributions towards a Knowledge of the Anatomy of the Lower Dicotyledons. II. The Anatomy of the Stem of the Berberidaceæ. *Earth Environ. Sci. Trans. R. Soc. Edinb.* **1920**, *52*, 501–515. [CrossRef]

86. Mattsson, J.; Ckurshumova, W.; Berleth, T. Auxin signaling in Arabidopsis leaf vascular development. *Plant Physiol.* **2003**, *131*, 1327–1339. [CrossRef]

87. Klee, H.J.; Horsch, R.B.; Hinchee, M.A.; Hein, M.B.; Hoffmann, N.L. The effects of overproduction of two *Agrobacterium tumefaciens* T-DNA auxin biosynthetic gene products in transgenic petunia plants. *Genes Dev.* **1987**, *1*, 86–96. [CrossRef]

88. Berleth, T.; Mattsson, J.; Hardtke, C.S. Vascular continuity and auxin signals. *Trends Plant Sci.* **2000**, *5*, 387–393. [CrossRef] [PubMed]

89. Zhai, N.; Xu, L. Pluripotency acquisition in the middle cell layer of callus is required for organ regeneration. *Nat. Plants* **2021**, *7*, 1453–1460. [CrossRef] [PubMed]

90. Sangwan, R.S.; Harada, H. Chemical regulation of callus growth, organogenesis, plant regeneration, and somatic embryogenesis in *Antirrhinum majus* tissue and cell cultures. *J. Exp. Bot.* **1975**, *26*, 868–881. [CrossRef]

91. Al-Mayahi, A.M.W. Effect of calcium and boron on growth and development of callus and shoot regeneration of date palm 'Barhee'. *Can J. Plant Sci.* **2019**, *100*, 357–364. [CrossRef]

92. Nable, R.O.; Bañuelos, G.S.; Paull, J.G. Boron toxicity. *Plant Soil* **1997**, *193*, 181–198. [CrossRef]

93. Day, S.; Aasim, M. Role of boron in growth and development of plant: Deficiency and toxicity perspective. In *Plant Micronutrients*; Aftab, T., Hakeem, K.R., Eds.; Springer: New York, NY, USA, 2020; pp. 435–453. [CrossRef]

94. Lewis, D.H. Boron, lignification and the origin of vascular plants—A unified hypothesis. *New Phytol.* **1980**, *84*, 209–229. [CrossRef]

95. Lovatt, C.J. Evolution of xylem resulted in a requirement for boron in the apical meristems of vascular plants. *New Phytol.* **1985**, *99*, 509–522. [CrossRef]

96. Miguel, J.F. Influence of High Concentrations of Copper Sulfate on In vitro Adventitious Organogenesis of *Cucumis sativus* L. *bioRxiv* **2021**, 1–13. [CrossRef]

97. Garcia, R.; Pacheco, G.; Falcao, E.; Borges, G.; Mansur, E. Influence of type of explant, plant growth regulators, salt composition of basal medium, and light on callogenesis and regeneration in *Passiflora suberosa* L. (Passifloraceae). *Plant Cell Tissue Organ Cult.* **2011**, *106*, 47–54. [CrossRef]

98. Webb, K.J. Growth and Morphogenesis of Tissue Cultures of *Pinus contorta* and *Picea sitchensis*. Doctoral Thesis, University of Leicester, Leicester, UK, 1978.

99. Simola, L.K. Structure of cell organelles and cell wall in tissue cultures of trees. In *Cell and Tissue Culture in Forestry. Forestry Sciences*; Bonga, J.M., Durzan, D.J., Eds.; Springer: Dordrecht, The Netherlands, 1987; Volume 24–26, pp. 389–418. [CrossRef]

100. Zhou, C.; Wang, S.; Zhou, H.; Yuan, Z.; Zhou, T.; Zhang, Y.; Xiang, S.; Yang, F.; Shen, X.; Zhang, D. Transcriptome sequencing analysis of sorghum callus with various regeneration capacities. *Planta* **2021**, *254*, 1–14. [CrossRef] [PubMed]

101. Bon, M.C.; Gendraud, M.; Franclet, A. Roles of phenolic compounds on micropropagation of juvenile and mature clones of *Sequoiadendron giganteum*: Influence of activated charcoal. *Sci. Hortic.* **1988**, *34*, 283–291. [CrossRef]

102. Ruffoni, B.; Massabò, F.; Costantino, C.; Arena, V.; Damiano, C. Micropropagation of Acacia "mimosa". *Acta Hortic.* **1992**, *300*, 95–102. [CrossRef]

103. Johansson, L. Effects of activated charcoal in anther cultures. *Physiol. Plant* **1983**, *59*, 397–403. [CrossRef]

104. Zaid, A. In vitro browning of tissues and media with special emphasis to date palm cultures—A review. Symposium on In vitro Problems Related to Mass Propagation of Horticultural Plants. *Acta Hortic.* **1987**, *212*, 561–566. [CrossRef]

105. Lorenzo, J.C.; de los Angeles Blanco, M.; Peláez, O.; González, A.; Cid, M.; Iglesias, A.; González, B.; Escalona, M.; Espinosa, P.; Borroto, C. Sugarcane micropropagation and phenolic excretion. *Plant Cell Tissue Organ Cult.* **2001**, *65*, 1–8. [CrossRef]

106. Alzubi, H.; Yepes, L.M.; Fuchs, M. Enhanced micropropagation and establishment of grapevine rootstock genotypes. *Int. J. Plant Dev. Biol.* **2012**, *6*, 9–14.

107. Sonibare, M.A.; Adeniran, A.A. Comparative micromorphological study of wild and micropropagated *Dioscorea bulbifera* Linn. *Asian Pac. J. Trop. Biomed.* **2014**, *4*, 176–183. [CrossRef]

108. Arce, J.P.; Medina, M.C. Micropropagation of Prosopis Species (Mesquites). In *High-Tech and Micropropagation V. Biotechnology in Agriculture and Forestry*; Bajaj, Y.P.S., Ed.; Springer: Berlin/Heidelberg, Germany, 1997; Volume 39, pp. 367–380. [CrossRef]

109. Allavena, A.; Rossetti, L. Micropropagation of bean (*Phaseolus vulgaris* L.); effect of genetic, epigenetic and environmental factors. *Sci. Hortic.* **1986**, *30*, 37–46. [CrossRef]

110. Lewandowski, I. Micropropagation of Miscanthus× giganteus. In *High-Tech and Micropropagation V. Biotechnology in Agriculture and Forestry*; Bajaj, Y.P.S., Ed.; Springer: Berlin/Heidelberg, Germany, 1997; Volume 39, pp. 239–255. [CrossRef]

111. Hajihashemi, S.; Jahantigh, O. Nitric Oxide Effect on Growth, Physiological and Biochemical Processes, Flowering, and Postharvest Performance of *Narcissus tazzeta*. *J. Plant Growth Regul.* **2022**, 1–16. [CrossRef]

112. Serk, H.; Gorzsás, A.; Tuominen, H.; Pesquet, E. Cooperative lignification of xylem tracheary elements. *Plant Signal. Behav.* **2015**, *10*, e1003753. [CrossRef] [PubMed]

113. Hajihashemi, S.; Noedoost, F.; Geuns, J.M.; Djalovic, I.; Siddique, K.H. Effect of cold stress on photosynthetic traits, carbohydrates, morphology, and anatomy in nine cultivars of *Stevia rebaudiana*. *Front. Plant Sci.* **2018**, *9*, 1430. [CrossRef] [PubMed]

114. Roberts, L.W. Physical factors, hormones, and differentiation. In *Vascular Differentiation and Plant Growth Regulators*; Springer Series in Wood Science; Springer: Berlin/Heidelberg, Germany, 1988; pp. 89–105. [CrossRef]

115. Lovisolo, C.; Schubert, A. Effects of water stress on vessel size and xylem hydraulic conductivity in *Vitis vinifera* L. *J. Exp. Bot.* **1998**, *49*, 693–700. [CrossRef]

116. Ben-Nissan, G.; Lee, J.Y.; Borohov, A.; Weiss, D. GIP, a Petunia hybrida GA-induced cysteine-rich protein: A possible role in shoot elongation and transition to flowering. *Plant J.* **2004**, *37*, 229–238. [CrossRef]

117. Wigoda, N.; Ben-Nissan, G.; Granot, D.; Schwartz, A.; Weiss, D. The gibberellin-induced, cysteine-rich protein GIP2 from *Petunia hybrida* exhibits in planta antioxidant activity. *Plant J.* **2006**, *48*, 796–805. [CrossRef] [PubMed]

118. Hirase, K.; Molin, W.T. Effect of inhibitors of pyridoxal-5′-phosphate-dependent enzymes on cysteine synthase in *Echinochloa crus-galli* L. *Pestic. Biochem. Phys.* **2001**, *70*, 180–188. [CrossRef]

119. Speiser, A.; Haberland, S.; Watanabe, M.; Wirtz, M.; Dietz, K.J.; Saito, K.; Hell, R. The significance of cysteine synthesis for acclimation to high light conditions. *Front. Plant Sci.* **2015**, *5*, 776. [CrossRef] [PubMed]
120. Ahmad, N.; Malagoli, M.; Wirtz, M.; Hell, R. Drought stress in maize causes differential acclimation responses of glutathione and sulfur metabolism in leaves and roots. *BMC Plant Biol.* **2016**, *16*, 1–15. [CrossRef]
121. Garcia-Molina, A.; Kleine, T.; Schneider, K.; Mühlhaus, T.; Lehmann, M.; Leister, D. Translational components contribute to acclimation responses to high light, heat, and cold in *Arabidopsis*. *iScience* **2020**, *23*, 101331. [CrossRef] [PubMed]

MDPI

St. Alban-Anlage 66

4052 Basel

Switzerland

www.mdpi.com

Horticulturae Editorial Office

E-mail: horticulturae@mdpi.com

www.mdpi.com/journal/horticulturae